Lecture Notes in Computer Science 9639

Commenced Publication in 1973
Founding and Former Series Editors:
Gerhard Goos, Juris Hartmanis, and Jan van Leeuwen

Juan Caballero · Eric Bodden
Elias Athanasopoulos (Eds.)

Engineering Secure Software and Systems

8th International Symposium, ESSoS 2016
London, UK, April 6–8, 2016
Proceedings

 Springer

Editors
Juan Caballero
IMDEA Software Institute
Madrid
Spain

Elias Athanasopoulos
VU University
Amsterdam
The Netherlands

Eric Bodden
Paderborn University & Fraunhofer IEM
Paderborn
Germany

ISSN 0302-9743   ISSN 1611-3349 (electronic)
Lecture Notes in Computer Science
ISBN 978-3-319-30805-0   ISBN 978-3-319-30806-7 (eBook)
DOI 10.1007/978-3-319-30806-7

Library of Congress Control Number: 2016932517

LNCS Sublibrary: SL4 – Security and Cryptology

Printed on acid-free paper

This Springer imprint is published by Springer Nature
The registered company is Springer International Publishing AG Switzerland

Preface

It is our pleasure to welcome you to the proceedings of the 8th International Symposium on Engineering Secure Software and Systems (ESSoS 2016). This event is part of a maturing series of symposia that attempts to bridge the gap between the software engineering and security scientific communities with the goal of supporting secure software development. The parallel technical sponsorship from ACM SIGSAC (the ACM interest group in security) and ACM SIGSOFT (the ACM interest group in software engineering) demonstrates the support from both communities and the need for providing such a bridge.

Security mechanisms and the act of software development usually go hand in hand. It is generally not enough to ensure correct functioning of the security mechanisms used. They cannot be blindly inserted into a security-critical system, but the overall system development must take security aspects into account in a coherent way. Building trustworthy components does not suffice, since the interconnections and interactions of components play a significant role in trustworthiness. Lastly, while functional requirements are generally analyzed carefully in systems development, security considerations often arise after the fact. Adding security as an afterthought, however, often leads to problems. Ad hoc development can lead to the deployment of systems that do not satisfy important security requirements. Thus, a sound methodology supporting secure systems development is needed. The presentations and associated publications at ESSoS 2016 contribute to this goal in several directions: First, by improving methodologies for secure software engineering (such as flow analysis and policy compliance). Second, with results for the detection and analysis of software vulnerabilities and the attacks they enable. Finally, for securing software for specific application domains (such as mobile devices and access control).

The conference program featured two keynotes by David Basin (ETH Zurich) and Karsten Nohl (Security Research Labs), as well as research and idea papers. In response to the call for papers, 50 papers were submitted. The Program Committee selected 13 full-paper contributions, presenting new research results on engineering secure software and systems. In addition, three idea papers were selected, giving a concise account of new ideas in the early stages of research. Overall, the acceptance rate was 32 %.

Many individuals and organizations have contributed to the success of this event. First of all, we would like to express our appreciation to the authors of the submitted papers and to the Program Committee members and external reviewers, who provided timely and relevant reviews. Many thanks go to the Steering Committee for supporting this series of symposia, and to all the members of the Organizing Committee for their tremendous work and for excelling in their respective tasks. The DistriNet research group of the KU Leuven did an excellent job with the website and the advertising for the conference. Finally, we owe gratitude to ACM SIGSAC/SIGSOFT and LNCS for continuing to support us in this series of symposia.

January 2016   Juan Caballero
                                                                                 Eric Bodden
                                                                  Elias Athanasopoulos

Organization

Program Committee

Javier Alonso	University of Leon, Spain
Eric Bodden	Fraunhofer, Germany
Michele Bugliesi	Università Ca' Foscari Venezia, Italy
Juan Caballero	IMDEA Software Institute, Spain
Werner Dietl	University of Waterloo, Canada
Michael Franz	University of California, Irvine, USA
Flavio D. Garcia	Radboud University Nijmegen, The Netherlands
Christian Hammer	Saarland University, Germany
Marieke Huisman	University of Twente, The Netherlands
Martin Johns	SAP Research, Germany
Stefan Katzenbeisser	TU Darmstadt, Germany
Johannes Kinder	Royal Holloway, University of London, UK
Andy King	University of Kent, UK
Jacques Klein	University of Luxembourg, Luxembourg
Andrea Lanzi	Università degli studi di Milano, Italy
Wenke Lee	Georgia Institute of Technology, USA
Zhenkai Liang	National University of Singapore, Singapore
Benjamin Livshits	Microsoft Research, USA
Heiko Mantel	TU Darmstadt, Germany
Nick Nikiforakis	Stony Brook University, USA
Martín Ochoa	Technische Universität München, Germany
Mathias Payer	Purdue University, USA
Frank Piessens	Katholieke Universiteit Leuven, Belgium
Alexander Pretschner	Technische Universität München, Germany
Awais Rashid	Lancaster University, UK
Mark Ryan	University of Birmingham, UK
Gianluca Stringhini	University College London, UK
Pierre-Yves Strub	IMDEA Software Institute, Spain
Helmut Veith	Vienna University of Technology, Austria
Santiago Zanella-Béguelin	Microsoft Research, UK

Additional Reviewers

Bai, Guangdong
Beckers, Kristian
Bissyande, Tegawende
Büscher, Niklas
Calzavara, Stefano
Chawdhary, Aziem
Chua, Zheng Leong
Denzel, Michael
Focardi, Riccardo
Grewal, Gurchetan
Gurov, Dilian
Kohnhäuser, Florian
Li, Li
Li, Xiaolei
Liu, Jia

Muehlberg, Jan Tobias
Noorman, Job
Oortwijn, Wytse
Ordean, Mihai
Oswald, David
Pani, Thomas
Radu, Andreea-Ina
Rizzo, Claudio
Robbins, Ed
Starostin, Artem
Thomas, Sam L.
Thomas, Susan
Ulbrich, Mattias
Weber, Alexandra

Contents

Security Testing Beyond Functional Tests

Mohammad Torabi Dashti and David Basin[✉]

Department of Computer Science, ETH Zurich, Zürich, Switzerland
basin@inf.ethz.ch

Abstract. We present a theory of security testing based on the basic distinction between system specifications and security requirements. Specifications describe a system's desired behavior over its interface. Security requirements, in contrast, specify desired properties of the world the system lives in. We propose the notion of a security rationale, which supports reductive security arguments for deriving a system specification and assumptions on the system's environment sufficient for fulfilling stated security requirements. These reductions give rise to two types of tests: those that test the system with respect to its specification and those that test the validity of the assumptions about the adversarial environment. It is the second type of tests that distinguishes security testing from functional testing and defies systematization and automation.

1 Introduction

Security testing plays an essential role in quality assurance for information technologies ranging from traditional software applications to cyber-physical control systems. Various security testing tools and techniques are available today, and a wide range of systems are regularly subjected to security tests. Yet, the literature lacks the necessary frame of reference to articulate and answer basic questions regarding security testing. For example, most practitioners would agree that security testing is harder than functional testing, measuring the adequacy of security tests is challenging, and some kinds of security testing, such as penetration testing, defy systematization and automation. However, there exists no coherent explanation for these phenomena.

We rationally reconstruct security testing around the notion of security requirements. Our starting point is the key distinction between system specifications and security requirements. A system specification describes how an artifact, or system, must behave in an environment. A security requirement, in contrast, expresses desired properties of the environment controlled by the system. Consider, for example, an office. A system specification for an electronic lock installed as part of the office's door might state that the lock opens the door if and only if a valid key is presented. A security requirement might be that access to the office is restricted to employees working there. Under certain assumptions, if the system (here, the lock) satisfies its specification, then the requirement is satisfied in the actual environment. In our example, these environmental assumptions include: the office has no entrance other than the door

© Springer International Publishing Switzerland 2016
J. Caballero et al. (Eds.): ESSoS 2016, LNCS 9639, pp. 1–19, 2016.
DOI: 10.1007/978-3-319-30806-7_1

controlled by the lock, and only those working in the office have a valid key. What distinguishes security requirements from other requirements is that they must hold in the presence of an adversary. In the office example, if the adversary can climb through an open window, then the requirement is violated, regardless of whether or not the deployed lock satisfies its specification. Accounting for the adversary's capabilities is therefore integral to testing security requirements.

We introduce the notion of a security rationale, which supports reductive security arguments for deriving a system specification and the assumptions on the system's environment sufficient for fulfilling stated security requirements. These reductions give rise to two types of tests: (1) those that test the system with respect to its specification and (2) those that test the validity of the assumptions about the adversarial environment. These types sharply distinguish security testing from functional testing. The purpose of functional tests is to refute the hypothesis that the system satisfies its (functional, security, or other) specification; this corresponds to just the first type. In contrast, security testing requires both types of tests. This distinction allows us to precisely explain in what sense security testing is harder than functional testing. It also provides a frame of a reference for delimiting the scope and reach of existing security testing techniques and procedures. We illustrate this point through examples from fuzz testing, fault injection, risk-based security testing, and vulnerability-driven security testing.

We describe why measuring the adequacy of security tests is challenging by demonstrating that security tests are inherently incomplete. This incompleteness, we argue, stems from the open-ended nature of the assumptions that are part of security rationales. It is therefore orthogonal to the incompleteness of functional tests, which is rooted in the infinite cardinality of the domains where test inputs are selected. The open-ended nature of the assumptions also explains why security testing intrinsically depends on the testers' creativity and resources, thereby defying automation and systematization. Finally, we clarify testing and vulnerability remediation procedures associated with our two test types.

The theory of security testing that we develop is novel. The most closely related work comes from the domain of requirements engineering, where one studies the relationship between systems and the domains in which they operate; see, for example, Jackson's world-machine model [13], which inspired our notion of security rationales. Security testing itself is a broad topic and an extended literature review is outside this paper's scope. Nevertheless, we discuss along the way various prominent security testing techniques and procedures, such as [11, 19, 24].

Structure of Paper. We define specifications and requirements in Sect. 2. We introduce the notion of a security rationale in Sect. 3, which relates specifications and security requirements. Security cases, introduced in Sect. 4, justify the conformance of a concrete system to a security rationale in the presence of a specific adversary. In Sect. 5, we define two types of security tests whose purpose is to refute the hypothesis that a security requirement is satisfied in an adversarial environment. We also discuss the role of these two types in practice and comment on vulnerability remediation. We draw conclusions in Sect. 6.

2 Specifications and Requirements

Our starting point is valuable **resources** worth protecting, such as data on a company's web server or documents in an office. We consider security **requirements** that pertain to these resources. For instance, if our resources are the books in a library, the security requirement might be that only those possessing a valid library card may borrow books. In general, such requirements reflect the constraints that stakeholders impose on access to the resources.

What distinguishes security requirements from other requirements, such as functional requirements, is that they must hold in the presence of an adversary. The **adversary** (or **threat agent**) is the entity against whom the resources must be protected. Examples of adversaries include disgruntled employees, curious trespassers, and nation-state attackers. A security requirement that is satisfied in the presence of one adversary might not hold in the presence of a more capable one. A resource's security is therefore not a meaningful property without fixing the adversary's capabilities, which is left implicit in the above library example. Risk analysis can for example be used to identify the adversary in whose presence a security requirement must be satisfied.

To satisfy a security requirement, we construct systems and deploy them in the (resource's) environment. A **system** is an artifact whose behaviors can be regulated and controlled. A system affects its environment by interacting with it through an **interface**. To control access to an office, we may for example install an electronic lock system. This system changes the environment by restricting who may enter through the office's door. A **specification** describes the desired behaviors of a system over its interface. For instance, for the lock system, described in more detail in Example 1 below, its specification relates input received by its sensors, e.g., a smart-card reader, with output to its actuators, which control the lock's cylinder. Note that specifications need not directly constrain a system's internal structure. Our electronic lock specification does not for instance express a preference for a particular memory layout for the lock's software.

There is a fundamental distinction between specifications and requirements. Specifications constrain a system's behaviors over its interface. Requirements, in contrast, constrain an adversary's access to resources in an environment. Therefore, requirements neither directly prohibit nor oblige any system behaviors, and systems do not directly guarantee a requirements' satisfaction. The following example illustrates this point.

Example 1. An R&D laboratory contains sensitive documents. To limit access to the documents, an electronic lock system is installed at the lab's door. A security requirement for the documents states that only those staff members working in the lab may read them. This does not prohibit (or oblige) any input-output behavior for the lock. In contrast, a specification for the lock states: the output signal open is produced only after receiving as input a key that belongs to the set validKeys. Here open is the signal that, say, triggers an actuator that opens the lock. The satisfaction of this specification, which describes the lock's desired input-output behavior over its interface, does not entail the requirement's satisfaction. The lab may have an open window.

As a side remark, the above requirement is rather weak. It does not, for example, prohibit information flow arising from a careless staff member leaking a document's content to an outsider. This requirement's satisfaction, therefore, does not entail the documents' confidentiality.   △

The fundamental distinction between specifications and requirements is at the heart of our development, and we explore its implications in detail. Two comments are however due here. First, while a specification applies to a system's (input-output) interface, a requirement cannot be attributed with an interface because resources and environments do not have definite interfaces. The following example illustrates this point.

Example 2. A publishing company's database stores data that is subject to the following integrity *requirement*: only copy editors may delete data. Dynamite that explodes in the database's vicinity constitutes an "input" that can delete the data, thereby affecting its integrity. Similarly, formatting the database's storage media or invoking the database's rollback operation are both "inputs" that can also delete the stored data. Clearly the integrity requirement above cannot be attributed to a specific interface.

Now suppose that the database's input-output interface is realized through an API. A *specification* for the database system, which applies to this interface, states: only the users who have the role `copy_editor` may execute the API's `delete` command. An input is a user's identity (or roles) together with the API command the user requests to execute. The system then either executes the command or denies the request.   △

Second, a system's interface may consist of multiple communication channels between the system and its environment. A **nominal** channel is a channel that has been anticipated in the system's specification. A trusted computing device, for example, has a nominal channel, realized through its API. A **side channel** is an unanticipated communication channel between the system and its environment, and by extension, the adversary. Measuring a trusted device's power consumption may for instance reveal a secret key stored on the device. Similarly, magnetic fields can degauss, and hence write to, the device's magnetic storage, and rowhammer attacks [16] can write to protected memory locations. These constitute side channels when the device's specification does not describe the device's behavior on these channels. Whether an adversary can exploit a nominal channel or a side channel to communicate with a system depends on the adversary's capabilities and the system's environment.

The relationship between specifications and requirements is central to security design and analysis. The problem security engineers solve is that of satisfying a security requirement in an adversarial environment. A solution to the problem consists of one or more systems that are deployed in the environment; cf. [13]. No single solution however solves all problems, because solutions are invariably contingent upon assumptions about the environment they address. We call these **environmental assumptions**. The following example illustrates this notion.

Example 3. Consider the scenario of Example 1. The lock system addresses the stated security requirement for the documents, provided various environmental assumptions are satisfied. These include that the only way to enter the lab is through the door controlled by the lock system. The adversary's capabilities affect this assumption's validity. Suppose that the lab has a window. If the adversary can climb in through the window, then this environmental assumption, and consequently the requirement, are violated.   △

The distinction between specifications and requirements is similar to the distinction between mechanisms and the policies they are intended to enforce [17]. A system is a mechanism that maps symbols to symbols, independently from the environment where it is deployed. This is of course desirable: a XACML policy enforcement point, an AES encryption module, and a lock system should do what their specifications promise, regardless of where they are deployed. In this sense, systems are just symbol manipulators, oblivious to their deployment environment. In contrast, a resource's security requirements impose certain (access) relations among actual entities in an environment. A symbol-manipulation entity can contribute to security only based on the environmental assumption that its input-output behavior is given an appropriate interpretation; cf. [22]. The following example illustrates this point.

Example 4. Consider the database scenario of Example 2. The database's specification (for its nominal channel) contributes to the security of the data under environmental assumptions, including: only copy editors have the role `copy_editor`, and there is no way to delete the data except by executing the API's `delete` command. These assumptions reflect the condition that the symbolic notions of a role and a command are interpreted appropriately in the context of the given scenario.   △

To capture the relationship between requirements, specifications, and environmental assumptions, we next introduce the notion of a security rationale.

3   Security Rationales

To address a security requirement RQ in an environment $\mathcal{E}$, we deploy a system in $\mathcal{E}$. The system's design, construction, and analysis are guided by a system specification SP. Moreover, deploying the system contributes to RQ if an environmental assumption EA holds true. Reducing RQ to SP and EA must clearly be justified: not every combination of SP and EA contributes to satisfying RQ. Security rationales embody such justifications. A **security rationale** for the four tuple $(\mathcal{E}, RQ, SP, EA)$ is a justification for the following condition: for any system S and adversary $\mathcal{A}$,

$$\mathcal{S} \models SP \ \wedge \ \mathcal{S}\|\mathcal{E}\|\mathcal{A} \models EA \quad \rightarrow \quad \mathcal{S}\|\mathcal{E}\|\mathcal{A} \models RQ \,. \tag{$\dagger$}$$

Here $\models$ and $\|$ represent satisfaction and composition. Moreover, SP, EA and RQ can each consist of multiple conjuncts, as illustrated subsequently in Example 5.

Intuitively, a rationale for $(\mathcal{E}, RQ, SP, EA)$ explains that, to address the requirement RQ in the environment $\mathcal{E}$ in the presence of the adversary $\mathcal{A}$, it suffices to deploy the system $\mathcal{S}$ provided that $\mathcal{S} \models SP$ and $\mathcal{S}\|\mathcal{E}\|\mathcal{A} \models EA$. The condition ($\dagger$) can be used from right to left to **reduce** a security requirement into a system specification and an environmental assumption sufficient for its establishment; see Fig. 1.

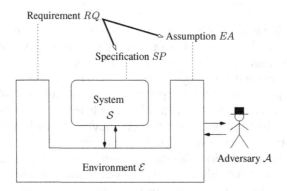

Fig. 1. A security rationale reduces (thick arrows) a security requirement to a specification and an environmental assumption. The validity of the environmental assumption depends on the adversary's capabilities. The adversary interacts (thin arrows) with the system over the environment.

In ($\dagger$), the premise $\mathcal{S} \models SP$ guides the design, construction, and analysis of systems, as mentioned above. The inclusion of the system $\mathcal{S}$ in the premise $\mathcal{S}\|\mathcal{E}\|\mathcal{A} \models EA$, which concerns the environmental assumptions, may appear counterintuitive. The adversary's role clarifies this point. The specification SP regulates the system's behaviors over its nominal channels. The adversary, against whom the requirement RQ must hold, may however interact with the system over side channels, i.e. channels not anticipated by SP. The system's interactions with the adversary over these channels must therefore be constrained as well. The environmental assumption EA includes such constraints.

In the following, we further clarify the condition ($\dagger$). First, there are numerous frameworks for formalizing, verifying, and testing the relation $\mathcal{S} \models SP$ in a precise manner. The two other relations in ($\dagger$) cannot however be readily formalized. In particular, the nebulous entities $\mathcal{E}$ and $\mathcal{A}$ often have no clear boundaries. This poses a major challenge to formalizing the notion of a security rationale. For the rest of this paper, we therefore treat the condition ($\dagger$) as an informal guideline and as a way to classify verification and refutation objectives.

Second, environmental assumptions and requirements have, in essence, the same type. In particular, (†) would be trivially satisfied if EA were RQ. The resulting reduction would however clearly not help with the requirement's analysis. Moreover, whether a statement is seen as a requirement or an assumption depends on the task at hand. For instance, in Example 3, the assumption that one cannot enter the lab through its window constitutes a requirement if we are interested in constructing the lab building. To satisfy this requirement we may, for example, install window bars; this would be preceded by a specification that would fix the window bars' construction in a way that is deemed sufficient to resist a given adversary.

Third, in the security literature, the environment is sometimes conflated with the adversary. To denote such an adversarial environment, let $\mathcal{E}^* = \mathcal{E}\|\mathcal{A}$. Then (†) boils down to $\mathcal{S} \models SP \wedge \mathcal{S}\|\mathcal{E}^* \models EA \rightarrow \mathcal{S}\|\mathcal{E}^* \models RQ$.

Finally, note that any security rationale can account for only a small set of entities and their interactions: we cannot reason about everything in the world. Therefore, any rationale inevitably relies upon the assumption that the excluded entities and interactions play no role in the requirement's satisfaction. This assumption in effect excludes certain adversarial actions. A prominent example is the assumption that the system has no side channels for communicating with the adversary; otherwise, its protection mechanisms can potentially be subverted. This further explains why we cannot dispense with $\mathcal{S}$ in $\mathcal{S}\|\mathcal{E}\|\mathcal{A} \models EA$ above.

The following example illustrates the above notions.

Example 5. Consider the R&D laboratory of Example 1. The requirement RQ states that only staff members may enter the lab. The lab has a door that is controlled by an electronic lock system. We reduce RQ to the requirement (SRQ): the lock opens the door only after a valid key is presented to it. The reduction relies on the following three environmental assumptions. (EA_1) Only staff members have a valid key. (EA_2) The door opens only after receiving the lock's signal[1]. (EA_3) The only way to enter the lab is through the door. Laws of logic justify the reduction.

$$
\begin{array}{ll}
(EA_1) & \text{hasValidKey}(X) \rightarrow \text{isStaff}(X) \\
(EA_2) & \text{doorOpensFor}(X) \rightarrow \text{signalFor}(X) \\
(EA_3) & \text{enterLab}(X) \rightarrow \text{doorOpensFor}(X) \\
(SRQ) & \text{signalFor}(X) \rightarrow \text{hasValidKey}(X) \\
\hline
(RQ) & \text{enterLab}(X) \rightarrow \text{isStaff}(X)
\end{array} \qquad (\star)
$$

The assumptions constrain the adversary's capabilities. The assumption EA_1, for instance, excludes numerous adversarial actions, both simple and elaborate. For example, according to EA_1, an adversary is not capable of bribing staff members to obtain a valid key. Similarly, the adversary cannot forge a valid key. The excluded adversarial actions clearly cannot be feasibly enumerated.

---

[1] We will abstract away from further temporal aspects in this example. For instance, once the door has been closed, it remains closed until the next signal arrives, and only one person can pass through the door while it is open.

In the final step, we reduce the requirement SRQ to the following specification for the lock system's nominal communication channel: (SP) the output signal open is produced only after receiving as input a key that belongs to the set validKeys. This reduction is justified by two assumptions EA_I and EA_S. The assumption EA_I states that the set validKeys, the input key, and the signal open are interpreted as expected, and that an entity cannot send a key to the lock system unless the entity has the key. The latter conjunct intuitively bridges the gap between the predicate hasValidKey(X) and the key the lock system receives from an entity X. The assumption EA_S states that all the communication channels between the lock system and the adversary are regulated by SP. It excludes for instance the possibility that the lock system has a hidden backdoor that bypasses its functionalities, or that disrupting the lock's electricity supply (which constitutes an "input" to the lock system) would leave the door open.

The arguments above constitute a security rationale for the tuple $(\mathcal{E}, RQ, SP, EA)$, where $\mathcal{E}$ is the lab's environment, RQ and SP are defined above, and EA is the conjunction of the assumptions EA_1, EA_2, EA_3, EA_I, and EA_S. Note that EA_I and EA_S cannot be expressed as assumptions on the environment alone: the lock system must be considered too.   △

The reduction steps carried out in a security rationale can be graphically represented as a **reduction tree**. Formally, a reduction tree is simply an and-or tree where the root denotes a security requirement, and the leaves are system specifications and environmental assumptions; see Fig. 2.

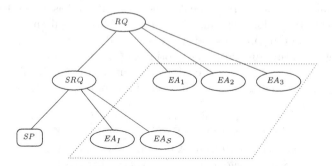

Fig. 2. A reduction tree for Example 5. Requirements and environmental assumptions are depicted as ovals, while specifications are depicted as rectangles. The dotted polygon contains the environmental assumptions. All branches here are and-branches.

Security rationales justify a reductive strategy for addressing security requirements. Such justifications can, in part, be formalized in a suitable proof system and justified using laws of logic, as (⋆) suggests. Laws of physics, such as nothing travels faster than the speed of light, can also be part of a security rationale. Formal models of the problem domain can assist security engineers with this task; cf. [1, 5, 14].

4 Security Cases

In this section, we introduce the notion of a security case. Intuitively, a security case explains why a rationale for a given security requirement is applicable to a concrete system in the presence of a specific adversary. Suppose we have a security rationale for the tuple $(\mathcal{E}, RQ, SP, EA)$. Then, deploying a system $\mathcal{S}$ in the environment $\mathcal{E}$ guarantees that RQ holds in the presence of an adversary $\mathcal{A}$ if the following condition holds:

$$\mathcal{S} \models SP \quad \wedge \quad \mathcal{S}\|\mathcal{E}\|\mathcal{A} \models EA. \qquad (\ddagger)$$

This statement is a direct consequence of (†). A **security case** is an argument for (‡)'s truth, for a concrete system $\mathcal{S}$ and a specific adversary $\mathcal{A}$. If the rationale's reduction steps are represented as a reduction tree, then a security case is an argument for the satisfaction of the tree's leaves.

Three remarks are due here. First, security cases are analogous both to safety cases, which argue for the safety of, say, vehicles (see for example ISO 26262-1:2011), and to dependability cases [12]. Security cases are (ideally) provided by security designers and analysts who explain why deploying $\mathcal{S}$ in the environment $\mathcal{E}$ solves the problem of addressing RQ in the presence of the adversary $\mathcal{A}$. For example, software verification techniques that demonstrate that a software system $\mathcal{S}$ satisfies a specification SP can contribute to a security case.

Second, the adversary's capabilities do not enter into a security rationale itself. Instead, once a specific adversary has been identified, for example, through risk analysis, the security case is given to justify the environmental assumptions' validity in the adversary's presence. The following example illustrates this point.

Example 6. Consider the security rationale of Example 5. This rationale does not depend on any particular adversary or system. However, the validity of the environmental assumptions critically depends on the adversary's capabilities, and the validity of the specification depends on the system's behaviors. For instance, the assumption EA_1 is violated if the adversary can threaten or bribe a staff member and thereby obtain a valid key. A security case here must argue that the given adversary, say, curious visitors, cannot violate this assumption. Similarly, the security case explains why a given lock system's behaviors over its nominal channels satisfy SP.   △

Third, whether or not a system satisfies a specification does not depend on the adversary's capabilities, as is evident in the condition (‡). This is a central point: systems can be designed, developed, and evaluated without knowledge about the environment where they will be deployed. That a system contributes to the security of protected resources in a given adversarial environment must be justified using security cases. This observation may seem counterintuitive as, for example, buffer overflow attacks and SQL injections, where an adversary takes control of a system by providing it with "malicious" inputs, are prevalent. We remark that these attacks exploit a system's inadequate handling of malformed inputs.

They can therefore be addressed by providing an adequate specification for the system's interface and requiring that the system satisfies it.

As mentioned in Sect. 3, the environmental assumptions always exclude certain adversarial actions. These exclusions cannot be justified without accounting for all interactions in the world, which is clearly infeasible. Therefore, to construct a manageable model of the environment, security cases invariably depend on **closed-world assumptions**, stating that what has not been considered plays no role in satisfying the given security requirement. Closed-world assumptions thus complete security cases in this merely formal sense [25]; see also Simon's *empty world hypothesis* [26]. The following example illustrates this point.

Example 7. Consider the security rationale of Example 5. The validity of EA_S, which states that the system has no side channels, depends on the adversary's capabilities and the system's behaviors. Suppose the lock system leaves the door open if its power is disrupted. The assumption EA_S is then not valid in the presence of an adversary who can cut off the system's power. It might however be valid for a weaker adversary. A security case here explains why a given system and adversary cannot communicate over this particular side channel in the environment $\mathcal{E}$. Alternatively, if the system leaves the door locked when the power is disrupted, then the security case can argue that although an adversary can affect the system over this side channel, the result does not adversely affect RQ's satisfaction. To complete the argument for EA_S's validity, all possible channels should be considered. These, however, cannot all be enumerated and argued for. The security case must therefore ultimately rely on the closed-world assumption that the considered side channels are the only ones relevant for RQ's satisfaction.   △

5   Security Testing

In this section, we define functional testing and security testing, and clarify their relationship. We then introduce two types of security tests, and illustrate them through examples from practice. Finally, we discuss vulnerabilities and their remediation, associated with these two test types.

By **functional testing** we refer to any process aimed at refuting the hypothesis that a system satisfies its (functional, security, or other) specifications. That is, given a system $\mathcal{S}$ and a specification SP, functional testing aims at refuting the hypothesis $\mathcal{S} \models SP$. Here we do not distinguish between black-box and white-box analysis. By **security testing** we refer to any process aimed at refuting the hypothesis $\mathcal{S}\|\mathcal{E}\|\mathcal{A} \models RQ$, for a system $\mathcal{S}$, environment $\mathcal{E}$, adversary $\mathcal{A}$, and security requirement RQ. Note that the purpose of both types of testing is to refute a hypothesis, rather than to verify it. This understanding, which is well-established in the literature [9, 21], sharply separates constructing security cases from security testing.

We remark that our notion of functional testing is more general than the term's conventional denotation in the literature, e.g., [2, 4, 21]. This is simply because, in our theory, a specification need not be confined to a system's desired functions, distilled, say, from its use cases. A bound on the system's delay in producing outputs, as well as a threshold on the system's electromagnetic radiation level are examples of system specifications. Tests aiming to refute these specifications therefore constitute functional tests in our theory. Conventionally, they are usually not deemed as functional tests because a system's delays and radiation levels are typically not considered to be part of a system's functionality; see also [10] for the murky boundary between functional and non-functional specifications. In our theory, the essence of a functional test is that it applies to the system's communication channels that are described in and constrained by the system's specification. To avoid confusion, we refer to the conventional forms of functional tests as **restricted functional tests**.

We now turn to security testing. Suppose that, in an environment $\mathcal{E}$, a requirement RQ is intended to be satisfied based on a rationale for $(\mathcal{E}, RQ, SP, EA)$. Let $\mathcal{S}$ be a system deployed in $\mathcal{E}$ that is intended to satisfy (‡), in the presence of an adversary $\mathcal{A}$. Perhaps surprisingly, refuting either conjunct of (‡) does *not* entail refuting $\mathcal{S}\|\mathcal{E}\|\mathcal{A} \models RQ$, which is the objective of security testing. However, the refutation of $\mathcal{S} \models SP$ or $\mathcal{S}\|\mathcal{E}\|\mathcal{A} \models EA$ does, of course, demonstrate that the intended rationale's premises are false for the system $\mathcal{S}$ and the adversary $\mathcal{A}$: the condition (†) is true due to the failure of its antecedent and one cannot construct a security case here. Therefore, the refutation of one of (‡)'s conjuncts *suggests* that the requirement RQ is violated because it is unlikely that RQ is satisfied due to unintended causes. This observation motivates the following hypothesis: If $\mathcal{S}\|\mathcal{E}\|\mathcal{A} \models RQ$, for a system $\mathcal{S}$ and an adversary $\mathcal{A}$, then $\mathcal{S} \models SP$ and $\mathcal{S}\|\mathcal{E}\|\mathcal{A} \models EA$. We call this the **intentional security hypothesis**, in short **H**.

Intuitively, **H** states that a security requirement is never satisfied unintentionally: a system addresses a security requirement by design, not by accident. Note that the hypothesis amounts to the condition (†)'s converse. This is expected: the condition (†) supports constructing security cases for verifying a security requirement's satisfaction. Security testing, whose goal is to refute the requirement's satisfaction, must rely on (†)'s converse, namely **H**. We show in Sect. 5.2 that **H** has been tacitly assumed in the literature.

5.1   S-Tests and E-Tests

Based on **H**, the tester can refute the hypothesis $\mathcal{S}\|\mathcal{E}\|\mathcal{A} \models RQ$ by refuting one of (‡)'s conjuncts. This results in the following two types of security tests.

S-Tests: Test the system with respect to its specification.

A test of this type, called an S-test, is intended to refute $\mathcal{S} \models SP$, which is an instance of functional testing. Tools and techniques for generating and automating functional tests can therefore be used here; see for example [2,4]. Note that S-tests pertain to symbol manipulating entities and are therefore independent of the adversary. Moreover, restricted functional tests are instances of S-tests. For example, suppose a radio transmission system must satisfy the specification that transmitted messages should be encrypted with 1024-bit keys. Restricted functional tests can be applied to this system because the specification describes a use case of the system.

E-Tests: Test the validity of the environmental assumptions.

A test of this type, called an E-test, is intended to refute the hypothesis $S\|\mathcal{E}\|\mathcal{A} \models EA$. Refuting this hypothesis is what distinguishes security testing from functional testing. Namely, functional tests pertain to a system's behaviors over its interface, described by a specification. In contrast, security E-tests apply not only to systems but also to a nebulous environment and an adversary with no interface (see Sect. 2). Therefore, testing the validity of environmental assumptions cannot be reduced to providing an input and observing an output over a definite interface. These tests are therefore not an instance of functional tests: they pertain to actual entities in the world. In particular, they depend on the adversary's capabilities. The diagram of Fig. 3 illustrates the relationship between these two types of tests.

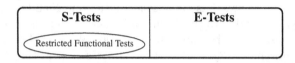

Fig. 3. S-tests, whose purpose is to refute the hypothesis that a system satisfies its specification, include restricted functional tests, which apply to the functionalities the system must offer. E-tests, in contrast, attempt to violate an environmental assumption in an adversarial environment.

Example 8. Consider the scenario of Example 5, with the reduction tree depicted in Fig. 2 for the requirement RQ. The purpose of security testing is to refute the hypothesis that RQ is satisfied in the presence of a given adversary $\mathcal{A}$. As previously explained, refuting the validity of the reduction tree's leaves (which is the goal of E-tests and S-tests) does *not* entail that RQ is violated, because RQ can be satisfied due to unanticipated reasons. It is only by **H** that design errors imply RQ's violation. We consider the task of violating some of Fig. 2's leaves in the following.

To violate the leaf SP, the tester tries to refute the hypothesis that the lock system satisfies the specification SP. The tester may, for instance, input very large keys into the lock system, where a key is a sequence of bits. If a buffer overflow is discovered, then the adversary might be able to take control of the lock and produce an **open** signal without possessing a valid key. Note that to violate SP the tester need not elicit the adversary's capabilities. The lock system must satisfy SP on its nominal channel for all possible inputs and outputs. This is an S-test. In contrast, the tests below are E-tests.

To violate EA_I, the tester checks if the lock system's local variables are misinterpreted, for example, the set `validKeys` might not actually consist of valid keys. If a staff member leaves the R&D team, then his key might still be stored in `validKeys`. The tester also checks whether the lock system is susceptible to replay attacks. If so, then EA_I is violated because the adversary can simply record the interaction between a valid key and the lock system and later send a valid key to the lock without legitimate possession of the key. Whether these scenarios refute EA_I's validity in the presence of a given adversary clearly depends on the adversary's capabilities.

To violate EA_2, the tester may try to intercept the communication between the lock and the door to inject an `open` signal. The tester may also assess the feasibility of breaking, or unhinging, the lab's door. To violate EA_3, the tester may try climbing through the window. The feasibility of these attacks naturally depends on the environment and the adversary's capabilities. If, for instance, the window is barred and the adversary neither has a metal saw nor is capable of squeezing through the bars, then climbing in through the window is infeasible, indicating that EA_3 is not refuted in these scenarios. $\triangle$

As the above example illustrates, when testing environmental assumptions and requirements, the tester must take the adversary's capabilities into account. For each goal the tester may ask whether the adversary can achieve it. Specific goals, such as unhinging a door, lead to specific questions regarding the adversary's capabilities. General goals, such as violating the assumption EA_2, which excludes a wide range of adversarial actions, lead to generic questions that cannot be directly answered. The tester must then elicit a list of attack scenarios and determine whether the adversary can realize them. This list can be developed by brainstorming and using experience with similar requirements. This can also be aided by consulting sources like [7], which go beyond enumerating common system vulnerabilities and consider malicious interaction from the environment. The investigated scenarios will however never be complete, because accounting for all possible interactions in the world is infeasible. Security testing is therefore an open-ended processes, hence inherently incomplete. Note that this incompleteness is orthogonal to the incompleteness of functional tests, which is rooted in the infinite cardinality of the domains where test inputs are selected. The difference is that in functional testing one picks inputs from a delimited, albeit infinite, domain, whereas E-tests come from a domain with no boundaries.

The following example illustrates the essentially unlimited creativity required by a security tester to anticipate all possible attack scenarios.

Example 9. A British secret operation, known as the Four Square Laundry affair, was carried out in Northern Ireland to collect information about the residents of a troubled neighborhood [20]. A rogue laundry service van visited the neighborhood regularly, and sent the collected laundry for various tests and inspections before washing it. The tests included checking for traces of explosive material or blood. The service also noted changes in the amount or kinds of clothing sent by each household for washing, which could indicate the presence of guests, and so forth. $\triangle$

The separation between S-tests and E-tests explains why security testing is harder than functional testing. A system specification describes the system's behaviors over its interface. It can therefore be used to construct functional tests, for example S-tests, independently from the adversary's capabilities and the environment in which the system is deployed. When it comes to security testing, the tester must also check the validity of requirements in an adversarial environment. Environments and adversaries are nebulous entities, with no clear interface. How, say, an environmental assumption can be violated depends on the adversary's capabilities, the environment's properties, and the system's behaviors. E-tests for checking an assumption's validity are only as thorough as the attack scenarios the tester anticipates.

5.2   S-Tests and E-Tests in Practice

Applying security testing in practice is challenging. If the security case (or the security rationale) intended to guarantee a resource's security is unavailable, then the tester must reconstruct, or approximate, it. This includes eliciting the adversary's capabilities and explicating specifications and environmental assumptions. These tasks are notoriously hard in practice; see for example [14,30]. Even when the security case and the security rationale are available, security E-tests amount to anticipating how the adversary can invalidate an environmental assumption or a requirement. This task defies prescriptive guidelines such as those available for functional testing. The effectiveness of E-tests therefore depends largely on the tester's creativity and resources; see the Four Square Laundry example above.

   These observations imply that security testing is largely a manual task that defies specific, thorough guidelines. It is therefore not surprising that existing methods fall short when it comes to E-tests. Below, we substantiate this claim by showing that existing security testing techniques have little to say in this regard.

Risk-Based Security Testing. Risk-based security testing [18,19,24] starts by explicating system specifications from risk analysis, misuse case diagrams, and other design and analysis documents. Roughly speaking, a risk corresponds to a security requirement that demands the risk's mitigation. The countermeasure that is intended to reduce or eliminate the risk can then be seen as a specification that defines how a system must implement the mechanisms that address the corresponding requirement. Afterward, risk-based security testing reduces security testing to S-tests applied to the mitigation mechanisms. E-tests are absent here because the environmental assumptions and the adversary that would make up a corresponding security rationale are not identified.

Fuzz Testing and Fault Injection. Fuzz testing [11,27] and fault injection techniques [29] aim at refuting generic system specifications such as: the system does not access unallocated memory areas. That is, they refute $\mathcal{S} \models SP_g$, for generic specifications SP_g and they therefore amount to S-tests. These techniques can be seen as generating S-tests guided by security-relevant fault models. For example, programs often fail to check their inputs length or format, and they have inadequate exception handling when dependency relations fail. Such fault models reflect how an adversarial environment may interact with the system. Consequently, they give rise to tests that are tailored to violate security-relevant specifications. E-tests are nonetheless absent here, simply because the resulting tests pertain to a system's nominal channels only; they do not analyze side channels and environmental assumptions.

Vulnerability-Driven Security Testing. Tests that try to identify a known, anticipated vulnerability in a particular system are sometimes said to be driven by that vulnerability. Since these tests are concerned with systems, they are clearly S-tests. OWASP's security test patterns fall under this class of security tests [24].

A more elaborate example of vulnerability-driven security testing is the NIST proposal [23] that associates security tests with security features of cryptographic modules. An example is that "environmental failure protection [...] features shall protect the cryptographic module against unusual environmental conditions or fluctuations (accidental or induced) outside of the module's normal operating range that can compromise the security of the module" [23]. The document associates a number of tests to this security feature, including "the tester shall extend the temperature and voltage outside of the specified normal range and determine that the module either shuts down to prevent further operations or zeroizes all plaintext secret and private keys and other unprotected [critical security parameters]". The NIST proposal is helpful in explicating how a module should behave in abnormal conditions, but it cannot describe under which assumptions on the adversary's capabilities and the environment a security requirement can be translated into the specifications subjected to functional tests. Note that although the NIST's suggested tests are not instances of restricted functional tests, they nevertheless apply to a system's communication channel that has been regulated by the system's specifications. They are therefore S-tests. E-tests are absent here as well.

In short, existing security testing methods and tools ignore E-tests. Since they all address security specifications, they tacitly assume that if the system violates its specification, then the security requirement is also violated. This amounts to the intentional security hypothesis, introduced in Sect. 5, about which the literature has not been explicit. The aforementioned shortcomings should not be construed as a criticism of the existing techniques' value. Rather, our security test types should be seen as a tool for delimiting their scope and reach. As mentioned before, E-tests depend on the adversary and target closed-world environmental assumptions that are impossible to delimit. It is therefore not surprising that, in contrast to S-tests, E-tests do not admit automation.

We conclude this section with two remarks. First, adversary models themselves are not subjected to E-tests (or S-tests). For example, discovering that a safe can be opened using standard office equipments demonstrates that the assumption that a curious co-worker could not open the safe has been false all along. It however does not help us decide whether a curious co-worker is a suitable adversary model for the documents protected by the safe. In general, E-tests and S-tests do not account for flaws rooted in unelicited requirements or weak attacker models. Requirements and the adversary are the parameters with respect to which these test types are defined. They are not themselves subject to these tests.

Second, the observations above shed light on the notion of adequacy for security tests. It is immediate that the adequacy of S-tests can be defined based on functional adequacy measures, such as coverage [31] and mutation analysis [8], and security-specific metrics such as [28]. The adequacy of E-tests, however, is an entirely different matter. Ideally, the validity of each environmental assumption must be "adequately" tested. These assumptions are however not only hard to explicate, but their validity also relies upon closed-world assumptions that can never be thoroughly tested. No finite set of security tests can therefore constitute an adequate set of E-tests. We return to this conundrum in Sect. 6.

5.3   Vulnerability Remediation

We can classify security vulnerabilities based on our test types. Let S be a system, $\mathcal{E}$ an environment, $\mathcal{A}$ an adversary, and RQ a security requirement. By a security **vulnerability** we refer to any cause for the violation of the security requirement, i.e., the violation of $S\|\mathcal{E}\|\mathcal{A} \models RQ$. Clearly this notion of a vulnerability is more general than, say, programming flaws.

We introduce two classes of vulnerabilities: S-vulnerabilities, and E-vulnerabilities. **S-vulnerabilities** are those vulnerabilities in the system S that lead to a violation of its specification SP. Due to **H**, these are indeed vulnerabilities as they lead to a violation of RQ. These vulnerabilities are revealed through S-tests, and remediating them amounts to fixing the system. **E-vulnerabilities** are those vulnerabilities that invalidate the environmental assumption EA. That is, vulnerabilities in this class cause the relation $S\|\mathcal{E}\|\mathcal{A} \models EA$ to fail. These too are vulnerabilities due to **H**, as they lead to a violation of RQ. To remediate an E-vulnerability, fixing the system alone is insufficient. The system must be re-engineered and the security rationale must be updated.

After fixing a system to address an S-vulnerability, only the system must be analyzed using S-tests; carrying out E-tests is unnecessary. Moreover, since the system's specification has not changed, these S-tests can be seen as regression tests. However, after re-engineering the design and updating the security rationale to address an E-vulnerability, both S-tests and E-tests must, in general, be carried out to analyze the security of the new design. Since these tests must address the new design's specification and environmental assumptions, they cannot be seen as simple regression tests. The following example illustrates these classes.

Example 10. Consider the scenario of Example 5, analyzed in part in Example 8. A window through which the adversary can enter the office is an E-vulnerability. To address it, new systems, such as window bars, can be installed in the environment. This system must then be tested with respect to its specification. As a second example, suppose that former staff members still have keys that are accepted by the lock system. This causes the environmental assumption EA_1 to fail. To address this E-vulnerability, the specification SP must be extended with the specification of a suitable key revocation mechanism. This likely entails changing the lock system entirely, installing a key revocation server, and so forth. These systems must then be tested with respect to the extended specification.   △

We have ignored flaws due to changes in the requirements or a mismatch between the stake-holder's expectations and the requirements. Although such cases are common in practice [15], they fall outside this paper's scope.

6   Concluding Remarks

Starting with the fundamental distinction between a system specification and a security requirement, we have provided a simple theory of security testing. Its ingredients — security rationales, security cases, the intentional security hypothesis, S-tests and E-tests — provide a basis for explaining the verification and refutation of security requirements in general, and security testing in particular. Our theory highlights the limitations of many testing and other quality assurance methods for reasoning about the security of systems: the vast majority of methods target the relationship between systems and their specifications, but not the assumptions made on their environments.

Targeting environmental assumptions is hard. One must ultimately resort to a closed-world assumption and posit that the adversary can only interact with the system and the environment in limited ways. As a result, the set of possible counter-examples is not only infinite, its domain cannot be precisely delimited. Hence, E-tests, which target environmental assumptions, defy automation and systematization.

The above difficulties raise the question of how practitioners can best approach E-testing and judge the quality of the resulting E-tests. We do not have the answer to this question. And any answer will certainly not be in terms of a logical method or formalism with conventional notions of completeness or coverage. Since testers' creativity and experience play a central role in refuting environmental assumptions, there is value in studying and learning from attacks [3,6]. We believe our theory can help in this regard as it suggests a frame of reference for documenting, classifying, and reusing the knowledge obtained through such studies. This includes explicating the assumptions that have been violated, associating common assumptions with attacks, and exploring possibilities for generalizations. Moreover, threats on different classes of systems and environments can be cataloged along with countermeasures; see, e.g., [7]. These catalogs can be analyzed using this frame of reference, highlighting cases where the attacks and mitigation methods refer to assumptions or specifications that are left implicit. Making these explicit can contribute to the body of knowledge developed around E-tests.

Security testing requires an open mind and a vivid imagination. It goes far beyond the well-charted territory of functional tests. One must raise one's sights to look beyond the machine and target the world as well.

Acknowledgment. We thank Peter Müller and Petar Tsankov for their comments on this paper.

References

1. Abrial, J.-R.: Modeling in Event-B: System and Software Engineering, 1st edn. Cambridge University Press, New York (2010)
2. Ammann, P., Offutt, J.: Introduction to Software Testing. Cambridge University Press, New York (2008)
3. Basin, D.A., Capkun, S.: The research value of publishing attacks. Commun. ACM **55**(11), 22–24 (2012)
4. Beizer, B.: Software Testing Techniques, 2nd edn. Van Nostrand Reinhold, New York (1990)
5. Bjorner, D.: Software Engineering 3: Domains, Requirements, and Software Design. Texts in Theoretical Computer Science. An EATCS Series. Springer, New York (2006)
6. BSI. A penetration testing model, The German Federal Office for Information Security(2003)
7. BSI. IT Grundschutz Kataloge, (Version: 14). The German Federal Office for Information Security (2014)
8. DeMillo, R.A., Lipton, R.J., Sayward, F.G.: Hints on test data selection: help for the practicing programmer. Comput. **11**(4), 34–41 (1978)
9. Dijkstra, E.W.: Notes on structured programming. Technical report T.H. Report 70-WSK-03, Technological University Eindhoven, April 1970
10. Glinz, M.: On non-functional requirements. In: 15th IEEE International Requirements Engineering Conference, RE, pp. 21–26. IEEE Computer Society (2007)
11. Godefroid, P., Levin, M.Y., Molnar, D.A.: SAGE: whitebox fuzzing for security testing. ACM Queue **10**(1), 20 (2012)
12. Jackson, D.: A direct path to dependable software. Commun. ACM **52**(4), 78–88 (2009)
13. Jackson, M.: The world and the machine. In: Proceedings of the 17th International Conference on Software Engineering, ICSE 1995, pp. 283–292. ACM, New York, NY, USA (1995)
14. Jackson, M.: Problem Frames. Addison-Wesley, Reading (2001)
15. Johnson, A.: Hitting the Brakes: Engineering Design and the Production of Knowledge. Duke University Press, London (2009)
16. Kim, Y., Daly, R., Kim, J., Fallin, C., Lee, J.-H., Lee, D., Wilkerson, C., Lai, K., Mutlu, O.: Flipping bits in memory without accessing them: an experimental study of DRAM disturbance errors. In: ACM/IEEE 41st International Symposium on Computer Architecture, ISCA, pp. 361–372. IEEE Computer Society (2014)
17. Levin, R., Cohen, E., Corwin, W., Pollack, F., Wulf, W.: Policy/mechanism separation in Hydra. SIGOPS Oper. Syst. Rev. **9**(5), 132–140 (1975)
18. McGraw, G.: Software Security: Building Security In. Addison-Wesley Professional, Boston (2006)
19. Michael, C.C., van Wyk, K., Radosevich, W.: Risk-based and functional security testing, Accessed 05 July 2013. https://buildsecurityin.us-cert.gov/
20. Moloney, E.: A Secret History of IRA. Penguin, Canada (2003)
21. Myers, G., Sandler, C., Badgett, T.: The Art of Software Testing, 3rd edn. Wiley, New York (2011)
22. Nelson, R.: What is a secret - and - what does that have to do with computer security? In: Proceedings of the Workshop on New Security Paradigms, pp. 74–79. IEEE (1994)

23. Derived test requirements for FIPS PUB 140–2, security requirements for crypto-graphic modules, NIST, CSEC and CMVP Laboratories Draft (2011)
24. OWASP. Testing guide v. 4, Accessed on 9 March 2014. https://www.owasp.org
25. Reiter, R.: On closed world data bases. In: Gallaire, H., Minke, J. (eds.) Logic and Data Bases, pp. 55–76. Plenum Press, New York (1978)
26. Herbert, A.: Simon.: The architecture of complexity. Proc. Am. Philos. Soc. **106**(6), 467–482 (1962)
27. Takanen, A., DeMott, J., Miller, C.: Fuzzing for Software Security Testing and Quality Assurance, 1st edn. Artech House Inc., Norwood (2008)
28. Tsankov, P., Dashti, M.T., Basin, D.A.: Semi-valid input coverage for fuzz testing. In: International Symposium on Software Testing and Analysis, ISSTA, pp. 56–66. ACM (2013)
29. Voas, J., McGraw, G.: Software Fault Injection. Wiley, New York (1998)
30. Wang, R., Zhou, Y., Chen, S., Qadeer, S., Evans, D., Gurevich, Y.: Explicating SDKs: Uncovering assumptions underlying secure authentication and authoriza-tion. In: Proceedings of the 22nd USENIX Conference on Security, pp. 399–414 (2013)
31. Zhu, H., Hall, P.A.V., May, J.H.R.: Software unit test coverage and adequacy. ACM Comput. Surv. **29**(4), 366–427 (1997)

Progress-Sensitive Security for SPARK

Willard Rafnsson[1]([✉]), Deepak Garg[2], and Andrei Sabelfeld[3]

[1] Carnegie Mellon University, Pittsburgh, USA
willardthor@cmu.edu
[2] Max Planck Institute for Software Systems, Kaiserslautern
and Saarbruecken, Germany
[3] Chalmers University of Technology, Gothenburg, Sweden

Abstract. SPARK 2014 is a safety critical language subset of Ada developed by Altran and used for developing safe and secure software by major industrial players in the aviation, commercial, medical, space, and military domains. This paper puts a spotlight on the SPARK flow analysis. Articulating the boundaries of what is achievable by the analysis, we spell out attacks to exploit termination, progress, resource exhaustion, and timing channels. We harden the analysis to achieve security against stronger attackers, with the focus on progress-sensitive security as our baseline. Instead of redesigning and reimplementing the enforcement, we leverage known flow analyses for weaker attackers by a transform on program dependence graphs. We establish the soundness of this approach for a core language and demonstrate that it can be applied as a source-to-source transform of SPARK code when modifying the compiler is undesirable. A case study, derived from publicly available code for a control unit of a missile, indicates the usefulness of the approach.

1 Introduction

SPARK is a safety critical language subset of Ada developed by Altran and used by industry in the aviation, commercial, medical, space, and military domains. Applications range from programming jet engines (Lockheed Martin) to military aviation (EuroFighter), UK's air traffic control system (Altran), cross-domain guards (Rockwell Collins), smart card OS (MULTOS), biometrics software (NSA), and multi-level security systems (Secunet) [42].

SPARK 2014. A recent major overhaul of SPARK has led to SPARK 2014 [44], a language and accompanying tools for developing safe and secure software. To aid security verification, a flow analysis is integrated in the compiler to track information flow in SPARK programs and is used in applications like separation kernels [29] and multi-level workstations [39].

Information Flow Security. The security model of SPARK programs draws on information flow tracking. The goal is to track the propagation of data from sources (inputs) to sinks (output) as information is manipulated by programs. For systems whose sources and sinks are classified into secret and public (or more

© Springer International Publishing Switzerland 2016
J. Caballero et al. (Eds.): ESSoS 2016, LNCS 9639, pp. 20–37, 2016.
DOI: 10.1007/978-3-319-30806-7_2

complex classifications [18]), the baseline policy is *noninterference* [16,21] that prevents secret inputs from affecting public outputs.

There are different ways in which noninterference can be broken, corresponding to different information flow channels. An *explicit* flow results from a data flow from the right-hand side to the left-hand side in an assignment. An *implicit* [19] flow is via control flow: for example, branching on a secret and outputting different public values in the branches is an implicit flow that leaks information about the secret without any explicit leaks. The *termination channel* [48] is another source of potential leaks: a program that loops on a secret and outputs a public value on exiting the loop reveals whether the loop has terminated and therefore leaks information about the secret guard. A generalization of this channel is the *progress channel* [7] that can be used to leak information about secrets via the progress of public outputs. In contrast to the one-bit termination channel, this channel allows leaking secrets in their entirety by brute force attacks [7]. Other channels of interest are *resource exhaustion* and *timing* [37], which allow the attacker to learn secret information by observing abnormal behavior and time variation, respectively.

SPARK Security Examined. Usage of the SPARK flow analysis in industry is encouraging. It makes the following questions important. What attacks does it prevent? How can it be extended to achieve security against more powerful attackers? Can it lead to a general methodology applicable to similar analyses?

This paper puts a spotlight on the flow analysis in SPARK GPL 2015. Released April 28 2015, it is, as of January 1 2016, the latest GPL edition of SPARK 2014. To articulate the boundaries of what it can achieve, we demonstrate that the analysis successfully tracks explicit and implicit flows and spell out attacks to exploit termination, progress, resource exhaustion, and timing channels.

SPARK Security Improved. With the goal to harden the analysis against stronger attackers, we set our baseline at the progress-sensitive security policy [8,12,31,32]. This policy is a natural generalization of noninterference to programs with output, in contrast to its progress-insensitive counterpart [7,8,12] that needs to carve out leaks due to computation progress. Further, as mentioned earlier, ignoring the progress channel implies opening up brute force leaks that may extract secrets in their entirety. Our key goal is to design a general approach that allows leveraging existing analysis and tools for explicit and implicit flows, such as SPARK flow analysis to enforce the stronger progress-sensitive security. This goal is particularly important given the state of the art where the vast majority of the information flow tools in addition to SPARK (e.g. FlowFox [22], JSFlow [24] and IFC4BC [10] for JavaScript, Jif [30], Paragon [14] and JOANA [23] for Java, FlowCaml [40] for Caml, all discussed in Sect. 8) are currently only able to enforce progress-insensitive security.

Achieving Progress-Sensitive Security. With this main goal at hand, the core idea for enforcement is as follows. We set out to leverage two independent components: graph-based analysis for explicit/implicit flows and termination analysis. There have been many successful efforts on developing such components, with the above-mentioned information flow tools for the former and much encouraging progress on the latter [17,27,45]. Facilitated by the latter, Moore et al. [28] show how to use *termination oracles* for termination-sensitive information flow analysis. Similarly, we parametrize our approach in the termination analysis to determine which loops terminate and perform a graph transform on the program dependence graph where we represent termination and progress flows by injecting additional edges going out of potentially diverging loops. This lets us reuse graph-based analyses, e.g. the one by Horwitz [26] that is behind the SPARK flow analysis, since we can simply apply it to the transformed graph. The elegance of this approach is that even if a trivial termination analysis ("all loops might diverge") is plugged into the framework, we get a sound and meaningful enforcement of progress-sensitive noninterference corresponding to Smith's and Boudol and Castellani's canonical restrictions for the termination channel [13,41].

We establish the soundness of this approach for a core language and demonstrate that it can be applied as a source-to-source transform of SPARK code when modifying the compiler is undesirable. We apply the source-to-source transformation on a case study with a control unit of a missile, loosely based on publicly available code by Hilton [25]. We formulate desired properties, such as "the orientation sensors may not affect self-destruction", in terms of information-flow policies and demonstrate how our enforcement verifies these properties.

Contributions. The paper's major contributions are (i) the attacks on the SPARK flow analysis to demarcate its boundaries, (ii) leveraging a progress-insensitive SPARK flow analysis (by changing the analyzer conservatively, or through source-to-source transformation) to enforce progress-sensitive noninterference, and (iii) a case study with a missile code controller to demonstrate the usefulness of the approach. While our work is motivated by improving the SPARK flow analysis, we believe the overall idea is portable to other approaches and tools. Thus, we present our results more generally. For example, our framework is graph-based, which opens up possibilities for natural adoption to other graph-based tools such as JOANA [23]. Combining the major and minor contributions, the paper contributes the following:

- Attacks illustrating the boundary of what SPARK's flow analysis can achieve, leaking via termination, progress, resource exhaustion, and timing (Sect. 2);
- A policy framework for expressing progress-(in)sensitive security conditions (Sect. 4) for an imperative language (Sect. 3) at the heart of SPARK;
- A general graph-based approach for dependency analysis using termination oracles to achieve progress-sensitive security (Sect. 5);
- A general graph-based framework for dependency analysis of reactive programs, also distinguishing output content from output presence (Sect. 5);
- Soundness of the graph-based enforcement for the core language (Sect. 5);
- Source-to-source transform leveraging existing graph-based flow analyses in a modular fashion (Sect. 6) to achieve progress-sensitive security; and

– Case study with a control unit of a missile that verifies desired security properties (Sect. 7).

Our code compiles with GNAT GPL 2015. Released April 28 2015, it is, as of January 1 2015, the latest GPL edition of the Ada 2012 compiler. Our code can be found online [49].

Scope. While resource exhaustion and timing channels are important, we leave their consideration and exploration of more sophisticated attacks on the SPARK security analysis for future work. Typically, attacks on these channels require more efforts from the attacker and result in attacks with lower bandwidth [37]. For similar reasons, we leave declassification [38] out of the scope of the present work. Although important and wished for by the SPARK developers [35], the flow analysis in SPARK is useful even without declassification, as indicated by its deployments by Secunet [29, 39] and as highlighted by our case study.

2   Attacks

We begin by providing evidence that SPARK's flow analysis is termination-, progress-, and timing-insensitive. We do this by providing minimal example programs which pass analysis yet leak information. Since SPARK's flow analysis implementation has no proof of soundness, this helps us identify the property it is meant to enforce, and thus how to improve it.

All our examples share the same structure: a *Main* file that reads a byte from standard input, and invokes a procedure **Leak** on said byte.

```
procedure Leak (H : in out Byte)
   with Global  => (In_Out => Standard_Output),
        Depends => (H => H, Standard_Output => Standard_Output);
```

This specification states that **Leak** performs I/O on its parameter H and the global **Standard_Output**, and that output on **Standard_Output** *only* depends on **Standard_Output**. That last bit is the *flow policy* of **Leak**. Our attacks, which differ only in how they implement **Leak**, aim to violate this flow policy while passing analysis, by making output on **Standard_Output** depend on H.

The source code for our attacks is in the appendix. In this section, we focus on the two attacks most relevant to our technical contributions (termination and progress), and summarize the other attacks when closing the section.

Termination. A flow analysis is *termination sensitive* when it tracks whether a value can affect termination behavior. To gauge whether SPARK's flow analysis is termination sensitive, we design **Leak** on the left of Fig. 1 such that the *presence of output* on standard output depends on whether the program enters an infinite loop, which depends on H. *Main* passes analysis with this implemetation of **Leak**. However, invoking **Leak** on input values 1 and 2 produces different observable behavior: with input 1, we see '!' on the standard output; with input 2, *Main* diverges. Thus, SPARK's flow analysis is *termination insenstive*.

```
 0 procedure Leak (H : in out Byte) is       0 procedure Leak (H : in out Byte) is
   begin                                         K : Byte := 0;
     H := H;                                    begin
                                                  H := H;
   -- if H is even: nontermination.              while True loop
 5 -- else terminate with output "!".      5        Write (Standard_Output , K);
   if H mod 2 = 0 then                             if K >= H then
      while True loop                                 while True loop
         H := H;                                         H := H;
      end loop;                                       end loop;
10 end if;                                  10    end if;
   Write (Standard_Output ,                       K := K + 1;
         Character 'Pos('!'));                   end loop;
   end Leak;                                   end Leak;
```

Fig. 1. Termination leaks (left) and progress leaks (right) in SPARK

Progress. A termination insensitive flow analysis permits one bit to leak through termination observations. Programs that pass such an analysis can leak much more when a value can affect the progress the program makes on producing its intermediate output [7]. A flow analysis that tracks such flows is *progress sensitive*. The right panel of Fig. 1 shows the brute force attack by Askarov et al. [7] modelled in SPARK. Here, Leak outputs on standard output all characters in their ASCII number order up to the character numbered H, and diverges, thus leaking all of H. Again, this program passes SPARK's flow analysis, indicating that the flow analysis is *progress insensitive*.

Summary. We studied SPARK's flow analysis under three additional attacks:

- *Resource Exhaustion:* We replace nontermination in the progress attack with abnormal termination (a stack overflow). We give two examples; one allocates an array too large to fit on the stack, the other creates infinitely many stack frames through infinite mutual recursion. Both pass analysis.
- *Timing:* A flow analysis is *timing sensitive* when it tracks whether a value can affect the time an effect occurs. We replace nontermination in the termination attack with a computation which takes considerable time (selection-sorting 2^{16} bytes). The attack passes analysis.
- *Explicit and Implicit flows:* As a sanity check, we provide two implementations of Leak: one creates an explicit flow from H to Standard_Output, the other an implicit flow. The attacks do *not* pass analysis.

Since SPARK detected the explicit and implicit flows, but failed to detect our other attacks, it appears that SPARK enforces progress-insensitive security. As demonstrated above, whole secrets can leak through progress. In this paper, we harden SPARK's flow analysis to detect progress leaks, to enforce progress sensitive security. Addressing the other attacks is out of scope of this paper.

3   Programs and Policies

We explain our ideas and results using a simple while language with flow annotations, inputs, outputs, and arrays, which is a stripped down version of SPARK. For a formal semantics and illustrative examples, see the appendix.

Programs. The syntax for our language is given in Fig. 2. Let p range over programs, b over blocks, x over array names, e over expressions, n over integers, c over channels, and $\odot$ over (total) binary integer operators. Here, $x[e]$ denotes index e in array x. To model non-array variables, we write x as syntactic sugar for $x[0]$. Statement $c \mathrel{<-} e$ outputs integer e to channel c, and $c \mathrel{->} x[e]$ inputs an integer on c and stores it in $x[e]$. The rest is standard.

$$
\begin{array}{ll}
p ::= \mathtt{skip} \quad b ::= \mathtt{skip} \\
\quad \mid\ b;\, p \qquad\ \mid\ x[e] := e \\
\quad\qquad\qquad\ \mid\ c \mathrel{<-} e \\
e ::= n \qquad\qquad \mid\ c \mathrel{->} x[e] \\
\quad \mid\ x[e] \qquad\ \mid\ \mathtt{if}\ e\ \{p\}\ \{p\} \\
\quad \mid\ e \odot e \qquad \mid\ \mathtt{while}\ e\ \{p\}
\end{array}
\qquad
\begin{array}{l}
\mathtt{if}\ e\ \{p_1\}\ \{p_0\};\ p \to p_1;\ p \\
\mathtt{if}\ e\ \{p_1\}\ \{p_0\};\ p \to p_0;\ p \\
\mathtt{while}\ e\ \{p_1\};\ p \to p_1;\ \mathtt{while}\ e\ \{p_1\};\ p \\
\mathtt{while}\ e\ \{p_1\};\ p \to p \\
\text{For } b \in \{\mathtt{skip},\ x[e] := e',\, c \mathrel{<-} e,\ c \mathrel{->} x[e]\}: \\
\quad b;\ p \to p
\end{array}
$$

Fig. 2. Program syntax (left) and CFG (right)

Control Flow Graphs. A control flow graph (CFG) represents a program as a directed graph. The CFG of a program p is defined by $\to$ in Fig. 2; p' is a node iff $p \to^* p'$, and (p', p'') is an edge iff p' and p'' are nodes and $p' \to p''$. We distinguish two nodes in the CFG of program p: the START node p and the END node $\mathtt{skip}$. START is defined as the root of the graph. END has no outgoing edges. Conventionally, CFG nodes are blocks, b. This representation is obtained by dropping p from nodes of the form $b;\, p$ and replacing $\mathtt{if}\ e\ \{\_\}\ \{\_\}$ and $\mathtt{while}\ e\ \{\_\}$ nodes with $\mathtt{branch}\ e$. See the appendix for an illustrative CFG.

Semantics. A program executes in a memory $m : \mathbb{X} \times \mathbb{N} \to \mathbb{Z}$, which provides a (mutable) binding for every location of every array (initially all set to 0 in the initial memory m_0), and an environment $e : \mathbb{C} \to \mathbb{Z}^\omega$, which provides an infinite stream of input values on every channel. We use a small-step reduction relation $(e, m, p) \xrightarrow{o} (e', m', p')$. Here, $o := \bullet \mid\ !cv$ is the *output* of the reduction step; if $p = c \mathrel{<-} e;\, p'$, then $o =\ !cv$ where v is the value e evaluates to; otherwise, $o = \bullet$. The full definition of $\xrightarrow{o}$ is shown in the appendix. Let $\bar{o} = o_1 \ldots o_n$ denote a finite sequence of outputs, and let $\xrightarrow{\bar{o}} = (\xrightarrow{\bullet})^* \xrightarrow{o_1} (\xrightarrow{\bullet})^* \ldots (\xrightarrow{\bullet})^* \xrightarrow{o_n} (\xrightarrow{\bullet})^*$.

Our environments are *total* [32], i.e., never block output, and always provide input on request. This is a natural fit for SPARK, as safety-critical systems typically perform nonblocking I/O (e.g. on files and POSIX shared memory using `read()` and `write()` from the Single UNIX Specification). The endpoints of channels thus, in general, form a collective store which can change independently of the program, and provide input that depends on past output. However,

Clark and Hunt [15] have shown that *when reasoning about security* of deterministic programs (as in our case), environments can be simplified to streams. We use this simplification here. Programs can be composed securely under these environments as long as their scheduler is secure and deterministic. For a more complete and general treatment of composition, see [32,33].

Flow Policies. A flow policy expresses permitted flows between input and output channels. We are interested in two kinds of dependencies: where input affects the *presence* (i.e. occurrence) resp. *content* (i.e.
value) of an output. The syntax of our flow policy language is given in Fig. 3. Let f range over flow policy specifications, and d over dependencies. The syntax c => c' (resp. c -> c') means that content (resp. presence) of output on c is

$$f ::= \text{null} \qquad d ::= c \Rightarrow c$$
$$\mid d; f \qquad \mid c \rightarrow c$$

Fig. 3. Syntax of flow policies

allowed to depend on input on c'. For instance, a flow policy stating that (only) the presence of output on StdErr (standard error) is allowed to depend on input on StdIn (standard input) can be written as StdErr -> StdIn; null. Every flow policy f straightforwardly yields a pair of functions (π, κ) where $\pi(c)$ (resp. $\kappa(c)$) is the set of input channels on which the presence (resp. content) of output on c may depend. We lift these functions to sets of channels: $\pi(C) = \bigcup_{c \in C} \pi(c)$ and $\kappa(C) = \bigcup_{c \in C} \kappa(c)$.

4   Security Property

Consider a fixed policy (π, κ). Our attackers observe all outputs on some output channels. An attacker or *observer* $\omega = (\omega_\pi, \omega_\kappa)$ is a pair where ω_π (resp. ω_κ) is the set of channels on which the presence (resp. content) of outputs is observed. If an observer sees the content of outputs on a channel, it can certainly detect the presence of outputs on the channel, so we require $\omega_\kappa \subseteq \omega_\pi$. Two environments are equivalent to an observer ω if the environments agree on all input channels that may flow to outputs visible to ω.

Definition 1 (ω-equivalence of e). *e and e' are ω-equivalent, $e \sim_\omega e'$, iff* $\forall c \in \pi(\omega_\pi) \cup \kappa(\omega_\kappa) \centerdot e(c) = e'(c)$.

The observables in an output are defined as follows: $!cv\rceil_\omega = {!cv}$ if $c \in \omega_\kappa$, $!cd$ if $c \in \omega_\pi \setminus \omega_\kappa$, and $\bullet$ otherwise (here d is a default output, like null or 0). We remove the unobservables of a sequence of outputs $\bar{o}$ follows: $\epsilon\rceil_\omega = \epsilon$, $(o.\bar{o})\rceil_\omega = \bar{o}\rceil_\omega$ if $o\rceil_\omega = \bullet$, and $(o\rceil_\omega).(\bar{o}\rceil_\omega)$ otherwise.

Definition 2 (ω-equivalence of $\bar{o}$). *$\bar{o}$ and $\bar{o}'$ are ω-equivalent, $\bar{o} \sim_\omega \bar{o}'$, iff* $\bar{o}\rceil_\omega = \bar{o}'\rceil_\omega$.

Our security property, *progress-sensitive noninterference* (PSNI), requires that under observably equivalent environments, a program must be able to componentwise observably-equivalently match observable outputs in its behaviors [8,12,31,32]. For an example involving PSNI, see the appendix.

Definition 3 (Progress-sensitive Noninterference). p *satisfies* PSNI *iff*
$$\forall \omega, e, e' \centerdot e \sim_\omega e' \implies \forall \bar{o} \centerdot (e, m_0, p) \overset{\bar{o}}{\rightarrow} \implies \exists \bar{o}' \centerdot (e', m_0, p) \overset{\bar{o}'}{\rightarrow} \wedge \bar{o} \sim_\omega \bar{o}'.$$

5 Enforcement

SPARK implements a dependency analysis on control flow graphs that prevents all explicit and implicit information leaks, but does not prevent leaks due to progress and termination. In this section, we explain how to augment such a dependency analysis with a loop termination oracle to enforce the stronger property progress-sensitive noninteference (PSNI, Definition 3). While loop termination oracles have been combined with type sytems to enforce PSNI in prior work (e.g., [28]), our technical development makes three novel contributions: (1) We use a graph-based analysis to enforce PSNI (2) Our dependency analysis handles reactive programs, and (3) Our dependency analysis accounts for the difference between output content and output presence. In the following, we describe our analysis for the core language from Sect. 3 and prove that it enforces PSNI. The core language captures all essential features of SPARK, so generalizing the analysis to all of SPARK should not be difficult.

Standard Data- and Control-Dependency Analysis. SPARK's flow analysis uses standard dependency analysis [20,26], which we review briefly. We say that a node b in a CFG reads array x if b contains x in at least one location other than $x[\ldots] := \ldots$. Dually, b writes to array x if $b = (x[e] := e')$. Node b reads a channel c if $b = (c \rightarrow \ldots [\ldots])$. Dependency analysis outputs all the nodes of the CFG on which a given node is *data dependent* or *control dependent*. Data dependence arises due to data flow. E.g., in x = 1; y = 3; z = x + 2; a = z, the statements z = x + 2 and a = z are data dependent on the statement x = 1, but not on y = 3. Similarly, in the example of Fig. 1 (right), the statements on lines 5 and 6 (output and branch K >= H, respectively) are data dependent on the statement K := K + 1 on line 11.

Definition 4 (Data Dependence). *A node b is data dependent on node b' in a CFG G, written $dd_G(b', b)$, if there is a path $b' \rightarrow^* b \in G$ and there is an array that b' writes and b reads, or there is a channel that both b and b' read.*

Note that the statement in b does not have to be an assignment; the definition implies a data dependence from x = y to c <- x in program x = y; c <- x. Also, as commonly assumed by flow analyses in prior work, e.g. Jif [30] and Paragon [14], our definition of data dependence is *flow-insensitive*. This means it ignores the effects of writes in nodes strictly between b' and b ; in program x = y; x = 0; z = x, node z = x is data dependent on the node x = y by our definition, even though x is overwritten by a constant between the nodes. (We use some lemmas from [1] in our proofs, but this difference does not impact those lemmas.) For clues on how to make this definition flow-sensitive, see [23].

Control dependence captures influence due to branches. In the program if (x > 0) { y = 1 } else { y = 2 }; z = 1, both the nodes y = 1 and y = 2 are control dependent on the branch node x > 0. However, the node z = 1 is *not* control dependent on x > 0 because it executes irrespective of the outcome of the test x > 0. There are many different definitions of control dependence in literature (see [34] for a survey). We define here the most standard notion of control dependence, which suffices for our purposes. We say that node b *post-dominates* b' if every path from b' to END passes through b.

Definition 5 (Control Dependence [1]). *A node b is control dependent on node b' in a CFG G, written $cd_G(b', b)$, if the following hold: (1) Either $b = b'$ or b does not post-dominate b' in G, and (2) There is a nontrivial path $b_1 \rightarrow \ldots \rightarrow b_k \in G$ with $b_1 = b'$, $b_k = b$ such that for all $i \in 2 \ldots k - 1$, b post-dominates b_i.*

For block-structured languages such as SPARK and the core calculus of Sect. 3, a node b is control dependent on node b' iff b is a branch or loop condition and b' lies within that branch or loop. However, control dependence is defined on arbitrary CFGs, even those without block structure (we exploit this generality later). Combining data- and control-dependency analysis, we define *dependence* as the reflexive-transitive closure of the data- and control-dependence relations. For example, in the program of Fig. 1 (right), the while loop on line 7 is dependent on the statement K := K + 1 on line 11 because the condition K >= H on line 6 is data-dependent on line 11, and line 7 is control-dependent on line 6. The set of all nodes on which a node b depends is called b's backward slice.

Definition 6 (Dependence and Backward Slice). *The dependence relation dep_G for CFG G is defined as $(dd_G \cup cd_G)^*$. The backward slice of node b, $BS_G(b) = \{b' \mid dep_G(b', b)\}$, is the set of all nodes on which b is dependent.*

Information Flow Control Using Dependency Analysis. The dependence relation dep_G captures all explicit and implicit flows, and, hence, can be used for enforcement of information flow policies. There are well-known algorithms to compute dependencies and backward slices efficiently, e.g., [26]. This analysis is already implemented in SPARK. However, noninterference enforced this way is progress-*in*sensitive because the dependency analysis described above does not take into account nonterminating loops. For instance, the program of Fig. 1 (right) passes SPARK's dependency analysis, even though it leaks H to a progress-sensitive adversary who can observe K. Additionally, the method so far has been limited to sequential programs where the adversary makes only one observation at the end of the program. We explain how the method can be adapted to enforce progress-sensitive noninterference on reactive programs, additionally accounting separately for output content and output presence.

Progress-sensitive Dependence. A leak due to progress happens when an attacker-visible output is pre-empted due to the nontermination of a branch with a secret branch condition. Our simple insight is that such leaks can be detected by a dependence analysis if we ensure the following:

Requirement 1. An output that can be reached *after* the end of a branch is dependent on the branch point if some loop in the branch may diverge.

To implement Requirement 1, we use a static termination analysis, often called a termination oracle [17,27,45]. This oracle determines which loops in the program may diverge. We add an edge from every node in such a loop to the END node of the CFG. It is easy to check that the modified CFG satisfies Requirement 1 if the termination oracle is *sound*, i.e., it flags all loops that diverge on some input. A trivial, sound termination oracle marks every loop as potentially non-terminating. The use of this oracle in our analysis causes every program that contains an attacker-visible output after a loop with a secret loop condition to be marked as leaky, irrespective of whether or not the loop diverges, which may result in false positives. This corresponds exactly to termination-sensitive analyses developed by Smith [41] and Boudol and Castellani [13]. False positives can be reduced using a more precise termination oracle. For example, the program while (h <> h) { }; l = 1 does not have a flow (via progress or otherwise) from the input variable h to the output variable l, but the trivial oracle above will cause this program to be marked leaky by the analysis. On the other hand, a slightly better oracle that uses symbolic analysis to infer that (h <> h) is always false will cause the program to be accepted. In general, fewer false positives in the termination oracle translate to fewer false positives in our dependence analysis. Consequently, we present our analysis *parametrically* in the termination oracle, leaving it to the specific implementation to decide how many resources to devote to the oracle (and, hence, how much precision to obtain).

Definition 7 (Termination Oracle). *A termination oracle T is a function that maps a CFG to a subset of the CFG's nodes. T is sound if for every CFG G and every node $b \in G$, $b \in T(G)$ if there is a memory m and environment e such that b appears infinitely often in the reduction sequence starting from the state (e, m, START).*

Definition 8 (Progress-sensitive Graph). *Given a control flow graph G and a termination oracle T, the progress-sensitive CFG $\mathrm{ps}_T(G)$ is defined by adding to G the edges $\{(b, \mathsf{END}) \mid b \in T(G)\}$.*

For the program of Fig. 1 (right), a sound termination oracle T will say that the while loop on line 7 is nonterminating and, hence, $\mathrm{ps}_T(G)$ will contain an edge from the branch condition of the loop to the end of the program. This makes the output statement on line 5 dependent on the branch condition K >= H and, hence, a dependency analysis will discover the progress leak in the program. Note that $\mathrm{ps}_T(G)$ may not correspond to any block-structured program.

Enforcing PSNI with Content and Presence Distinction. Our analysis takes as input a policy f and a program p. It works as follows. Let G be p's CFG. We compute the progress-sensitive CFG $G' = \mathrm{ps}_T(G)$. Then, for each node $b \in G'$ that outputs to some channel c, we compute the backward slice of b in G', and check that the policy relation κ allows a flow to c from any channel c' on which an input is made in the backward slice. This ensures that information flows to the *content* of messages on c only in accordance with the policy. To account for flows due to *presence* of outputs on c, we compute a second backward slice from the same node b, but after erasing the payload of the output in b. We check that the policy relation π (not κ) allows a flow to c from any channel c' on which an input is made in this backward slice. Thus, by computing two backward slices per output node, we capture separate observations of content and presence. In the sequel, we assume a fixed policy $f = (\kappa, \pi)$.

Definition 9 (Enforcement of PSNI). *Let p be a program with CFG G. Let $G' = ps_T(G)$. We say p passes the PSNI enforcement, written $\mathsf{check}_T(p)$, if the following hold for any node b of the form c <- e in G':*

1. *If c' -> x [e'] $\in BS_{G'}(b)$ then $c' \in \kappa(c)$.*
2. *If c' -> x [e'] $\in BS_{\hat{G}'}(\hat{b})$ then $c' \in \pi(c)$, where $\hat{G}'$ is obtained by replacing b with $\hat{b} = c$ <- d in G'.*

Our main theorem is that the enforcement above is sound: If $\mathsf{check}_T(p)$, then p satisfies PSNI. We prove the theorem using bisimulations on backward slices [1]. Our proof is inspired by a related proof for enforcement of progress-insensitive noninterference in a sequential language [50]. In contrast to that proof, our proof captures progress-sensitive noninterference for a reactive language. Handling reactivity is quite involved: With multiple outputs, we have to argue that the *order* of observable outputs (at different program points) is independent of secret inputs. To do this, we construct a hypothetical slice that is the union of slices from all outputs visible to a given adversary. See the appendix for details.

Theorem 1 (Soundness of Enforcement). *If T is a sound termination oracle and $\mathsf{check}_T(p)$, then p satisfies PSNI.*

6   Source-to-Source Transform

The previous section describes a CFG transformation which ensures Requirement 1 – that any outputs after a potentially divergent branch depend on the branch's condition, which is used to enforce PSNI. In this section, we describe a source-to-source transform that also implies Requirement 1. The transform can be used to enforce PSNI using a standard, *unmodified* (and, hence, closed-source) dependency analysis of the kind that exists for SPARK.

The goal of our source-to-source transform, like the CFG transform, is to add a direct path from every potentially infinite loop to the end of the program. If the existing dependency analysis supports programmatic exceptions, then the transform is trivial: Just before every potentially infinite loop (identified by the oracle T), we add a statement to raise an unhandled exception, conditional on an unsatisfiable predicate. This has the effect of simulating an edge from the loop to the end of the program because the exception is not handled anywhere. It is quite easy to see that this has the same effect as the CFG transform. For example, the program in the right of Fig. 1 would be transformed to the program in the left of Fig. 4. Observe the new line 12 with the `raise` statement.

```
 0 procedure Leak (H : in out Byte) is       0 procedure Leak (H : in out Byte) is
     E : exception; -- new exception            E : Byte := 0; -- E := 1 when an
     X : Byte := 0;                                            -- exception is raised
     K : Byte := 0;                             K : Byte := 0;
     O : File_Type := Standard_Output;          O : File_Type := Standard_Output;
 5 begin                                      5 begin
     H := H;                                     if E = 0 then H := H; end if;
     if X = 1 then raise E; end if;              if E > 0 then E := 1; end if;
     while True loop                             while E = 0 and then True loop
                                                   E := E;
10     Write (O, K);                       10     if E = 0 then Write (O, K); end if;
       if K >= H then                             if E = 0 and then K >= H then
         if X = 1 then raise E; end if;             if E > 0 then E := 1; end if;
         while True loop                            while E = 0 and then True loop
                                                      E := E;
15         H := H;                          15         if E = 0 then H := H; end if;
         end loop;                                  end loop;
       end if;                                    end if;
       K := K + 1;                                if E = 0 then K := K + 1; end if;
     end loop;                                   end loop;
20 end Leak;                               20 end Leak;
```

Fig. 4. Source-to-source transformation of the program in the left of Fig. 1, using exceptions (left) and using an emulation of exceptions (right)

If the flow analysis, like SPARK, does not track flows through exceptions [1], then the source-to-source transform can emulate an exception by adding a new boolean variable, say E, initially set to 0. E is set to 1 where the exception is to be raised, and the program is transformed to check that E is still 0 before executing any statement or entering any branch of the original program. This ensures that once the exception is "raised" (E is set to 1), no statement from the original program executes and control propagates to the end of the program silently. This transform can be defined formally, but we only illustrate it for the progress leak in Fig. 1 in the right panel of Fig. 4. Observe that there is now a dependency between the branch condition K >= H and the output statement on line 10.

We note that to enforce PSNI, the dependency analysis should be applied directly to the output of either of the two transforms described above, without any intervening compiler optimizations. Such optimizations can negate the effects of our transforms. For instance, constant propagation followed by dead code elimination would remove the two **raise** statements on lines 7 and 12 in the left panel of Fig. 4 and, hence, also remove the control dependencies introduced deliberately by the transform.

7   Case Study

We demonstrate the usefulness of our approach on a nontrivial application by implementing a control system for a cruise missile (derived from publicly available code by Hilton [25]), and applying our approach on the code to prove desired

---

[1] The SPARK 2014 documentation states that SPARK programs are allowed to raise exceptions, but may not handle them. However, in our experiments with SPARK GPL 2015, we found that the flow analysis did not track flows through exceptions.

properties. The code steers the missile towards a target coordinate, and deto-
nates a nuclear warhead once within range, or self-destructs in the event that
a device fails. The code is intended to be an illustrative model native to the
domain of SPARK. The code makes several simplifications (e.g. the missile flies
in 2D space), and there are many safety- and mission-critical considerations for
more realistic missile control systems that we have not considered. For details
on such considerations, see Hilton [25]. We give an overview of our case study
(all our code is online [49]).

The missile has three sensors: a *failure detector*, which reports when a device
has failed; an *intertial navigation system*, which provides spatial orientation and
displacement readings (via accelerometers and a ring laser gyroscope) for navigation by dead reckoning; and a *clock*, used to calculate orientation and displacement from accelerometer readings through integration. Using these readings,
the code controls three actuators: a *watchdog*, which, if not actuated at regular intervals, triggers self-destruction (to avoid unwanted consequences of device
or software failure); a *nuclear warhead*, which is detonated when the missile
reaches its target; and *steering*, consisting of aerodynamic fins which the code
actuates for trajectory corrections. Architecturally, our code draws inspiration
from an existing case study on implementing a controller for a water boiler [43,
Section 7]. The *Main* module consists of a sense-control-actuate loop, in which
it commands the sensor modules to read from their device, uses these readings
to compute values to control the actuators, and invokes the actuator modules to
actuate their device. The inter-module information flows are given in Fig. 5.

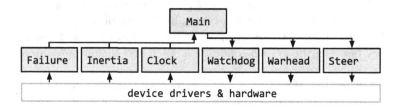

Fig. 5. Inter-module information flows in the missile control system

To illustrate our approach, we aim to prove that orientation does not affect
self-destruction. The body of the *Main* procedure, in the left of Fig. 6, is of
primary concern. Without our approach, since SPARK assumes loops terminate,
reaching "Watchdog.Actuate" in each iteration is deemed inevitable by SPARK,
so SPARK (incorrectly) claims the presence of a destruct event does not depend
on *any* input. However, if we instead apply the SPARK analysis on the code
resulting from applying our source-to-source analysis from Sect. 6 on the missile
control system source code, we get a different result: SPARK (correctly) infers
that the presence of a destruct event depends *only* on device failure. This can
be seen by inspecting the result of the transformation of the main loop, in the
right of Fig. 6. Since both loops are of the form "while True loop", any sound

termination oracle would flag them both as possibly diverging. Hence we emulate a raised exception before both loops, and add a check on variable E to each branch. SPARK no longer deems that reaching "Watchdog.Actuate" is inevitable; it now depends on the value of E. SPARK deems that the value of E depends on Destruct, since there is an assignment to E under a branch on Destruct. Since Destruct depends *only* on device failure, and since the only other assignment to E branches only on E, self-destruction depends only on device failure.

8 Related Work

We focus on the three most closely related areas of work: information flow tools, progress-sensitive security, and information flow analysis for SPARK.

Information Flow Tools. As mentioned before, much progress has been made on enforcement of increasingly rich policies for increasingly expressive programming languages. This has resulted in tools for mainstream programming languages as FlowFox [22], JSFlow [24] and IFC4BC [10] for JavaScript, Jif [30], Paragon [14] and JOANA [23] for Java, FlowCaml [40] for Caml, LIO [47] for Haskell, and SPARK flow analysis [9] for SPARK. With the exception of the latest versions of LIO, these tools target *progress-insensitive noninterference* [7,8,12], allowing secrets to affect progress of public computation. With the focus on the termination and timing channels, Stefan et al. [46] introduced restrictions in LIO on side effects that follow secret branching, which help enforce stronger policies.

```
0                                           0  if E > 0 then E := 1; end if;
   while True loop                             while E = 0 loop
     -- [...] (sense, control)                   E := E; -- [...] (sense, control)
     Steer.Actuate;                              if E = 0 then Steer.Actuate; end if;
     if Destruct then                            if E = 0 and then Destruct then
5      -- block watchdog.                    5     if E > 0 then E := 1; end if;
       while True loop                              while E = 0 loop
         null;                                        E := E; null;
       end loop;                                    end loop;
     end if;                                     end if;
10   Watchdog.Actuate;                      10  if E = 0 then Watchdog.Actuate; end if;
     if Detonate then                           if E = 0 and then Detonate then
       Warhead.Actuate;                           Warhead.Actuate;
     end if; -- [...]                            end if; -- [...]
   end loop;                                   end loop;
```

Fig. 6. Main loop, before (left) and after (right) transformation

Progress-sensitive Security. *Progress-sensitive noninterference* [6,8,12,31,32] (PSNI) disallows progress leaks. PSNI is not susceptible to laundering secrets by brute-force attacks [7] or re-running programs [11]. A typical approach to enforcing PSNI is to disallow loops with secret guards, going back to Volpano and Smith's technique to deal with termination leaks [48], or to allow loops with secret guards but prohibit assignments to public variables that follow such loops [13,41]. While the theory of progress-sensitive security has been explored [6, 8, 12, 31, 32], our work connects the theory with tools, showing how we can leverage a progress-insensitive tool (SPARK's flow analysis) to achieve PSNI. Related to our source-to-source transform, Russo et al. [36] discuss *magnification patterns* in the context of distinguishing flows in malicious and nonmalicious code. A magnification pattern in a control-flow graph consists of a branching on a secret guard inside of a loop. We note that in the absence of such patterns (as is sometimes the case in non-malicious code [36]), progress-sensitive security and progress-insensitive security coincide. Moore et al. [28] use *termination oracles* for termination-sensitive tracking. Their prototype implementation utilizes an SMT solver to analyze examples in a simple imperative language. While related, there are several distinguishing features of our work: we focus on practical information flow control in SPARK and push the approach to the full SPARK language; our case study goes beyond code snippets to a suite for a missile controller; on the theoretical side, our framework is graph-based, which opens up possibilities for natural adoption to other graph-based tools such as JOANA [23].

Information flow Analysis in SPARK. A line of work by Amtoft et al. shares with our work the motivation to improve SPARK's information flow analysis. Based on an expressive information logic [2], they enhance the information flow contract language to support compositional policies and conditional information flows [5]. They improve the precision of the analysis by breaking out of a limitation of the original analysis that treats arrays as indivisible entities and evaluate the approach on a collection of SPARK programs [4]. They extend the logical framework to produce machine-checkable formal certificates of correctness for verified code [3]. Extending the results by Amtoft et al. to guarantee progress-sensitive security is a promising direction for future work.

9 Conclusion

This paper puts a spotlight on the SPARK flow analysis. Articulating the boundaries of what is achievable by the analysis, we spell out the attacks to exploit such channels as termination, progress, resource exhaustion, and timing channels. We suggest how to harden the analysis to achieve security against stronger attackers, with the focus on progress-sensitive security as our baseline. Instead of redesigning and reimplementing the enforcement, we show how to leverage known flow analyses for weaker attackers by a transform on program dependence graphs. The graph transform represents termination and progress flows by injecting additional edges. We establish the soundness of this approach for a core language and demonstrate that it can be applied as a source-to-source transform of SPARK code when modifying the compiler is undesirable. A case study with a control unit of a missile written in SPARK 2014 indicates the usefulness of the approach. Future work is focused on enriching the policy and enforcement mechanisms with possibilities for *declassification* [38], a feature on the wish list of the SPARK developers [35]. We are also interested in extending the framework with treating resource exhaustion and timing leaks and exploring more sophisticated attacks.

Acknowledgments. Thanks are due to Angela Wallenburg for inspiration and regular updates about developments on SPARK. This work was funded by the European Community under the ProSecuToR and WebSand projects, the Swedish research agencies SSF and VR and the German DFG priority program "Reliably Secure Software Systems" (RS3). This research was supported in part by US Navy grant N000141310156; NSF grants 1320470.

References

1. Amtoft, T.: Slicing for modern program structures: A theory for eliminating irrelevant loops. Inf. Process. Lett. **106**(2), 45–51 (2008)
2. Amtoft, T., Bandhakavi, S., Banerjee, A.: A logic for information flow in object-oriented programs. In: POPL, pp. 91–102 (2006)
3. Amtoft, T., Dodds, J., Zhang, Z., Appel, A., Beringer, L., Hatcliff, J., Ou, X., Cousino, A.: A certificate infrastructure for machine-checked proofs of conditional information flow. In: Degano, P., Guttman, J.D. (eds.) Principles of Security and Trust. LNCS, vol. 7215, pp. 369–389. Springer, Heidelberg (2012)
4. Amtoft, T., Hatcliff, J., Rodríguez, E.: Precise and automated contract-based reasoning for verification and certification of information flow properties of programs with arrays. In: Gordon, A.D. (ed.) ESOP 2010. LNCS, vol. 6012, pp. 43–63. Springer, Heidelberg (2010)
5. Amtoft, T., Hatcliff, J., Rodríguez, E., Robby, E., Hoag, J., Greve, D.: Specification and checking of software contracts for conditional information flow. In: Cuellar, J., Maibaum, T., Sere, K. (eds.) FM 2008. LNCS, vol. 5014, pp. 229–245. Springer, Heidelberg (2008)
6. Askarov, A., Chong, S., Mantel, H.: Hybrid monitors for concurrent noninterference. In: CSF, July 2015
7. Askarov, A., Hunt, S., Sabelfeld, A., Sands, D.: Termination-insensitive noninterference leaks more than just a bit. In: Jajodia, S., Lopez, J. (eds.) ESORICS 2008. LNCS, vol. 5283, pp. 333–348. Springer, Heidelberg (2008)
8. Askarov, A., Sabelfeld, A.: Tight enforcement of information-release policies for dynamic languages. In: Proceeding of the IEEE Computer Security Foundations Symposium, July (2009)
9. Barnes, J.: High Integrity Software: The SPARK Approach to Safety and Security. Addison-Wesley Longman Publishing Co., Inc., Boston, MA, USA (2003)
10. Bichhawat, A., Rajani, V., Garg, D., Hammer, C.: Information flow control in webkit's javascript bytecode. In: Abadi, M., Kremer, S. (eds.) POST 2014 (ETAPS 2014). LNCS, vol. 8414, pp. 159–178. Springer, Heidelberg (2014)
11. Birgisson, A., Sabelfeld, A.: Multi-run security. In: Atluri, V., Diaz, C. (eds.) ESORICS 2011. LNCS, vol. 6879, pp. 372–391. Springer, Heidelberg (2011)
12. Bohannon, A., Pierce, B., Sjöberg, V., Weirich, S., Zdancewic, S.: Reactive noninterference. In: ACM Conference on Computer and Communications Security, pp. 79–90, November 2009
13. Boudol, G., Castellani, I.: Non-interference for concurrent programs and thread systems. Theor. Comput. Sci. **281**(1), 109–130 (2002)

14. Broberg, N., van Delft, B., Sands, D.: Paragon for practical programming with information-flow control. In: Shan, C. (ed.) APLAS 2013. LNCS, vol. 8301, pp. 217–232. Springer, Heidelberg (2013)
15. Clark, D., Hunt, S.: Noninterference for deterministic interactive programs. In: Workshop on Formal Aspects in Security and Trust (FAST 2008), October 2008
16. Cohen, E.S.: Information transmission in sequential programs. In: DeMillo, R.A., Dobkin, D.P., Jones, A.K., Lipton, R.J. (eds.) Foundations of Secure Computation, Academic Press, pp. 297–335 (1978)
17. Cook, B., Podelski, A., Rybalchenko, A.: Termination proofs for systems code. In: PLDI, pp. 415–426 (2006)
18. Denning, D.E.: A lattice model of secure information flow. Comm. ACM **19**(5), 236–243 (1976)
19. Denning, D.E., Denning, P.J.: Certification of programs for secure information flow. Comm. ACM **20**(7), 504–513 (1977)
20. Ferrante, J., Ottenstein, K.J., Warren, J.D.: The program dependence graph and its use in optimization. ACM Trans. Program. Lang. Syst. **9**(3), 319–349 (1987)
21. Goguen, J.A., Meseguer, J.: Security policies and security models. In: Proceedings of the IEEE Symposium on Security and Privacy, pp. 11–20, April 1982
22. Groef, W.D., Devriese, D., Nikiforakis, N., Piessens, F.: Flowfox: A web browser with flexible and precise information flow control. In: ACM Conference on Computer and Communications Security (2012)
23. Hammer, C., Snelting, G.: Flow-sensitive, context-sensitive, and object-sensitive information flow control based on program dependence graphs. Int. J. Inf. Secur. **8**(6), 399–422 (2009)
24. Hedin, D., Birgisson, A., Bello, L., Sabelfeld, A.: JSFlow: Tracking information flow in JavaScript and its APIs. In: Proceeding of the 29th ACM Symposium on Applied Computing (2014)
25. Hilton, A.J.: High Integrity Hardware-Software Codesign. Ph.D. thesis, The Open University, April 2004
26. Horwitz, S., Reps, T.W., Binkley, D.: Interprocedural slicing using dependence graphs. In: PLDI, pp. 35–46 (1988)
27. Kroening, D., Sharygina, N., Tsitovich, A., Wintersteiger, C.M.: Termination analysis with compositional transition invariants. In: Touili, T., Cook, B., Jackson, P. (eds.) CAV 2010. LNCS, vol. 6174, pp. 89–103. Springer, Heidelberg (2010)
28. Moore, S., Askarov, A., Chong, S.: Precise enforcement of progress-sensitive security. In: ACM Conference on Computer and Communications Security, pp. 881–893(2012)
29. The Muen Separation Kernel. http://muen.codelabs.ch/
30. Myers, A.C., Zheng, L., Zdancewic, S., Chong, S., Nystrom, N.: Jif: Java Information Flow. Software release. Located at, July 2001 http://www.cs.cornell.edu/jif
31. O'Neill, K., Clarkson, M., Chong, S.: Information-flow security for interactive programs. In: Proceedings of the IEEE Computer Security Foundations Workshop, pp. 190–201, July 2006
32. Rafnsson, W., Hedin, D., Sabelfeld, A.: Securing interactive programs. In: Proceedings of the IEEE Computer Security Foundations Symposium, June 2012
33. Rafnsson, W., Sabelfeld, A.: Compositional security for interactive systems. In: CSF, pp. 277–292 (2014)
34. Ranganath, V.P., Amtoft, T., Banerjee, A., Hatcliff, J., Dwyer, M.B.: A new foundation for control dependence and slicing for modern program structures. ACM Trans. Program. Lang. Syst. **29**, 5 (2007)

35. Refined Information Flow Requirement. http://lists.forge.open-do.org/pipermail/spark2014-discuss/2012-December/000683.html
36. Russo, A., Sabelfeld, A., Li, K.: Implicit flows in malicious and nonmalicious code. 2009 Marktoberdorf Summer School (IOS Press) (2009)
37. Sabelfeld, A., Myers, A.C.: Language-based information-flow security. IEEE J. Sel. Areas Commun. **21**(1), 5–19 (2003)
38. Sabelfeld, A., Sands, D.: Declassification: Dimensions and principles. J. Comput. Secur. **17**(5), 517–548 (2009)
39. Workstation, M.: High-Security Framework, Pilot, and Formalization Architecture. http://www.secunet.com/fileadmin//sina_downloads/Produktinfo_englisch/SINA-Multilevel_Brochure_en.pdf
40. Simonet, V.: The Flow Caml system. Software release. Located at, July 2003. http://cristal.inria.fr/~simonet/soft/flowcaml
41. Smith, G.: A new type system for secure information flow. In: Proceedings of the IEEE Computer Security Foundations Workshop, pp. 115–125, June 2001
42. SPARK (programming language). http://en.wikipedia.org/wiki/SPARK_%28programming_language%29
43. Development, T.: Support: INFORMED Design Method for SPARK. http://docs.adacore.com/sparkdocs-docs/Informed.htm
44. SPARK (2014). http://www.spark-2014.org/
45. Spoto, F., Mesnard, F., Payet, É.: A termination analyzer for java bytecode based on path-length. ACM Trans. Program. Lang. Syst. **32**(3), Article no. 8, 70 (2010)
46. Stefan, D., Russo, A., Buiras, P., Levy, A., Mitchell, J.C., Maziéres, D.: Addressing covert termination and timing channels in concurrent information flow systems. In: ICFP, pp. 201–214 (2012)
47. Stefan, D., Russo, A., Mitchell, J., Mazières, D.: Flexible dynamic information flow control in haskell. In Proceedings of the Haskell Symposium, pp. 95–106. ACM (2011)
48. Volpano, D., Smith, G.: Eliminating covert flows with minimum typings. In: Proceedings of the IEEE Computer Security Foundations Workshop, pp. 156–168, June 1997
49. Rafnsson, W., Garg, D., Sabelfeld, A.: Progress-Sensitive Security forSPARK. Full version: http://research.precise.li/pub/2016essos
50. Wasserrab, D., Lohner, D., Snelting, G.: On PDG-based noninterference and its modular proof. In: PLAS, pp. 31–44 (2009)

Sound and Precise Cross-Layer Data Flow Tracking

Enrico Lovat[1], Martín Ochoa[2]([✉]), and Alexander Pretschner[1]

[1] Technische Universität München, Munich, Germany
{enrico.lovat,alexander.pretschner}@in.tum.de
[2] Singapore University of Technology and Design, Singapore, Singapore
martin_ochoa@sutd.edu.sg

Abstract. We connect runtime monitors for data flow tracking at different abstraction layers (a browser, a mail client, an operating system) and prove the soundness of this generic model w.r.t. a formal notion of explicit information flow. This allows us to (1) increase the precision of the analysis by exploiting the high-level semantics of events at higher levels of abstraction and (2) provide system-wide guarantees at the same time. For instance, using our model, we can soundly reason about the flow of a picture from the network through a browser into a cache file or a window on the screen by combining analyses at multiple layers.

1 Introduction

Research in data flow tracking [4,22] tackles the problem of monitoring flows of data from sources (i.e. input parameters to methods, sockets, files) to sinks (i.e. outputs to sockets, files). Data flow analysis systems can answer the question if data has (potentially) flowed, or will (potentially) flow, from a source to a sink.

Dynamic approaches for data-flow tracking implement reference monitors at various levels of abstraction: binary code, Java bytecode, operating systems, and dedicated applications. Dynamic analyses can exploit layer-specific semantic information and be precise in the presence of reflection or call-backs. They cannot, by definition, detect flows that are a consequence of non-executed branches and they do impose a non-negligible runtime overhead. In the absence of OS-layer monitoring and if monitoring is not done at the binary level, dynamic analysis results are confined to the considered layer. In this paper, we elaborate on the idea of using multiple monitors at different layers, with the goal of improving the precision of the single layers by exploiting known relations between them.

As an example, consider Fig. 1 where an application loads two files from the OS and then saves one of them with a different name. Data d, contained in the first file *file f*, enters the application via container *src1* (1), is propagated through the application internals (2) and finally leaves the application (3). If dynamic monitoring was performed solely at the OS layer, the analysis would report data e to have flowed to *file i* as well— which is sound but over-approximating. If monitoring was solely performed at the application level, data flows at the OS layer could not be observed, e.g., the flow of data d from *file i* to *file h*.

© Springer International Publishing Switzerland 2016
J. Caballero et al. (Eds.): ESSoS 2016, LNCS 9639, pp. 38–55, 2016.
DOI: 10.1007/978-3-319-30806-7_3

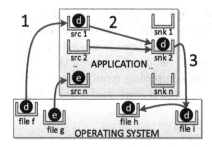

Fig. 1. Intra-app data flow example

Using monitors at multiple layers, two in this example, allows us to increase precision of single-layer analyses, to detect and control system-wide (rather than program-wide) data flows, and to exploit semantic information at various layers.

It has been shown [7,26,27] that dynamic analyses can be performed at the level of every piece of binary code that is executed. This obviates the need for multiple layers. However, in addition to overhead issues of purely dynamic solutions, at low levels it is also hard to detect and model semantics of high-level events (like "take a screenshot") or objects (like "a mail"). While implementations of such system-wide information-flow trackers exist [7,26,27], we are not aware of any formal description of the guarantees that can be provided. Using monitors at levels different from the binary code, we decrease the number of events that need to be intercepted and exploit semantic information of the single levels.

Information flow researchers consider both explicit and implicit flows (as in non-interference [9]). Our work focuses on *explicit* data flows only, with the well-known limitations that this choice entails. The reason is twofold: first, we want to take advantage of the *semantic* information gained by monitoring high-level events (e.g. "play" or "forward") and those events typically involve only explicit data flows (such as copying data from a file to another). Second, many implicit flows are harmless in practice (e.g., the famous password-checking example [21]), such that by enforcing strict non-interference one is likely to severely hamper the intended functionality of a system. Enhancing our cross-layer analysis to include some implicit information flows while preserving functionality requires non-trivial declassification policies and is left to future work.

The **problem** that we tackle is the following. We assume that there are dynamic analyses at two or more levels, all of them different from that of CPU-level instructions, including operating system, application level, database, programming language and window manager. How can we connect the analysis results of the different layers, and what guarantees can we give?

Our **contributions** are **(a)** a formal definition of soundness of data flow tracking for single levels different from that of the CPU; **(b)** a formal definition of soundness of data flow tracking when multiple layers are combined; **(c)** a generic schema to compose data flow analyses at various levels and that thus enables us to detect system-wide data flows; and **(d)** a proof that this generic schema returns sound data flow results, provided that the single layers are correct and given some partial information about shared resources at both layers.

We do not discuss implementations of this model due to space restrictions, but we point to examples in Sect. 7, where some strategies for cross-layer composition described here have been applied without a proper formal justification and with implicit assumptions that we now make explicit. See [15] for more details.

The rest of the paper is organized as follows: Sect. 2 introduces fundamental concepts and sets the notation for the rest of the paper. Section 3 discusses the security guarantees for single layers. Section 4 defines layer composition and extends the soundness notion to composed systems. Section 5 presents the main result of the paper, i.e. an algorithm for soundly composing monitor results at different abstraction layers. In Sect. 6 we review related work and we conclude in Sect. 7.

2   Background and Roadmap

We consider flows in systems described as tuples $(\mathcal{E}, \mathcal{D}, \mathcal{C}, \Sigma, \sigma_i, \mathcal{R})$ for system events $\mathcal{E}$, data items $\mathcal{D}$ (e.g., "a picture"), and containers $\mathcal{C}$ (these are representations: a pixmap, a file, a memory region, a set of network packets, and so on). For the time being, these systems can be understood as single layers. In the following we assume that the alphabet of system events $\mathcal{E}$, the set of data items $\mathcal{D}$ and the set of containers $\mathcal{C}$ can be of arbitrary but finite size.

States Σ are defined by a *storage function* of type $\mathcal{C} \rightarrow \mathbb{P}(\mathcal{D})$ that describes which set of data is potentially stored in which container. Data items will be often referred to as *labels* in the following.

The transition function $\mathcal{R} : \Sigma \times \mathcal{E} \rightarrow \Sigma$ is the core of the data-flow tracking model, encoding how the execution of events affects the dissemination of data in the system (and therefore also referred to as a *monitor* in the following). At runtime, events are intercepted and the data state is updated according to $\mathcal{R}$. $\mathcal{R}$ applied to a sequence of events is the recursive application to each event in the sequence (i.e. $\mathcal{R}(\sigma, \langle e_1, \langle ... \rangle \rangle) = \mathcal{R}(\mathcal{R}(\sigma, e_1), \langle ... \rangle)$).

Abstraction Layers: We will show desirable properties by relating the model for one layer A to a very low level model $\perp$ with intuitive completeness and correctness properties. Layer A can be an operating system, a data base, a windowing system, an application, etc. $\perp$ is the level of the CPU and volatile as well as persistent memory cells, and represents the *real* execution of the system. Let $\mathcal{V}$ be the set of all total functions of type $\mathcal{C}_\perp \rightarrow \mathbb{N}$ that map containers to actual *values* stored in memory. We provide level $\perp$ with a function $v \in \mathcal{V}$, that indicates the current state of memory, and a trace execution semantics eval : $\mathcal{V} \times seq(\mathcal{E}_\perp) \rightarrow \mathcal{V}$ that describes the state after executing a trace, such that the system at $\perp$ is given by $(\mathcal{E}_\perp, \mathcal{D}_\perp, \mathcal{C}_\perp, \Sigma_\perp, \sigma_i, \mathcal{R}_\perp, \mathcal{V}, \text{eval})$.

A in contrast is some distinct higher layer. Set $\mathcal{L}$ denotes the set of all these high levels, while $\mathcal{L}^\perp = \mathcal{L} \cup \{\perp\}$. Data $\mathcal{D}$ is layer-independent. For $\dagger \in \mathcal{L}$, $\mathcal{C}_\dagger$ denotes the set of representations of some data item at layer $\dagger$. To relate two layers, we assume pairs of functions γ and α that relate events and containers as follows. The idea is that an A-level container corresponds to a set of $\perp$-level containers (volatile and persistent memory cells) and an A-level action to a sequence of CPU-level instructions (machine instructions such as MOV, BNE, ADD, LEQ). For a layer $\dagger \in \mathcal{L}^\perp$, each state $\sigma_\dagger \in \Sigma_\dagger$ is defined by the respective storage function. Additionally, for $\perp$, $v \in \mathcal{V}$ encodes the memory state.

Relating Events and States: In the following we will introduce abstraction and concretization functions that relate events and states at a higher abstraction level with $\bot$. This notation will allow us to define formal soundness properties. Note that our goal is to reason about what happened at the $\bot$ level, *without monitoring it*, by assuming availability of partial information on the abstraction/concretization functions. This will be made operational in practice by *oracles*, as we will discuss in Sect. 5. For the moment, concretization and abstraction functions are *ideal*: they relate to "what has really happened" in a monitored system, and where for instance scheduling of concurrent processes has already been fixed.

Traces are sequences of events that reflect the execution of some functionality at a specified layer. A trace $t \in seq(\mathcal{E}_\dagger)$ at layer $\dagger \in \mathcal{L}$ is thus the sequence of events generated by the system during a specific run (e.g. "*print x*" or "*mov AX,BX*"). At the $\bot$-level, traces are sequences of CPU instructions.

We assume events to be unique to exactly one level and to contain an implicit timestamp and duration, yielding a natural order on a trace's events. Each event at a higher layer corresponds to a sequence of CPU-level instructions. For simplicity's sake, we assume that we can bijectively map an abstract sequence of events to a concrete sequence of $\bot$-events. This embodies the fundamental assumption of a *single-core system*: all traces can be uniquely sequentialized. While a single-core can simulate parallelism alternating execution of different tasks, we assume that events at high levels happen *sequentially* (e.g. first save $file_1$ and then load $file_2$). This allows us to use a simpler notation in these introductory sections. In Sect. 5 we relax this assumption, because serializable traces can capture also concurrent executions.

Moreover, we deliberately *discard events at $\bot$ that do not correspond to an event at a higher layer*, e.g. those generated by an application for which there is no explicit monitor. The implication is that our approach can only be sound w.r.t. those CPU-level instructions for which a monitor at some level exists.

In the following, we define abstraction and concretization functions for events and states. For this purpose, we redefine α and γ as follows. Let $\dagger \in \mathcal{L}$:

$$\begin{aligned}
\textit{Events}: &\quad \gamma_\dagger : seq(\mathcal{E}_\dagger) \rightarrow seq(\mathcal{E}_\bot), \quad \alpha_\dagger : seq(\mathcal{E}_\bot) \rightarrow seq(\mathcal{E}_\dagger) \\
\textit{States}: &\quad \gamma_\dagger : \Sigma_\dagger \rightarrow \Sigma_\bot, \quad\quad\quad \alpha_\dagger : \Sigma_\bot \rightarrow \Sigma_\dagger \\
\textit{Containers}: &\quad \gamma_\dagger : \mathcal{C}_\dagger \rightarrow \mathbb{P}(\mathcal{C}_\bot), \quad\quad \alpha_\dagger : \mathcal{C}_\bot \rightarrow \mathbb{P}(\mathcal{C}_\dagger) \text{ such that}
\end{aligned}$$

$$\begin{aligned}
\gamma_\dagger(\sigma_\dagger) &= \{(c_\bot, \sigma_\dagger(c_\dagger)) : c_\dagger \in \mathrm{dom}(\sigma_\dagger) \wedge c_\bot \in \gamma_\dagger(c_\dagger)\} \\
\alpha_\dagger(\sigma_\bot) &= \{(c_\dagger, \sigma_\bot(c_\bot)) : c_\bot \in \mathrm{dom}(\sigma_\bot) \wedge c_\dagger \in \alpha_\dagger(c_\bot)\}.
\end{aligned}$$

Additionally, $\forall \mathcal{C} \subseteq \mathcal{C}_\dagger : \gamma_\dagger(\mathcal{C}) = \bigcup_{c \in \mathcal{C}} \gamma_\dagger(c)$ and $\forall \mathcal{C} \subseteq \mathcal{C}_\bot : \alpha_\dagger(\mathcal{C}) = \bigcup_{c \in \mathcal{C}} \alpha_\dagger(c)$. For each layer $\dagger \in \mathcal{L}$ we assume the existence of a special container $c_\dagger^U$ that represents the abstraction of all those $\bot$-level containers not observable at the $\dagger$-level ($\forall c_\bot \in \mathcal{C}_\bot : (\forall c_\dagger \in \mathcal{C}_\dagger \setminus \{c_\dagger^U\} : \alpha_\dagger(c_\bot) \neq c_\dagger) \implies (\alpha_\dagger(c_\bot) = c_\dagger^U)$). By definition $\sigma(c_\dagger^U) = \mathcal{D}$ for any state σ.

State and Trace Union: Given two states σ_1 and σ_2 at the same abstraction level, let $\sigma_1 \bowtie_\sigma \sigma_2 = \{(c, D) \mid D = \sigma_1(c) \cup \sigma_2(c)\}$. Recall that events at any level are assumed to be unique and to contain an implicit timestamp, yielding a natural order on a trace's events. We denote by $t_1 \bowtie_t t_2$ the time-ordered trace consisting of unique elements of t_1 and t_2.

Roadmap. In the following we give a high-level account on the strategy used in the rest of the paper to justify the proposed cross-layer analysis algorithm. **(1)** We relate the notion of taint propagation at the lowest abstraction layer ($\perp$) with that of *weak secrecy*. **(2)** We define a notion of *soundness* for a single layer A with respect to $\perp$. The intuition behind this definition is that the taint propagation in A (specified by $\mathcal{R}_A$) must be coherent with respect to taint propagation happening at $\perp$. This definition offers a semantic characterization for single monitors at any level. **(3)** We then define composed states $\sigma_{A \otimes B}$ as pairs of states (σ_A, σ_B) at different layers and give a notion of sound composed monitor. **(4)** We construct and prove the soundness of a composed monitor $\hat{\mathcal{R}}_{A \otimes B}$ that relies on the soundness of monitors $\mathcal{R}_A$ and $\mathcal{R}_B$ at the single layers and on partial information on γ. **(5)** We show an example where additional information about A and B can lead to a more precise cross-layer tracking and **(6)** we use it to motivate the usefulness of further *oracles* that encode partial information about γ and α. **(7)** We construct a composed monitor $\hat{\mathcal{R}}_{A \otimes B}$ that relies on the soundness of single layer monitors $\mathcal{R}_A$ and $\mathcal{R}_B$ and on the information from the oracles and show its soundness.

Thanks to the above construction, we can connect existing data-flow tracking analyses for different layers of abstraction (e.g. [13,16,25]) to capture data flows across layers, and show overall soundness, or weak secrecy, respectively. All proofs are provided in a technical report [17].

3   Security Guarantees at Single Layers

In the following, we define the notion of information flow which will be the fundamental security property guaranteed by our framework. We use this notion to show soundness of the propagating data flow monitors at various layers.

3.1   Step 1: Security Property at the $\perp$ Layer

Data-flow tracking estimates which containers are "dependent" from the data stored in some other containers after a system run. The strongest guarantees in this sense are given by Non-Interference [9], which relates inputs and outputs in terms of pairs of executions (or state of variables before and after executing a program [24]).

Definition 1 (Non-interference). *Let* $\mathcal{C}_H^i, \mathcal{C}_H^o \subseteq \mathcal{C}_\perp$ *be sets of containers at* $\perp$ *and* $\mathcal{C}_L^i, \mathcal{C}_L^o$ *their complements. A trace* $t_\perp \in seq(\mathcal{E}_\perp)$ *respects* non-interference *w.r.t. this partition of the containers if*

$$\forall v, v' \in \mathcal{V} : \bigwedge_{c \in \mathcal{C}_L^i} v(c) = v'(c) \Longrightarrow \bigwedge_{c \in \mathcal{C}_L^o} \mathsf{eval}(v, t_\perp)(c) = \mathsf{eval}(v', t_\perp)(c)$$

In other words, the values of a certain memory region (represented by *low* containers) are independent from its complement (the *high* containers) after execution of a trace at $\bot$. This represents the notion of *absence of flows* from high to low containers. As discussed in the introduction, in this work we focus on explicit information flows. An information flow property which captures such flows is *weak-secrecy* [23].

To formally define this property in our context, consider a trace $t_\bot \in seq(\mathcal{E}_\bot)$. We say that its *branch-free* version $bf(t_\bot)$ consists of the same assembly-level instructions in the same order, except for branch statements such as BNE (branch-non-equal), which are removed from the observed trace. Of course there are no *branches* in one actually executed trace. There are, however, conditional jumps like *branch-not-equal* that may lead to *implicit, or control-flow-based* information flows. In order to cater to explicit flows only, these instructions are ignored, i.e. removed. The resulting trace then corresponds to one *path* through the CFG of the original program where all conditional nodes are replaced by empty statements. By doing so, our notion of security becomes the verification of non-interference on a sequence of explicit data flows only.

Definition 2 (Weak Secrecy). *Let* $\mathcal{C}_H^i, \mathcal{C}_H^o \subseteq \mathcal{C}_\bot$ *be sets of containers at* $\bot$. *A trace* $t_\bot \in seq(\mathcal{E}_\bot)$ *respects* weak-secrecy *w.r.t.* $\mathcal{C}_H^i, \mathcal{C}_H^o$ *if its branch-free version* $bf(t)$ *respects non-interference w.r.t.* $\mathcal{C}_H^i, \mathcal{C}_H^o$.

Note that non-interference in general, for arbitrary program languages, does not imply weak secrecy. Moreover, our construction is not intended to guarantee non-interference: we need it to define weak secrecy only.

A monitor $\mathcal{R}_\bot$ propagates labels (i.e. data items) in-between containers as the consequence of the execution of a trace.

Definition 3. *A monitor* $\mathcal{R}_\bot$ *is sound w.r.t. weak-secrecy if given an initial state* σ_i, *for all data items* $d \in \mathcal{D}$, *all traces* $t_\bot \in seq(\mathcal{E}_\bot)$ *respect weak-secrecy for the initial partition of the containers as induced by* d: $\mathcal{C}_H^i = \{c \in \mathcal{C}_\bot \mid d \in \sigma_i(c)\}$ *and, at the end of trace* $t_\bot$, *the resulting partition of the containers as computed by the monitor:* $\mathcal{C}_H^o = \{c \in \mathcal{C}_\bot \mid d \in \mathcal{R}_\bot(\sigma_i, t_\bot)(c)\}$.

In other words, if $\mathcal{R}_\bot$ claims a container c does not hold data d after the execution of a trace, then the values of c are independent from the values of d in the weak-secrecy sense. In the following, $\mathcal{R}_\bot^\#$ indicates the (virtual) most precise sound monitor at level $\bot$, i.e. for all $d \in \mathcal{D}$ and $t_\bot \in seq(\mathcal{E}_\bot)$, the output partition $\mathcal{C}_H^o$ induced by any sound monitor includes that induced by $\mathcal{R}_\bot^\#$.

Sources and Destinations. From the point of view of $\mathcal{R}_\bot^\#$, events move data from a container to another: an instruction typically reads from a certain memory region and writes to another. For any given event e and a transition function $\mathcal{R}_\bot^\#$, the functions $\mathbb{S}_{\mathcal{R}_\bot^\#}, \mathbb{D}_{\mathcal{R}_\bot^\#} : \mathcal{E}_\bot \to 2^{\mathcal{C}_\bot}$ denote the sets of source destination containers of the events. We assume the two functions to be given as an oracle of the event, such that for all $\perp$-containers c, states σ and data items d,

$$d \in \mathcal{R}^{\#}_{\perp}(\sigma, e)(c) \implies d \in \sigma(c) \vee (\exists c' \in \mathbb{S}_{\mathcal{R}^{\#}_{\perp}}(e) : d \in \sigma(c') \wedge c \in \mathbb{D}_{\mathcal{R}^{\#}_{\perp}}(e)),$$

i.e., if after executing e a container c holds d, then this was already present before the execution of e, or there was a flow from a container in the sources of e (making c a destination).

Note that there could be coarse partitions that fulfill this property. In the following we assume that the oracle provides the most precise ones in the sense that e respects weak secrecy w.r.t. $\mathbb{S}_{\mathcal{R}^{\#}_{\perp}}(e)$, $\mathcal{C}_{\perp} \setminus \mathbb{D}_{\mathcal{R}^{\#}_{\perp}}(e)$, and w.r.t $\mathcal{C}_{\perp} \setminus \mathbb{S}_{\mathcal{R}^{\#}_{\perp}}(e), \mathbb{D}_{\mathcal{R}^{\#}_{\perp}}(e)$. In other words, there is non-interference between the partitions induced by sources and destinations and their dual complement. Intuitively, this ensures that all relevant sources and all relevant destinations are captured and no more.

We overload the notation of $\mathbb{S}$ and $\mathbb{D}$ for traces of events $t \in seq(\mathcal{E}_{\perp})$ as $\mathbb{S}_{\mathcal{R}_{\perp}}(t) = \bigcup_{e \in t} \mathbb{S}_{\mathcal{R}^{\#}_{\perp}}(e)$ and $\mathbb{D}_{\mathcal{R}^{\#}_{\perp}}(t) = \bigcup_{e \in t} \mathbb{D}_{\mathcal{R}^{\#}_{\perp}}(e)$. A similar overloading applies to sets of events. We also extend the notation for events and monitors at higher layers of abstraction, $\mathbb{S}_{\mathcal{R}_{\dagger}}$ and $\mathbb{D}_{\mathcal{R}_{\dagger}}$ for $\dagger \in \mathcal{L}$, such that the same relation between $\mathcal{R}_{\dagger}$, containers and data holds at level $\dagger$.

3.2   Step 2: Soundness at a Single Layer

An A-level state of the system, σ_A, is sound if, for every container c_A, the set of data stored in c_A is a superset of the data "actually" stored in it, i.e. of the data stored in the concretization of c_A. For this reason, soundness is defined w.r.t. a $\perp$-state. In the following, we assume a fixed pair of γ_A/α_A w.r.t. which soundness is defined.

Definition 4. *A state σ_A is* sound *w.r.t.* $\sigma_{\perp}$, *written* $\sigma_{\perp} \vdash \sigma_A$, *iff*

$$\forall c_A \in \mathcal{C}_A : \sigma_A(c_A) \supseteq \bigcup_{c_{\perp} \in \gamma_A(c_A)} \sigma_{\perp}(c_{\perp}).$$

This implies that $\forall \sigma_A \in \Sigma_A : \gamma_A(\sigma_A) \vdash \sigma_A$ and that $\forall \sigma_{\perp} \in \Sigma_{\perp} : \sigma_{\perp} \vdash \alpha_A(\sigma_{\perp})$. The data flow analysis for A is sound w.r.t. $\perp$ (i.e., it respects weak-secrecy) if $\mathcal{R}_A$ preserves the soundness of the state (w.r.t. the canonical $\mathcal{R}^{\#}_{\perp}$ of Definition 3).

Definition 5 (Soundness of Single Layer Monitor). *A monitor $\mathcal{R}_A$ at a level A is* sound *w.r.t.* $\perp$, *written* $\mathcal{R}^{\#}_{\perp} \vdash \mathcal{R}_A$, *if given an initial state $\sigma^i_{\perp} \vdash \sigma^i_A$, modeling any trace of events $t_A \in seq(\mathcal{E}_A)$ results in a state σ_A which is sound with respect to the state reached by the canonical $\mathcal{R}^{\#}_{\perp}$ at $\perp$ for $\gamma_A(t_A)$. Formally, $\forall t_A \in seq(\mathcal{E}_A), \sigma^i_A \in \Sigma_A, \sigma^i_{\perp} \in \Sigma_{\perp} : \mathcal{R}^{\#}_{\perp} \vdash \mathcal{R}_A \iff \sigma^i_{\perp} \vdash \sigma^i_A \wedge \mathcal{R}^{\#}_{\perp}(\sigma^i_{\perp}, \gamma_A(t_A)) \vdash \mathcal{R}_A(\sigma^i_A, t_A)$.*

As direct corollary, $\mathbb{S}_{\mathcal{R}^{\#}_{\perp}}(\gamma_A(e)) \subseteq \gamma_A(\mathbb{S}_{\mathcal{R}_A}(e))$ and $\mathbb{D}_{\mathcal{R}^{\#}_{\perp}}(\gamma_A(e)) \subseteq \gamma_A(\mathbb{D}_{\mathcal{R}_A}(e))$.

4 Guarantees for Multiple Layers

A monitoring infrastructure is unsound if there exists a container at a higher abstraction layer that ignores the presence of data in its concretization. Unless one performs tracking at the level of single machine instructions, this situation is likely, and we cannot expect to achieve system-wide soundness in practice. What we can do, however, is to show *cross-layer soundness*: under the strong assumption of having sound models of two (or n) layers, we can show that if data moves exclusively within or in-between these layers, and we have information about their shared resources, our cross-layer model captures all cross-layer flows.

4.1 Step 3: Layer Composition

We proceed to define a notion of layer composition and discuss possible ways in which events observable at one layer may interfere with another layer. We then show a first overly-conservative way to model composition and prove its soundness. In the following, we focus on a system composed by two layers only; n-layered systems can be modeled by applying the same concepts recursively to each further layer. Without loss of generality we assume that $C_A \cap C_B = \emptyset$ for each pair of distinct $A, B \in \mathcal{L}^\perp$. Given two sound models for two layers of abstractions A and B in a system, our goal is to define a sound model for the system composed by A and B, denoted $A \otimes B$.

We begin by defining the composed system using the abstraction and concretization functions to compose the observations of monitors at the single layers. Let $C_{A \otimes B} = C_A \cup C_B$ be the set of containers in the composed system and $T_{A \otimes B} \subseteq seq(\mathcal{E}_A) \times seq(\mathcal{E}_B)$ the set of event traces, given by pairs of traces in A and B. We denote the composed state $\sigma_{A \otimes B} \in \Sigma_{A \otimes B} \subseteq \Sigma_A \times \Sigma_B$ as a pair of states in layers A and B respectively. For this notion of states, we derive an ideal (w.r.t $\perp$) composed monitor given by concretization and abstraction functions.

When talking about the state of the system or about traces in a multilayered system $A \otimes B$, we use the notation $|_\dagger$ to denote the projection to layer $\dagger$.

Mathematically speaking, it is simple to compose two monitors as follows.

Definition 6 (Ideal Composed Monitor). *Let t_A and t_B be traces at layers A and B, respectively, and σ_A^i and σ_B^i initial sound states. Let $\sigma_{A \otimes B}^\perp = \gamma_A(\sigma_A^i) \bowtie_\sigma \gamma_B(\sigma_B^i)$, $t_{A \otimes B}^\perp = \gamma_A(t_A) \bowtie_t \gamma_B(t_B)$ and $\sigma_{A \otimes B}'^\perp = \mathcal{R}_\perp^\#(\sigma_{A \otimes B}^\perp, t_{A \otimes B}^\perp)$. The function $\mathcal{R}_{A \otimes B}^\# : \Sigma_{A \otimes B} \times T_{A \otimes B} \to \Sigma_{A \otimes B}$ is defined as:*

$$\mathcal{R}_{A \otimes B}^\#((\sigma_A^i, \sigma_B^i), (t_A, t_B)) = (\alpha_A(\sigma_{A \otimes B}'^\perp), \alpha_B(\sigma_{A \otimes B}'^\perp)).$$

Practically speaking, we usually do not have access to the particular sequence of events occurring at $\perp$, i.e., to the ideal $\mathcal{R}_\perp^\#$ monitor and to precise concretization/abstraction functions for the containers. However, as we did for the single layers, we can characterize sound approximations of composed monitors.

Definition 7 (Soundness of Composing Monitor). *A monitor $\mathcal{R}_{A\otimes B}$ is sound w.r.t $\bot$, written $\mathcal{R}_{A\otimes B}^{\#} \vdash \mathcal{R}_{A\otimes B}$ if for all $\sigma_A, \sigma_B, t_A, t_B$ with $\sigma' = \mathcal{R}_{A\otimes B}((\sigma_A, \sigma_B), (t_A, t_B))$ the projections to A-level containers $\sigma'|_A$ and B-level containers $\sigma'|_B$ are sound w.r.t. $\mathcal{R}_{\bot}^{\#}(\gamma_A(\sigma_A) \bowtie_\sigma \gamma_B(\sigma_B), \gamma_A(t_A) \bowtie_t \gamma_B(t_B))$.*

4.2 Step 4: Sound Monitor Based on the State Relation

Let two containers at different layers c_A and c_B be *related*, written $c_A \sim c_B$, if their $\bot$-concretizations overlap, $\gamma_A(c_A) \cap \gamma_B(c_B) \neq \emptyset$. Without any additional information about related containers in A and B, the only sound approximation for $\sigma' = \mathcal{R}_{A\otimes B}(\sigma, (t_A, t_B))$ is $\forall c \in \mathcal{C}_{A\otimes B} : \sigma'(c) = \mathcal{D}$, i.e. every container possibly contains any data. This is because some data d may be transferred to a container c_B by some event $e_B \in \mathcal{E}_B$, and if $c_A \sim c_B$, d would also be stored in $\gamma_A(c_A)$ because of the non-empty intersection of the concretizations. Unless d is stored in c_A, this is a violation of the soundness of $A \otimes B$ (cf. Definition 4).

However, assuming information about related containers to be known (see Sect. 5.2) it is easy to build a monitor $\hat{\mathcal{R}}_{A\otimes B}$ that approximates the data-flows induced by a trace of events by propagating the data from every source of the trace to any destination of the trace and to any container related to the destinations:

$$\forall c \in \mathcal{C}_{A\otimes B},$$
$$\hat{\mathcal{R}}_{A\otimes B}(\sigma, (t_A, t_B))(c) = \begin{cases} \sigma(c) \cup \bigcup_{c' \in \mathbb{S}_{A\otimes B}} \sigma(c') & \text{if } c \in \mathbb{D}_{A\otimes B} \vee \exists\, \tilde{c} \in \mathbb{D}_{A\otimes B} : c \sim \tilde{c} \\ \sigma(c) & \text{otherwise} \end{cases}$$

with $\mathbb{S}_{A\otimes B} = \mathbb{S}_{\mathcal{R}_A}(t_A) \cup \mathbb{S}_{\mathcal{R}_B}(t_B)$ and $\mathbb{D}_{A\otimes B} = \mathbb{D}_{\mathcal{R}_A}(t_A) \cup \mathbb{D}_{\mathcal{R}_B}(t_B)$.

$\hat{\mathcal{R}}_{A\otimes B}$ is sound (see [17] for a proof) but overly conservative. The next section shows how to leverage additional cross-layer information to increase precision.

5 Cross-Layer Models

Although the complete definition of γ and α may not be available, it is often the case that, in some contexts, partial information about it is known by domain experts (e.g. the set of related containers). The goal of this section is to model such partial information in form of *oracles* and to formalize a refined data flow tracking model that, leveraging these oracles, provides more precise results. The key idea is that if more information about the relation between layers A and B is available, a more precise sound model can be constructed.

After extending our notation to capture the duration of events, we illustrate an example of using additional information to improve tracking precision; afterward, we abstractly define *properties* for the oracles (operationally, the oracles are implementation-specific and have to be instantiated by experts); and finally, we show an algorithm that, given two instances of the model and of the oracles, soundly approximates their composition.

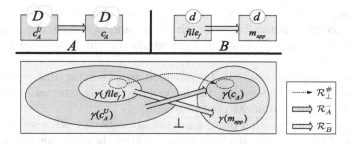

Fig. 2. Example of application loading a file, according to single layer monitors. Dotted sets represent actual $\mathbb{S}_{\mathcal{R}_\perp^\#}$ and $\mathbb{D}_{\mathcal{R}_\perp^\#}$ sets.

Events in $A \otimes B$: In the case of single layers, events are assumed to be instantaneous. However, in a multi-layer context the duration of an event at one layer may span several timesteps at the other layer. For instance, an event like LOAD() can be considered atomic at an application layer while it corresponds to many system call events at the operating system layer. For this reason, it is useful to distinguish between the moments in time when an event e begins and ends when reasoning about multiple layers.

Without loss of generality, the following assumes that every monitor for a layer † is defined over events in $\mathcal{E}_†^- \subseteq \mathcal{E}_† \times \{S, E\}$, where the suffixes S and E for an event e indicate, respectively, the beginning and the end of the execution of e. $\mathcal{R}_†^-$ denotes a monitor for such traces. While for simplicity's sake, we assume high level events to happen sequentially, serialized traces could also capture interleaving of events, e.g. $\langle e_S, e_S', e_E, e_E' \rangle$. Note that at level $\perp$, events are still serialized (single-core assumption, Sect. 2). To simplify notation, whenever a trace contains a certain event e_S directly followed by e_E, we write both events as e, e.g. instead of $\langle \text{LOAD}_s(), \text{READ}_s(), \text{READ}_e(), \text{LOAD}_e() \rangle$ we write $\langle \text{LOAD}_s(), \text{READ}(), \text{LOAD}_e() \rangle$. A precise formalization of all these concepts can be found in Appendix A.

5.1   Step 5: Increasing Precision — Example

Consider an application loading file f and two monitors, one for the application (A) and one for the operating system (B), both sound w.r.t. $\perp$. This generates the trace $t = \langle \text{LOAD}_s(f), \text{OPEN}(f, fd), \text{READ}(fd), \text{CLOSE}(fd), \text{LOAD}_e(f) \rangle$ where the first and last events happen at layer A and all the others at layer B Fig. (2).

Because files are not properly modeled in A, the source of the transfer in A is given by c_A^U (see definition of c^U in Sect. 2). Because the file is unknown to the application, it could possibly carry any data. This explains why $\forall \sigma_† \in \Sigma_† : \sigma_†(c_†^U) = \mathcal{D}$. The execution of $t|_A$ induces then a flow of all data $\mathcal{D}$ from c_A^U to c_A, where c_A is an internal container of the application, e.g. a document.

At the OS level, the file has a proper abstraction. Let *file$_f$* be such a container and d the data item stored in it. The execution of $t|_B$ is then modeled in B as a flow from *file$_f$* to container m_{app} representing the memory of the application.

If A and B were considered in isolation, the storage of c_A and m_{app} after the execution of t would be, respectively, $\mathcal{D}$ and d. Using the model presented in Step 4, instead, both containers would contain $\mathcal{D}$, a sound but coarse approximation.

A better approximation can be provided by observing that $\gamma_B(\textit{file } f) \subseteq \gamma_A(c_A^U)$ and $\gamma_A(c_A) \subseteq \gamma_B(m_{app})$, the latter because any internal object of the application is stored within its process memory.

Assuming the application process has not accessed any other sensitive data, the content of all the $\perp$-containers in $\gamma_B(m_{app})$ after the execution of t, including those in $\gamma_A(c_A)$, is at most d, as reported by $\mathcal{R}_B^-$ and because of its soundness. Therefore, a more precise monitor for the combined system would model t as a flow from $\gamma_B(\textit{file } f)$ to $\gamma_A(c_A)$, thus estimating that after the execution both c_A and m_{app} contain d. Note that this result is more precise than $\mathcal{R}_A^-$'s estimation.

What this scenario illustrates is that one layer has a more precise knowledge than the other about *the sources* of a certain event (e.g. the content of the file), while the other layer has a finer-grained understanding of the *destination* of the transfer (e.g. the app-specific container c_A). Let the term *cross actions* indicate those high-level operations, like "'`Application x loading file f`'" in the example (cf. Fig. 3), that correspond to traces of events at both layers in which this intuition holds.

In the following, we characterize events in this kind of traces, by referring to `IN`,`OUT` and `INTRA` as *behaviors* of events. If a certain cross action generates two events $e_A \in \mathcal{E}_A^-$ and $e_B \in \mathcal{E}_B^-$ such that $\gamma_A(\mathbb{S}_{\mathcal{R}_A^-}(e_A)) \subseteq \gamma_B(\mathbb{S}_{\mathcal{R}_B^-}(e_B))$ and $\gamma_B(\mathbb{D}_{\mathcal{R}_B^-}(e_B)) \subseteq \gamma_A(\mathbb{D}_{\mathcal{R}_A^-}(e_A))$, we say that e_A is an `OUT` event and that e_B is an `IN` event. If an event is neither `IN` nor `OUT` then it is an `INTRA` event. In completely independent layers or when a layer is considered in isolation, every event is an `INTRA` event. In a multi-layer context an `INTRA` event at layer † propagates data within † according to $\mathcal{R}_†^-$ and, in turn, to any other layer via related containers.

Hence, in addition to the dependency between layers generated by related containers and discussed in Sect. 4.2, we consider also a second class of cross-layer flows, i.e. those due to `IN` and `OUT` events.

Definition 8. *A* cross-layer flow *of data is generated by either: (1) the result of executing an event that transfers data to a container at one layer that is related with a container at the other layer, or (2) a cross action generating a sequence of events at both layers that includes at least one* IN *event at one layer and at least one respective* OUT *event at the other layer.*

The intuition behind `IN` and `OUT` events is that, in spite of what the single layer monitors may estimate, the only data flowed to the destinations of a certain `IN` event (e.g. `LOAD()`) is at most the same data read by the respective `OUT` events (e.g. `READ()`). In the next subsections we capture the two kinds of cross-layer dependencies described in Definition 8 in form of two oracles, which describe the relation between two layers. Provided an instantiation of these oracles, and the models for the layers, it is possible to automatically generate a sound precise model for the whole system composed by both layers (Sect. 5.3).

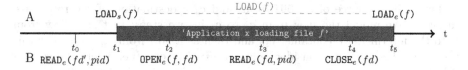

Fig. 3. Example of cross action. The X_B oracle applied to generic $\mathtt{READ}_e(..)$ events at time t_0 returns $(\mathtt{INTRA}, \emptyset)$. The $\mathtt{READ}_e(..)$ event at time t_3 instead, being part of the loading cross action, corresponds to $(\mathtt{OUT}, \text{'Application x loading file } f\text{'})$. The respective $\mathtt{IN}$ event at layer A is the $\mathtt{LOAD}_e(..)$ event at time t_5.

5.2 Step 6: Definition of Oracles

$\underline{X_A \text{ oracle}}$: Information about related containers, as needed by the model described in Step 4, is captured by oracle $X_A : \mathcal{C}_{A \otimes B} \to \mathbb{P}(\mathcal{C}_{A \otimes B})$, which maps each container c to the set of all the containers related to c at other layers.

Oracle Property 1. $\forall c \in \mathcal{C}_{A \otimes B} : X_A(c) = \{c' \in \mathcal{C}_{A \otimes B} \mid \exists l \in \mathcal{L} : c' \in \mathcal{C}_l \wedge c \notin \mathcal{C}_l \wedge c \sim c'\}$.

Leveraging X_A, it is also possible to model the *sync* operator, which will be useful in the following. Given a state of the system, $sync : \Sigma_{A \otimes B} \to \Sigma_{A \otimes B}$ returns a new state in which all the data stored in each container have been propagated to all the related containers at other layers, i.e. $\forall c \in \mathcal{C}_{A \otimes B}, \sigma \in \Sigma_{A \otimes B} \bullet sync(\sigma)(c) = \sigma(c) \cup \bigcup_{c' \in X_A(c)} \sigma(c')$. Because the *sync* operator only adds data to containers, it is easy to prove that if σ is a sound state (cf. Sect. 3.2), then $\sigma' = sync(\sigma)$ is also a sound state.

$\underline{X_B \text{ oracle}}$: In a multi-layer system, the behavior of a given event may differ in different contexts. For instance, a $\mathtt{READ}()$ event signaled by the operating system is related to a $\mathtt{LOAD}()$ event at the application layer, only if the process that invoked the system call is the application's one and if the target file of the system call is the same file being loaded by the application. Similarly, if the application is loading two files at the same time, then a sound and precise modeling needs to associate each $\mathtt{LOAD}()$ with the respective $\mathtt{READ}()$ events only.

To model this distinction, we use a unique identifier, called *scope id*, for each distinct instance of a cross action. All the $\mathtt{IN}$ and $\mathtt{OUT}$ events at both layers that pertain to a certain cross action are associated to that cross action's scope id.

This is captured by oracle $X_B : \mathcal{E}_{A \otimes B}^- \times \Sigma \to \{\mathtt{IN}, \mathtt{OUT}, \mathtt{INTRA}\} \times SCOPE$, where $SCOPE$ is the set of scope ids, like 'Application x loading file f'. X_B maps each event to its respective behavior in the context of a cross action.

It is also important to aggregate and store the content of the data being transferred by the $\mathtt{OUT}$ events in a way that is usable by the next corresponding $\mathtt{IN}$ event, because multiple $\mathtt{IN}(\mathtt{OUT})$ events may correspond to the same $\mathtt{OUT}(\mathtt{IN})$ event, e.g. one $\mathtt{LOAD}()$ event may correspond to multiple $\mathtt{READ}()$ system calls.

For each scope id sc, we model the existence of an *intermediate* container c_{sc} for the cross layer flow. Storage information for the intermediate containers $(\mathcal{C}_{sc})$ must be part of the system state in form of storage function $s_{sc} : \mathcal{C}_{sc} \to \mathbb{P}(\mathcal{D})$.

Let c_s be a source of an OUT event and c_d a destination of the respective IN event. We model the flow from c_s to c_d in two steps: first as a flow from c_s to the intermediate container c_{sc} and then as a flow from c_{sc} to c_d. For this reason, in this work we consider only *serialized* traces, (i.e. where the sorting of indexed events by timestamp is unique, cf. Definition 9 in Appendix A), and where IN events take place *after* the respective OUT events. This assumption is not restrictive in practice and always held in concrete instantiations [15,16].

In summary, augmenting the set of states for the composed system $\Sigma_{A\otimes B} \subseteq \Sigma_A \times \Sigma_B \times (C_{sc} \to \mathbb{P}(\mathcal{D}))$, we can encode the relation between two given layers A and B by using the oracles $X_A : C_{A\otimes B} \to 2^{C_{A\otimes B}}$ and $X_B : \mathcal{E}_{A\otimes B}^- \times \Sigma_{A\otimes B} \to \{IN, OUT, INTRA\} \times SCOPE$, which, by definition, guarantee the following:

Oracle Property 2. *Let $t \in seq(\mathcal{E}_{A\otimes B}^-)$ be a trace of events terminating with the event e^I, identified as IN by the oracle X_B. Let $E^O \subseteq \mathcal{E}_{A\otimes B}^-$ be a set of respective (i.e. w.r.t. the same scope) events in t identified as OUT by the oracle X_B. Then, in an ideal monitoring ($\mathcal{R}_{A\otimes B}^{\#-}$) of t, the destinations of e^i contains at most the content of the sources of all the events in E^O at the time of their execution. Formally,*

$$\left(\begin{array}{l} X_B(\sigma^{e^I}, e^I) = (IN, sc) \wedge \\ \forall e \in E^O : X_B(\sigma^e, e) = (OUT, sc) \end{array} \right) \implies \sigma^{e^I}(\mathbb{D}_{\mathcal{R}^\#}(e^I)) \subseteq \bigcup_{e \in E^O} \sigma^e(\mathbb{S}_{\mathcal{R}^\#}(e))$$

where $\mathcal{R}^\#$ stands for $\mathcal{R}_{A\otimes B}^{\#-}$, t_e denotes the subtrace of events in t from the beginning until event e included, and σ^e is the state reached by the ideal monitor $\mathcal{R}_{A\otimes B}^{\#-}$ after executing t_e from the initial state, i.e. $\sigma^e = \mathcal{R}_{A\otimes B}^{\#-}(\sigma_i, t_e)$

The intuition is that if the oracle X_B states that a certain event e is an IN event in a trace, then the execution of e will transfer to e's destination containers at most the data stored in the sources of the respective OUT events in the past trace. This is the key behind the refined precision offered by $\dot{\mathcal{R}}_{A\otimes B}$ defined in Algorithm 1 in comparison with $\hat{\mathcal{R}}_{A\otimes B}$.

5.3   Step 7: Algorithm for Sound Composition

We now come to the main result of this paper. Our goal is to show that, given an instantiation of the oracles for which the two properties defined in Sect. 5.2 hold, a composition algorithm considering such oracles is sound w.r.t. an ideal monitor at $\bot$, and thus ensures weak-secrecy.

Let $\gamma_{A\otimes B}$ be the overloading of γ for $C_{A\otimes B}$, $\Sigma_{A\otimes B}$, $\mathcal{E}_{A\otimes B}^-$ and traces of events in $\mathcal{E}_{A\otimes B}^-$. Given the models for A and B and these two oracles, the model $A \otimes B$ for the composed system is specified as follows: First, the set of containers in the system $C_{A\otimes B}$ is given by $C_A \cup C_B \cup C_{sc}$, where C_{sc} is the set of intermediate containers (which represent no real container in the system, i.e. $\forall c \in C_{sc} \bullet \gamma_{A\otimes B}(c) = \emptyset$). Secondly, a state of the system $\sigma_{A\otimes B} \in \Sigma_{A\otimes B}$ corresponds to the state of the two layers A and B and the storage function for intermediate containers s_{sc}, $\sigma = (\sigma_A, \sigma_B, s_{sc})$.

ALGORITHM 1. $\dot{\mathcal{R}}_{A\otimes B}((\sigma_A, \sigma_B, s_{sc}), e)$

---

1   $s_{sc_{RET}} \longleftarrow s_{sc}; \sigma_{A_{RET}} \longleftarrow \sigma_A; \sigma_{B_{RET}} \longleftarrow \sigma_B;$

2   $(beh, sc) \longleftarrow X_B((\sigma_A, \sigma_B, s_{sc}), e);$

3   **switch** beh **do**

4   **case** INTRA

5   **if** $e \in \mathcal{E}_A^-$ **then** $\sigma_{A_{RET}} \longleftarrow \mathcal{R}_A^-(\sigma_A, e);$

6   **else** $\sigma_{B_{RET}} \longleftarrow \mathcal{R}_B^-(\sigma_B, e);$

7   **case** IN

8   **if** $e \in \mathcal{E}_A^-$ **then** $\sigma_{A_{RET}} \longleftarrow (\sigma_A[t \leftarrow \sigma_A(t) \cup s_{sc}(c_{sc})]_{t \in \mathbb{D}_{\mathcal{R}_A^-}(e)});$

9   **else** $\sigma_{B_{RET}} \longleftarrow (\sigma_B[t \leftarrow \sigma(t) \cup s_{sc}(c_{sc})]_{t \in \mathbb{D}_{\mathcal{R}_A^-}(e)});$

10   **case** OUT

11   **if** $e \in \mathcal{E}_A^-$ **then**

12        $s_{sc_{RET}} \longleftarrow s_{sc}[c_{sc} \leftarrow \sigma_A(t)]_{t \in \mathbb{S}_{\mathcal{R}_A^-}(e)};$

13        $\sigma_{A_{RET}} \longleftarrow \mathcal{R}_A^-(\sigma_A, e);$

14   **else**

15        $s_{sc_{RET}} \longleftarrow s_{sc}[c_{sc} \leftarrow \sigma_B(t)]_{t \in \mathbb{S}_{\mathcal{R}_A^-}(e)};$

16        $\sigma_{B_{RET}} \longleftarrow \mathcal{R}_B^-(\sigma_B, e);$

17 **return** $sync(\sigma_{A_{RET}}, \sigma_{B_{RET}}, s_{sc_{RET}})$

Given two sound instantiations of the model for A and B and the two oracles defined above, a sound and precise model of the data flows within and across these two layers is captured by $\dot{\mathcal{R}}_{A\otimes B}$ defined in Algorithm 1[1].

Theorem 1. *Given two oracles X_A and X_B, for which properties 1 and 2 hold, two monitors for two layers $\mathcal{R}_A^-$, $\mathcal{R}_B^-$, an initial state $\sigma_{A\otimes B} = (\sigma_A, \sigma_B)$ and a serializable trace of events $t \in seq(\mathcal{E}_{A\otimes B}^-)$, if $\sigma_\perp \vdash \sigma_A$, $\sigma_\perp \vdash \sigma_B$, $\mathcal{R}_\perp \vdash \mathcal{R}_A^-$ and $\mathcal{R}_\perp \vdash \mathcal{R}_B^-$, then $\dot{\mathcal{R}}_{A\otimes B}((\sigma_A, \sigma_B), (t_A, t_B))$ is sound, i.e. $\mathcal{R}_{A\otimes B}^\# \vdash \dot{\mathcal{R}}_{A\otimes B}$.*

A detailed proof is provided in [17]. The intuition is that, for INTRA events, $\dot{\mathcal{R}}_{A\otimes B}$ behaves similarly to $\hat{\mathcal{R}}_{A\otimes B}$, and therefore it is sound, and for OUT events related to a scope sc, the content of the sources is also stored in a container c_{sc}, from where it can be "read" by the corresponding IN events and transferred to their destinations. The soundness then comes from Oracle Property 2.

6   Related Work

In terms of system-wide data flow tracking, we distinguish three classes of solutions in the literature. The first class includes solutions that focus on a single layer, like the operating system [8,10,12], the hypervisor [27] or the hardware level [4]. With respect to our model, hardware level solutions could be seen as the $\perp$ layer. Despite recent improvements in efficiency both at the software [1]

---

[1] Let m be a function of type $S \to T$ and $X \subseteq S$. $m' = m[x \leftarrow expr]_{x \in X}$ indicates a function $S \to T$ such that $m'(y) = expr$ for any $y \in X$ and $m'(y) = m(y)$ otherwise.

and hardware level [6], solutions in this class fail to capture the high-level semantics of events and objects (e.g. "forward a mail").

The second class of solutions includes those approaches that consider multiple instantiations of the same solution for *one* specific level of abstraction, usually the application layer. This class of work includes solutions like [11,20], where the inter-application flow tracking relies on the simultaneous execution of the sender and the receiver events, both at the application layer. None of them can model a flow of data toward resources at different layers, e.g. toward a file; given a monitor for the second layer, this is instead possible with our model.

The third class of related work includes approaches that consider multiple layers of abstractions at the same time. [18] is a work from the area of provenance aware storage systems, where representations of data are considered at three system layers at the same time (network, file system, workflow engine). Depending on the type of the content being handled, this work relies on tracking solutions that interact with each other and exchange taint results across different layers. Similarly, the *Garm* tool [7], aims at tracking data provenance information across multiple applications and machines. Garm instruments application binaries to track and store the data flow within and across applications, and to monitor interactions with the OS. Although both [7,18] address multiple layers of abstraction at the same time, none describes a general model applicable to a different number or type of layers, but each rather focuses on hard-coded solutions for the specific layers of abstraction considered.

[19] addresses multiple layers of abstraction generically by integrating a basic data flow tracking schema with a usage control framework. Here, the specification of the cross-layer dependencies is performed ad-hoc and all the monitors are executed in parallel in an independent fashion. Step 6 of [19] defines the meaning of cross-layer flows at the semantics model, but does not provide any notion of soundness, nor any operationalized way to monitor such flows at runtime.

A work more related to ours is *Shrift* [16], a solution for system-wide hybrid information flow tracking. Shrift replaces the runtime monitoring of an application with a statically computed mapping between its inputs and outputs, which is used at runtime by an operating system layer monitor to model data flows through the application. While using the model presented here, [16] does not describe cross layer flows in general.

7   Conclusions

In this paper we presented a formal definition of soundness, in terms of a notion related to information flow (weak-secrecy), for system-wide data flow tracking at and across different layers of abstraction. This semantic characterization of soundness is the first of its kind and represents the paper's first contribution.

We also proposed a generic schema to compose data flow analyses at various levels. Our schema relies on the existence of partial oracles that spell out the relation between the different levels in an actual system. The operationalization of the composition as an algorithm for runtime monitoring and the proof of its soundness represent the second major contribution of this research.

It is crucial to make the oracle assumptions explicit, even though in practice it is challenging to prove that single layer monitors and oracles are accurate, due for instance to non-deterministic low level interleavings and implementation details such as temporary variables and files. Such assumptions are usually reasonable, given that domain experts can accurately model the data-flow propagation of single high-level events, and whenever relations between layers are well known.

We have instantiated the framework described in this work to connect instantiations for different layers of abstraction, including a mail client [14], X86 binaries [2], Java Bytecode [16], and different operating systems [10,25], proving its feasibility. We argue that the genericity of the approach makes it possible to capture other solutions for data flow tracking from the literature, e.g. [3,5,28], as single-layer monitor instances, and to connect them to trackers at other layers in a sound manner. We do not discuss implementation details and experiments here because of space restrictions, but refer to [15] for more information.

In sum, our proposed cross-layer algorithm $\mathcal{R}_{A\otimes B}$ conservatively estimates and synchronizes the data propagation state between layers or inside one layer given that monitors are sound in isolation, that oracles are accurate, and that traces are serializable. Our implementation experiments show that these conditions are met often in practice, allowing for a sound and precise analysis. If, however, some of these conditions are not met, then one is forced to use a more conservative analysis (like $\hat{\mathcal{R}}_{A\otimes B}$) which propagates data from all sources to all destinations of a trace. Ultimately, if sources and destinations are unknown, the only possible sound analysis is to propagate all data to all containers.

Appendix

A   Serialized Events

Let $t^S(e) : \mathcal{E} \to \mathbb{N}$ and $t^E(e) : \mathcal{E} \to \mathbb{N}$ be two functions that return, respectively, the time at which a certain event e starts and ends. In the context of multiple layers, we assume that for any event $e_\dagger \in \mathcal{E}_\dagger$ it holds that $e_\dagger$ terminates only after starting $(t^S(e_\dagger) < t^E(e_\dagger))$ and that for every event e observed, the single layer monitors report an event e^S at time $t^S(e)$ to notify the beginning of e and an event e^E at time $t^E(e)$ to notify its end. In concrete implementations it is usually possible to observe or approximate these two aspects of any event.

For $\dagger \in \mathcal{L}$, let $\mathcal{E}_\dagger^- \subseteq \mathcal{E}_\dagger \times \{S, E\}$ be the set of such *indexed* events that denote when events in $\mathcal{E}_\dagger$ start and end. Let $ser : seq(\mathcal{E}_\dagger) \to seq(\mathcal{E}_\dagger^-)$ the operator that converts a trace of events $t_\dagger \in seq(\mathcal{E}_\dagger)$ into its indexed equivalent $t_\dagger^- \in seq(\mathcal{E}_\dagger^-)$ by replacing every event $e_\dagger \in t_\dagger$ with the sequence $\langle e_\dagger^S, e_\dagger^E \rangle$.

Lemma 1. *For each monitor $\mathcal{R}_\dagger$ ($\dagger \in \mathcal{L}$), there always exists a monitor $\mathcal{R}_\dagger^-$: $\Sigma_\dagger \times \mathcal{E}_\dagger^- \to \Sigma_\dagger$ such that $\forall \sigma_\dagger \in \Sigma_\dagger, \forall t_\dagger \in seq(\mathcal{E}_\dagger) : \mathcal{R}_\dagger(\sigma, t_\dagger) = \mathcal{R}_\dagger^-(\sigma, ser(t_\dagger))$.*

Proof. Given $\mathcal{R}_\dagger$, the monitor $\mathcal{R}_\dagger^-$, defined as $\mathcal{R}_\dagger^-(\sigma, (e_\dagger, i)) = \sigma$ if $i = S$ and $\mathcal{R}_\dagger^-(\sigma, (e_\dagger, i)) = \mathcal{R}_\dagger(\sigma, e_\dagger)$ if $i = E$, respects the property. $\square$

It is hence safe to assume, without loss of generality, that every monitor for a layer $\dagger$ is defined over events in $\mathcal{E}_{\dagger}^{-}$. We denote such a monitor $\mathcal{R}_{\dagger}^{-}$.

Definition 9 (Serializable Trace). *A trace $t = (t_A, t_B)$ is serializable if for every pair of events $e_A \in t_A, e_B \in t_B$, $t^S(e_A) \neq t^S(e_B)$ and $t^E(e_A) \neq t^E(e_B)$.*

Let $\mathcal{E}_{A \otimes B} = \mathcal{E}_A \cup \mathcal{E}_B$ and $\mathcal{E}_{A \otimes B}^{-} = \mathcal{E}_{A \otimes B} \times \{S, E\}$. If a trace $t = (t_A, t_B) \in seq(\mathcal{E}_A) \times seq(\mathcal{E}_B)$ is serializable, then it is possible to construct a trace $t^{-} \in seq(\mathcal{E}_{A \otimes B}^{-})$ that is equivalent to t, in the sense that it is possible to reconstruct each one given the other. t^{-} is given by the events in $ser(t_A) \bowtie_t ser(t_B)$ sorted by timestamp. The monitor for the composed system $\dot{\mathcal{R}}_{A \otimes B}$ described in step 7 of this work assumes the trace of input events $t = (t_A, t_B)$ to be serializable and provided as a sequence of events in $\mathcal{E}_{A \otimes B}^{-}$ ($\dot{\mathcal{R}}_{A \otimes B} : \Sigma_{A \otimes B} \times \mathcal{E}_{A \otimes B}^{-} \to \Sigma_{A \otimes B}$).

Note that we can relax the assumption on the serializable traces because any trace of events $t_{A \otimes B} = (t_A, t_B)$ in $A \otimes B$ can be seen as longest possible concatenation of subtraces $t_i = (t_{iA}, t_{iB})$, such that any event starting in t_i also terminates within t_i and viceversa and such that $(t_{1A} :: t_{2A} :: .. :: t_{nA}) = t_A$ and $(t_{1B} :: t_{2B} :: .. :: t_{nB}) = t_B$. Then, for each t_i,

$$\mathcal{R}_{A \otimes B}(\sigma, t_i) = \begin{cases} \dot{\mathcal{R}}_{(\sigma, t_i)} & \text{if } t_i \text{ is serializable} \\ \hat{\mathcal{R}}_{(\sigma, t_i)} & \text{otherwise} \end{cases}$$

$\mathcal{R}_{A \otimes B}$ is a sound monitor that is no less precise than $\hat{\mathcal{R}}_{(\sigma, t)}$ and does not require t to be serializable.

References

1. Austin, T.H., Flanagan, C.: Efficient purely-dynamic information flow analysis. ACM Sigplan Not. **44**(8), 20–31 (2009)
2. Biswas, A.K.: Towards improving data driven usage control precision with intra-process data flow tracking. Master's thesis, Technische Universität München (2014)
3. Chin, E., Wagner, D.: Efficient character-level taint tracking for java. In Proceedings of the ACM Workshop on Secure Web Services, pp. 3–12 (2009)
4. Chow, J., Pfaff, B., Garfinkel, T., Christopher, K., Rosenblum, M.: Understanding data lifetime via whole system simulation. In: USENIX Security (2004)
5. Crandall, J.R., Chong, F.T.: Minos: control data attack prevention orthogonal to memory model. In: Proceedings MICRO37, pp. 221–232. IEEE (2004)
6. de Amorim, A.A., Dénes, M., Giannarakis, N., Hritcu, C., Pierce, B.C., Spector-Zabusky, A., Tolmach, A.: Micro-policies (2015)
7. Demsky, B.: Cross-application data provenance and policy enforcement. ACM Trans. Inf. Syst. Secur. **14**(1), 1–22 (2011)
8. Enck, W., Gilbert, P., Chun, B.-G., Cox, L.P., Jung, J., McDaniel, P., Sheth, A.N.: TaintDroid: an information-flow tracking system for realtime privacy monitoring on smartphones. In: USENIX OSDI (2010)
9. Goguen, J.A., Meseguer, J.: Security policies and security models. In: IEEE Symposium on Security and Privacy (1982)
10. Harvan, M., Pretschner, A.: State-based usage control enforcement with data flow tracking using system call interposition. In: NSS (2009)

11. Kim, H.C., Keromytis, A.D., Covington, M., Sahita, R.: Capturing information flow with concatenated dynamic taint analysis. In: ARES (2009)
12. Krohn, M., Yip, A., Brodsky, M., Cliffer, N., Kaashoek, M.F., Kohler, E., Morris, R.: Information flow control for standard OS abstractions. In: SOSP (2007)
13. Kumari, P., Pretschner, A., Peschla, J., Kuhn, J.-M.: Distributed data usage control for web applications: A social network implementation. In: Proceedings of the First ACM Conference on Data and Application Security and Privacy, CODASPY 2011, pp. 85–96. ACM (2011)
14. Lörscher, M.: Usage Control for a Mail Client. Master thesis, TU Kaiserslautern (2012)
15. Lovat, E.: Cross-layer Data-centric Usage Control. Ph.D. thesis, Technische Univesität München (2015)
16. Lovat, E., Fromm, A., Mohr, M., Pretschner, A.: SHRIFT system-wide hybrid information flow tracking. In: Federrath, H., Gollmann, D., Chakravarthy, S.R. (eds.) SEC 2015. IFIP AICT, vol. 455, pp. 371–385. Springer, Heidelberg (2015). doi:10.1007/978-3-319-18467-8_25
17. Lovat, E., Ochoa, M., Pretschner, A.: Sound and precise cross-layer data flow tracking. Technical Report TUM-I1629, Technische Universität München, January 2016. https://mediatum.ub.tum.de/node?id=1289467
18. Muniswamy-Reddy, K., Braun, U., Holland, D.A., Macko, P., Maclean, D., Margo, D., Seltzer, M., Smogor, R.: Layering in provenance systems. In: USENIX (2009)
19. Pretschner, A., Lovat, E., Büchler, M.: Representation-independent data usage control. In: Garcia-Alfaro, J., Navarro-Arribas, G., Cuppens-Boulahia, N., de Capitani di Vimercati, S. (eds.) DPM 2011 and SETOP 2011. LNCS, vol. 7122, pp. 122–140. Springer, Heidelberg (2012)
20. Rasthofer, S., Arzt, S., Lovat, E., Bodden, E.: Droidforce: Enforcing complex, data-centric, system-wide policies in android. In: ARES (2014)
21. Smith, G.: On the foundations of quantitative information flow. In: Alfaro, L. (ed.) FOSSACS 2009. LNCS, vol. 5504, pp. 288–302. Springer, Heidelberg (2009)
22. Suh, G.E., Lee, J.W., Zhang, D., Devadas, S.: Secure program execution via dynamic information flow tracking. In: ACM SIGARCH (2004)
23. Volpano, D.: Safety versus secrecy. In: Cortesi, A., Filé, G. (eds.) SAS 1999. LNCS, vol. 1694, p. 303. Springer, Heidelberg (1999)
24. Volpano, D., Smith, G.: A type-based approach to program security. In: Bidoit, M., Dauchet, M. (eds.) CAAP 1997, FASE 1997, and TAPSOFT 1997. LNCS, vol. 1214, pp. 607–621. Springer, Heidelberg (1997)
25. Wüchner, T., Pretschner, A.: Data loss prevention based on data-driven usage control. In: 23rd IEEE International Symposium on Software Reliability Engineering (ISSRE), pp. 151-160, November 2012
26. Yin, H., Song, D., Egele, M., Kruegel, C., Kirda, E.: Panorama: Capturing system-wide information flow for malware detection and analysis. In: CCS (2007)
27. Zhang, Q., McCullough, J., Ma, J., Schear, N., Vrable, M., Vahdat, A., Snoeren, A.C., Voelker, G.M., Savage, S.: Neon: System support for derived data management. SIGPLAN Not. **45**(7), 63–74 (2010)
28. Zhu, Y., Jung, J., Song, D., Kohno, T., Wetherall, D.: Privacy scope: A precise information flow tracking system for finding application leaks. Technical Report UCB/EECS-2009-145, EECS Department, University of California, Berkeley, October 2009

Automatically Extracting Threats
from Extended Data Flow Diagrams

Bernhard J. Berger[✉], Karsten Sohr, and Rainer Koschke

Center for Computing Technologies (TZI), Universität Bremen, Bremen, Germany
{berber,sohr,koschke}@tzi.de

Abstract. Architectural risk analysis is an important aspect of developing software that is free of security flaws. Knowledge on architectural flaws, however, is sparse, in particular in small or medium-sized enterprises. In this paper, we propose a practical approach to architectural risk analysis that leverages Microsoft's threat modeling. Our technique decouples the creation of a system's architecture from the process of detecting and collecting architectural flaws. This way, our approach allows an software architect to automatically detect vulnerabilities in software architectures by using a security knowledge base. We evaluated our approach with real-world case studies, focusing on logistics applications. The evaluation uncovered several flaws with a major impact on the security of the software.

Keywords: Architectural risk analysis · Threat modeling · Automatic flaw detection

1 Introduction

Software security is an important topic for software vendors. There are complementary measures to assess the security status of software systems. First, one can assess the security at the architectural level to ensure that there are no security flaws. This approach is based on software models and attempts to identify conceptual problems, such as missing encryption of confidential data. Second, static analyzers are available that can detect implementation-level bugs, such as SQL injection and Cross-Site Scripting vulnerabilities.

Industry tends to make use of static security analyzers because they are easy to use and do not require deep security knowledge to employ them. Finding architecture-level security flaws using modeling techniques requires a deeper understanding of security, typical security problems, security measures, and their implications. In academia, other more formal approaches have been established, notably, language-based security [20], model-driven development for security [6], and stepwise refinement [15]. Nevertheless, it will take time until these approaches have a practical impact on industry.

In this paper, we present a tool-supported, practical approach to architectural risk analysis based on Microsoft's Threat Modeling [9,24]. A manually-crafted

© Springer International Publishing Switzerland 2016
J. Caballero et al. (Eds.): ESSoS 2016, LNCS 9639, pp. 56–71, 2016.
DOI: 10.1007/978-3-319-30806-7_4

model of a security architecture can be automatically analyzed with the help of security rules defined in a knowledge base. These rules identify well-known architecture-level security flaws and existing countermeasures. The analysis leads to a list of tackled security problems and a list of not handled security flaws. The tool support speeds up the process of architectural risk analysis and hence reduces the monetary effort.

In particular, our contributions are as follows:

1. introduction of *extended data flow diagrams*, a refinement of data flow diagrams, which are a representation of the system architecture and are used by Microsoft's Threat Modeling to identify security flaws [24],
2. provision of a catalog of threats based on well-known resources, such as CAPEC and CWE,
3. a tool for automatically finding these threats in extended data flow diagrams,
4. an evaluation of the catalog with manually crafted data flow diagrams in the context of three real-world applications.

2   Background

Threat Modeling and architectural risk analysis can be used to detect architecture-level security flaws. These techniques target fundamental security flaws in the software architecture rather than detecting low-level security bugs. Consequently, it is expected that the impact of such flaws is higher than low-level bugs [16].

Threat Modeling has been introduced by Microsoft [9,24]. It is part of the design phase in Microsoft's Security Development Lifecycle [17] and therefore is employed every time the design phase is executed. In the first step of Threat Modeling, data flow diagrams (DFDs) for the system in question are created. In the second step, the STRIDE approach is used to identify security flaws based on the data flow diagrams created in the first step. The STRIDE approach is an attacker-centric approach where the analyst tries to find points of the software where an attacker can breach the protection goals of information security. The identified threats are a target for risk analysis to identify the most important threats that must be addressed. Data flow diagrams are kept simple to ease their usage. Different publications propose extensions of these diagrams to capture more information of the system (see [3,8]).

Architectural Risk Analysis (ARA) is described by McGraw in *Software Security: Building Security In* [16]. In the first step of ARA, an architecture overview is created. How it is represented is not defined. In this way, the approach fits into every software development process. This step is followed by three different analysis steps with different priorities, namely, attack resistance analysis, ambiguity analysis, and weakness analysis. Attack resistance analysis focuses on finding well-known security flaws in the architecture, whereas the ambiguity analysis targets security flaws that are specific to the analyzed system. The weakness analysis searches for problems in external software components, such as used frameworks.

3   Analysis Approach

The goal of our analysis is to automatically identify threats (architectural flaws) and mitigations. Since Threat Modeling employs DFDs to model an application's system architecture [24], our approach is also based on these diagrams. We follow the Threat Modeling approach, which is used by many large software vendors such as Microsoft, SAP, and EMC. Alternatively, we could have used a UML-based approach to architectural risk analysis [2,12], but we employed Threat Modeling due its relevance in industry.

To better support the automated analysis for security flaws, we introduce Extended Data Flow Diagrams (EDFDs). EDFDs cover all concepts of traditional data flow diagrams, but use enhancements allowing us to add additional semantics. Furthermore, we capture possible threats in a knowledge base that is applicable to previously created software models. This results in a Threat Model process linking threats to elements of a diagram. Here, the threats are noted in the diagram as well as possible mitigations.

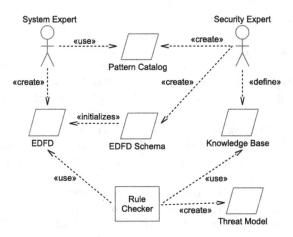

Fig. 1. Usage scenario

Figure 1 depicts the usage scenario and the responsibilities. Security experts define a knowledge base with rules about common architectural security flaws (see Sect. 3.4) and create a catalog with predefined patterns (see Sect. 3.3) as well as an EDFD schema (see Sect. 3.2). The EDFD schema is used to initialize EDFDs (see Sect. 3.2), which are then created by system experts (who are not necessarily security experts). To create these EDFDs, the system expert consults the given pattern catalog. The resulting concrete EDFD and the knowledge base are used as input by a rule checker, which automatically creates a threat model (see Sect. 3.6).

3.1   Dataflow Diagrams

DFDs consist of five different modeling elements. Processes are active components that process data. Data stores are components such as databases or files that store data. Interactors are external systems or users who interact with the analyzed system. All these elements can interact with each other using data flows. Data flows may cross trust boundaries. A trust boundary symbolizes different trust levels within a system. Figure 2a shows all available DFD elements.

3.2   Introducing EDFDs

We use EDFDs since we witnessed several shortcomings with the existing data flow diagrams while we applied Threat Modeling to real industry systems. First of all, DFDs do not capture knowledge on existing security measures or requirements. Therefore, an automated analysis of an already-defined security architecture against security rules is not possible. Second, data flows are unidirectional channels between arbitrary elements. Data that are transported in this channel are specified by simple labels that are attached to data flow edges. For this reason it is hard to identify data that flow through the whole system. Furthermore, trust boundaries are used to indicate trust areas. There is no information which components belong to a trust area making it impossible to identify dangerous data flows or access paths. Therefore, we decided to model aspects, such as data, communication, and security measures, in a more explicit way to allow an automated analysis of DFDs.

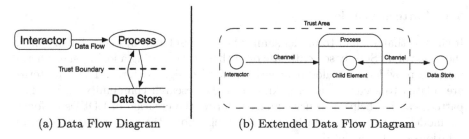

(a) Data Flow Diagram   (b) Extended Data Flow Diagram

Fig. 2. Exemplary Diagrams

EDFDs comprise four central concepts, namely *Elements*, *Channels*, *Trust Areas*, and *Data* (see Fig. 2b). Each of these concepts is typable and allows us to add additional information called annotations. These annotations help us add information about security measures or requirements to these components. Types and annotations are designed hierarchically allowing us to refine them. An HTTP connection, for instance, is a special kind of interprocess data flow. Each defined type may imply annotations to the component. The set of defined types and annotation types is called EDFD schema, which allows us to adapt EDFDs to the needs of new systems easily. Nevertheless, EDFDs come with a predefined schema to provide a starting point for modeling systems.

Elements: EDFDs abstract the entities *Data Store*, *Process*, and *Interactor* from traditional DFDs to the *Element* concept. We use corresponding types to model these entities. A *Java Process*, for instance, is an element type and inherits from *Process* implying the *Java* annotation.

Elements can also be structured hierarchically, e.g., processes can consist of subprocesses. In this case we can define a parent-child relationship (see Fig. 2b).

Channels: We use channels to model data flows. This allows us to model different kinds of communications, such as one-to-one, one-to-many, and many-to-many communications. The predefined schema contains three root-channel types *InterProcessConnection, IntraProcessConnection* and *ManualInput.*

Trust Areas: In software systems, several circumstances exist that group elements to trust areas. Therefore, there are quite different kinds of trust areas, such as *Network, Machine Boundary,* or *Software Area.*

Data: The existing data instances depend on the current software system to be modelled. Therefore, we introduce a small number of predefined data types, such as *User Data* and its refinement *Credentials.* A frequently used annotation for data is the flag *IsConfidential.*

In summary, we use types and implied annotations and not mere annotations for the sake of usability. Since knowledge on architectural vulnerabilities is sparse (in particular, in small and medium-sized enterprises) and the implications of using certain security mechanisms are often unclear, we introduced the typing mechanism. This typing mechanism takes the burden from the analyst.

3.3 Pattern Catalog

It showed that we had to model similar facts for different systems, for instance, the usage of an SQL-based database or authentication using a login page. Therefore, we provided a catalog of security-related design patterns. These patterns are related to security patterns known in the research community [4,23]. The pattern catalog helps a security analyst during the creation of EDFDs to focus on modeling the system instead of thinking about the way certain technical solutions can be modeled.

3.4 Knowledge Base Rules

Our knowledge base captures descriptions of common security flaws in a machine readable form. A rule consists of a *name*, a *description* that explains the details of the flaw, and an estimation of the *severity* and the *likelihood* of the flaw. The rules contain queries to identify possible flaws and corresponding mitigations. The queries are defined using a graph query language called *Cypher Query Language* [10]. It is designed to describe subgraphs that should be matched in an arbitrary graph. Therefore, it is possible to define properties of nodes and paths within the subgraph. It is even possible to describe paths of infinite length.

3.5 Rule Checker

The detection of flaws consists of three steps. First the EDFD is lowered to a labeled and attributed graph representation. Then all rules are applied to the graph to identify possible threats. In the last step the threat model is generated based on the findings. During the lowering all elements that can be found in the EDFD are mapped to nodes or edges of a normal graph (see Fig. 3). A trust area is mapped to a node and an *include* edge to each contained element. Attributes and types are mapped to attributes in the resulting graph. For matching the rules are all matching subgraphs for each flaw pattern are identified. If a match is found, a corresponding entry in the threat model is created. Then the matching engine searches for possible mitigations starting with the identified subgraph.

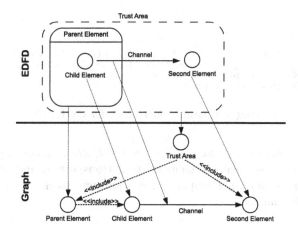

Fig. 3. Lowering of EDFDs to simple graphs

We implemented a graphical EDFD editor on top of Eclipse's Graphiti. It is able to create new EDFDs, initialized with a predefined schema, and modify existing ones. The visualization can be customized by type-dependent icons, and arrow styles in case of channels. Hierarchical elements can be folded and expanded. If channels exist where one of their endpoint elements is hidden because its parent has been folded, we automatically lift this edge to the visible parent [22].

3.6 Threat Model

The result of our automatic detection process is a threat model. Each threat references a *rule* from our knowledge base. Furthermore, it links at least one entity from an EDFD that is the target of the threat. Additionally, the threat can link a set of mitigations in the EDFD.

4 Knowledge Base

We populated our knowledge base with threats from the Common Weakness Enumeration (CWE) and the Common Attack Pattern Enumeration and

Classification (CAPEC). The most common problems are aggregated in the *Top 25 Most Dangerous Software Errors* [18]. Table 1 gives an overview of the supported CWE entries[1]. If a rule belongs to the Top 25, the corresponding ranking is given. The most interesting rules are those where we can give mitigation rules, for instance, *CWE-288: Authentication Bypass Using an Alternate Path or Channel*, or *CWE-319: Cleartext Transmission of Sensitive Information*. We distinguish between threats where we are able to detect possible mitigations and those where we are not able to do so. For the latter group a manual code review or additional static analyses are necessary to check whether the threat has been addressed.

```
MATCH ( src  :  Element )  −[flow  :  Channel]−>  ( tgt  :  Element )
WHERE flow . type . subtypeof ( " InterProcessCommunication " )
      AND ANY ( d IN  flow . data WHERE d . IsConfidential )
      AND NOT  flow . IsEncrypted
```

Listing 1.1. Graph Query Rule for CWE-319

Listing 1.1 shows our rule for CWE entry *CWE-319* as an example. The rule looks for a channel between two elements that transports confidential information. If the channel does not employ encryption, an attacker may capture the transported information.

```
1    MATCH ( src  :  Element )  −[p1: Channel *]−>  ( n1  :  Element )
2                        −[entry  :  Channel]−>  ( tgt  :  Element )
3                        <− [:  include ] − ( area  :  TrustArea {
                              Authentication_Required  :  true } )
4    WHERE NONE ( d in  entry . data WHERE d . IsCredential OR  d . IsSessionToken )
5         AND NOT ( src ) <−[:include]− ( area )
6         AND NOT ( n1 ) <−[:include]− ( area )
7         AND NONE ( n IN  nodes ( p1 ) WHERE ( area )  −[:includes]−> ( n ) )
```

Listing 1.2. Graph Query Rule for CWE-306

As a second example, we give the rule for *CWE-306: Missing Authentication for Critical Function* in Listing 1.2. The corresponding pattern graph in Fig. 4. We look for a path *p1* from a node *src* to a node *n1*. *n1* in turn is directly connected to the second node *tgt*. At last, there is a trust area node *area* that has an *include* relation to *tgt*. This part of the query searches for a path from *src* to *tgt* where *tgt* is contained in a trust area that requires authentication according to line 3 of the query. In line 4, we check that no session token or credential exists that is sent to *tgt*. Lines 5 to 7 ensure that none of the nodes on the *p1* are contained in the same trust area *area* including *src* and *n1*. With the help of these checks, we can find a flow to a component that is not authenticated.

---

[1] For the sake of presentation, we only give the complete names of CWE as well as CAPEC entries in the appendix.

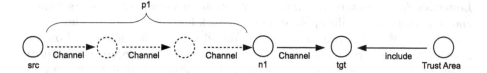

Fig. 4. Pattern Graph for CWE-306

Table 1. Excerpt of Supported Rules

C(W)E C(A)PEC	Rule	Top 25	(T)hreat (M)itigation	C(W)E C(A)PEC	Rule	Top 25	(T)hreat (M)itigation
W	89	1	T	W	190	24	T
W	78	2	T	W	759	25	T
W	120	3	T	W	288		M
W	79	4	T	W	319		M
W	306	5	M	W	602		M
W	311	8	M	A	108		T
W	352	12	T	A	16		T
W	22	13	T	A	22		T (M)
W	327	19	M	A	66		T
W	134	23	T	A	94		(T) (M)

The currently supported CAPEC rules are listed in Table 1. Supporting them is more complex than supporting CWE rules since they subsume complete classes of attack patterns. Sometimes it is impossible to detect all possible variants of an attack pattern. Therefore, we note some threats and mitigations in brackets to show that we support just some special cases of this attack pattern.

We support **additional rules** that do not match CWE or CAPEC entries. For instance, we are checking information flow policies with the help of our rule set. This is useful for broadcast channels where sensitive data may flow to processes that are not trustworthy enough to process this information. In total, we currently support 25 security-related rules that are not application-specific. In future we will add further rules depending on projects and use cases.

5   Evaluation

We evaluated our approach with the help of three industrial case studies from different vendors. The effort to construct these EDFDs summed up to half a day of work for each one. Two of the applications are from the logistics domain having a similar purpose and the third one is from e-government. In particular, we investigated the following questions.

1. Can we automatically identify threats and security flaws with EDFDs?
2. What is the impact of the identified security flaws?
3. Can we find similarities between the logistics applications?

Logistics Application A. Our first application comes from the logistics domain and is a company-specific application framework based on JavaEE technologies. The vendor offers a large number of domain-relevant applications based on this framework and wanted to identify framework-based flaws to improve the security status of their complete product portfolio. The software is mainly offered on a software-as-a-service (SaaS) base and has therefore a number of security requirements beyond the functional requirements. The applications help customers with customs clearance and fleet and port management. Figure 5 shows a simplified EDFD that we created during a workshop together with three system experts who are responsible for the architecture and the main development of the framework components. No-one of them had deep knowledge in the area of software security at that time.

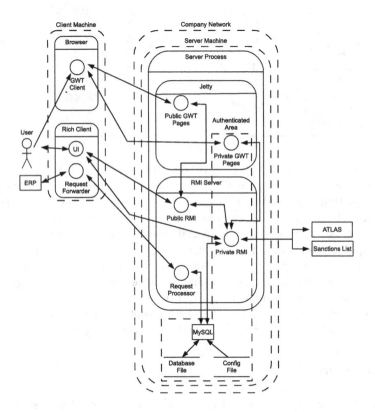

Fig. 5. Simplified EDFD for Logistics Application A

There are two possible clients. First, a browser-based client that is accessible by the end user. Second, a Java-based rich client supporting a user interface and an integrated *request forwarder* forwarding requests made by external customer systems, such as enterprise resource planning software, to the server. The server process and a related SQL-based database are running in the trust area of the software vendor's network. The server implements two interfaces for the clients. One of them is a GWT-based web interface and the second is an RMI-based interface for the rich client. Both interfaces consist of public and private components that require authentication. The clients send *credentials* to the public interfaces to authenticate the user and receive a *session object* for identifying the user for subsequent requests in return. Please note that we omit the visualization of transferred data and the annotations for the sake of clarity.

Our analysis approach identified several possible threats for this system that we discussed with the system experts as well as detected mitigations. An overview of the matching CWE entries can be found in Table 2. The rules identified correctly that there were confidential data that flowed between the clients and the server process. Hence, an attacker can try to capture these data during transmission. Our approach also found the mitigation that the channels were using TLS for transport encryption and therefore secured the data.

The rule checker found an instance of CWE entry 306 (see Fig. 4). The external *ERP* system sends requests to the *Request Forwarder* contained in the *Rich Client*. The requests are then forwarded to the *Request Processor* on the server-side. The forwarder in turn communicates directly with the *MySQL database*. This communication path does not transport authentication data which indicates that an authentication check is missing. The communication path was added afterwards to the software since some clients wanted to integrate the logistics application directly into their ERP software. Moreover, a possible SQL injection was detected for the private RMI interface, but a review showed that they used hand-written SQL statements in conjunction with a company-specific API that ensured that an SQL injection could not occur. Consequently, this finding is a false positive. Another problem we detected was the circumstance that the client implemented authorization checks rather than the server. Hence, this is an instance of the CWE entry 602. Our process detected more threats, which were then examined in manual reviews and revealed additional security flaws in this application framework.

Table 2. Detected Threats for *Logistics Applications*

CWE Top 25	(D)etected/(V)ulnerable	App A	App B	CWE Top 25	(D)etected/(V)ulnerable	App A	App B
89	1	D	V	327	19	V	
78	2	D	V	134	23		
120	3			190	24		
79	4	V	V	759	25	V	V
306	5	V		288		V	
311	8		V	319			V
352	12	V	V	602		V	V
22	13	D					

The consequences and the impact of these findings are high if one considers the sensitivity of the data and systems involved in port logistics. For example, an attacker can circumvent the client-side security checks and access all data and functions available. This can be achieved by using the request forwarder or simply by modifying the rich client. Since these problems occur in the application framework, each of the applications based on the framework is vulnerable. The consequence would be that seaports, international forwarding agents, and parcel services are not able to do their job. In the end, this would result in a major financial loss for the software vendor due to contractual penalties and for economy due to outstanding delivery of goods.

Logistics Application B. The second application from the logistics domain is similar to the first one. It helps manufacturing industry with customs declarations, sanction lists checks, and commissioning. It is implemented based on the JavaEE specifications and provides a web-based interface to the customers. The software is distributed and sold on a SaaS basis as well. The data are stored in a single database making multi-tenancy an important topic for the application's security. The application is divided into several products, such as import, export, and sanction lists. In total there are seven different products that can be bought by users (Fig. 6).

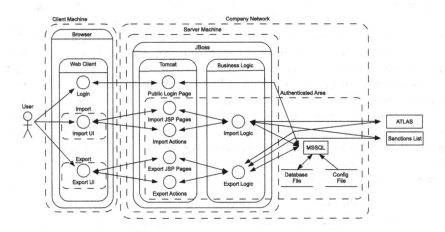

Fig. 6. EDFD for Logistics Application B

The software is structured like a typical Java Enterprise application. The client of the web application is a browser and is divided into an public *Login* page and one component for each aforementioned product. Each product on the client side is independent of other products. We modeled this as different trust areas. On the server side in the company's network, a JBoss application container runs a Tomcat web container and all product-specific *Business Logic* components implementing the view-independent algorithms and the persistence. Tomcat serves different Java Servlets (dynamic web content), a *Login Page* and JSP pages as well as Struts actions for each product. Struts is a framework that allows web programmers to implement the model view controller pattern

for dynamic web pages. For persistence purposes, the *Business Logic* components talk to a Microsoft SQL database running on a different machine. Where
necessary, the *Business Logic* components communicate with external systems,
such as online sanction lists and the German electronic customs interface Atlas.

Our automated analysis identified several existing flaws. An overview of
detected threats is given in Table 2. First of all it contains injection-based vulnerabilities, such as XSS and SQLi. The application was vulnerable to these
kinds of attacks since the programmers neglected input validation and did not
use an SQL abstraction layer such as the Java Persistence API—also an architectural flaw. Furthermore, we identified a threat based on *CWE-602: Client-Side
Enforcement of Server-Side Security.*

Similar to the findings in *Logistics Application A* the impact is crucial for
the security of the software system. It may be easy for attackers to bypass most
of the security measures if she has a valid account for the system. Furthermore,
if the communication between client and server is not protected by appropriate
transport encryption, an attacker has the possibility to steal the credentials of
an arbitrary user. Depending on the motivation of an attacker the vulnerabilities
can have different consequences. On the one hand, the attacker can be interested
in harming a specific customer and steal sensitive information, such as the list of
customers or exported goods. Furthermore, there can be a financial impact for
customers since some taxes are collected automatically on import or export in
advance. On the other hand, the attacker can be interested in compromising the
software provider and delete customer data or disrupt the functionality. Since
the contracts contain penalties for delays caused by the software, this finally can
lead to financial damage and may have legal consequences.

E-Government Application. The third case study, Governikus Service Components, belongs to the e-government domain and is part of a service-oriented
architecture. Its purpose is to create qualified digital signatures that are legally
binding in Germany. These signatures are created with the help of signature cards and are used by authorities for signing documents, such as birth
certificates.

Therefore, security is an immanent requirement and is taken into account
during development. Hence, the software is being evaluated according to the
Common Criteria (CC) [7].

For reasons of space we cannot give an EDFD here, but briefly describe it.
An external system communicates with a public web service using an HTTPS
connection. The web service runs in a Tomcat instance and sends the signing
requests to a worker application using Java Message Service, an asynchronous
message bus. The worker application in turn dispatches the requests to different
signature cards according to their purpose. The signature is then returned to the
calling process. The automatic analysis process did not detect any threats, which
was not surprising due to the efforts spent into the CC evaluation. Nevertheless,
the EDFD produced for this application helped us find a serious security flaw
manually. One important security measure is to enforce proper access control
for the signature card. The implemented checks ruled out that an outsider could

access the card. However, the case was not checked that a legitimate user tries to access a card of another user, i.e., the user identity was not compared with the identity required to access the card. Interestingly, the flaw was not detected before, although an EAL 4+ evaluation according to the CC was carried out on this system—an evaluation level with relatively high effort.

Results. Regarding the aforementioned questions, we can conclude from these three case studies:

1. We were able to automatically identify threats and security flaws in EDFDs. The detected flaws are known from collections such as the CWE or CAPEC.
2. In general we can see that these flaws lead to a major security breach, enabling an attacker to circumvent most parts of the existing security measures.
3. The investigated logistics applications have major security problems. It is obvious that security was not a concern that was addressed during development. This is problematic given the purpose of the systems.

6   Discussion

Our work is not the first one that attempts to identify security flaws in software architectures. Nevertheless, many companies try to avoid this step for budget reasons. Therefore, we split the responsibilities and define the different roles shown in Fig. 1. This way, we can reuse gathered knowledge and reduce the effort necessary to conduct architectural risk analysis. The pattern catalog additionally helps one decrease the time necessary to create an accurate EDFD. The approach can be implemented using other modeling techniques such as UML.

Since the knowledge base is not complete, our approach produces false negatives. A source of false positives and false negatives is the accuracy of the checked system model. Specifically, application-specific flaws cannot be detected automatically as we have seen in the context of the e-government case study. Since the number of found threats is within the double-digit range for all case studies, it is still possible to discuss the identified threats.

We are aware of the fact that our approach of finding possible threats and mitigations is known as subgraph isomorphism problem that is NP-complete in general. Nevertheless, our rule base is processed within a few minutes because most of the rules are local or the search space can be reduced due to the attribute and type constraints.

7   Related Work

In this paper, we focus on architectural security analysis and hence we discuss related work from this perspective. Our technique can be differentiated from static code analyzers [5]. Static code analyzers attempt to detect low-level programming bugs, such as SQLi vulnerabilities, at code level. In contrast, our approach works at the architectural level and aims to identify flaws, e.g., an application basically does not carry out input validation to avoid SQLi vulnerabilities.

Microsoft provides a tool that supports the Threat Modeling process [9]. This tool, in essence, makes available a catalog of questions that an analyst can apply to a given DFD rather than providing an analysis engine. Schaad and Borozdin applied STRIDE to block diagrams and identified possible threats and vulnerabilities introduced by the usage of third-party standard software components [21]. They also support a question-based assessment.

There are also works based on the UML that allow one to analyze security architectures [2,12,13]. The UML-based approaches let a software architect formulate security requirements that a software architecture must satisfy, e.g., access control or confidentiality requirements. In general, the introduction of UML and its constraint language OCL was meant to allow an architect to specify positive system requirements rather than anti-requirements (things that can go wrong), although UML/OCL can also be used for this purpose. In contrast to the aforementioned approaches, we utilize security knowledge and experience as provided by CWE and CAPEC. Consequently, our technique complements common UML-based approaches to security.

Almorsy et al. use *formalized vulnerability signatures* defined in OCL to automatically detect different kinds of security issues in C#, C++, C and VB.Net applications. Their approach allows them to detect implementation-level vulnerabilities, such as SQL-Injection and Cross-Site Scripting vulnerabilities [1]. Furthermore, they calculate security metrics, e.g. the attack surface metric (see Manadhata and Wing [14]). Currently, they do not aim at detecting architecture-level security flaws.

Jung et al. present a technique to check a service-oriented architecture implemented using the Apache Tuscany Framework. Their security rules are decomposed using a tree structure and therefore resemble the Security Goal Indicator Tree approach [19]. This approach, however, does not consider existing threats of a system [11].

8   Conclusion and Outlook

In this paper, we proposed an approach to the automated security analysis of software architectures. This analysis technique allows organizations to conduct architectural risk analysis more cost-effectively. We employed extended data flow diagrams as well as a knowledge base that contains information on architectural weaknesses and possible mitigations. We applied this technique to three real-world case studies and detected critical security flaws, in particular in the context of a port logistics system.

In the future, we will extend the knowledge base including the supported rule set with the help of further case studies from different domains. We can also combine the knowledge base with a reverse engineering approach that automatically extracts the EDFDs from legacy code. This allows us to integrate our analysis technique in later steps of the Security Development Lifecycle.

Acknowledgement. This work was supported by the German Federal Ministry of Education and Research (BMBF) under the grant 16KIS0069K (ZertApps project).

A   CWE and CAPEC Rules

CWE-22 Improper Limitation of a Pathname to a Restricted Directory ('Path Traversal')

CWE-78 Improper Neutralization of Special Elements used in an OS Command ('OS Command Injection')

CWE-79 Improper Neutralization of Input During Web Page Generation ('Cross-site Scripting')

CWE-89 Improper Neutralization of Special Elements used in an SQL Command ('SQL Injection')

CWE-120 Buffer Copy without Checking Size of Input ('Classic Buffer Overflow')

CWE-134 Uncontrolled Format String

CWE-190 Integer Overflow or Wraparound

CWE-288 Authentication Bypass Using an Alternate Path or Channel

CWE-306 Missing Authentication for Critical Function

CWE-311 Missing Encryption of Sensitive Data

CWE-319 Cleartext Transmission of Sensitive Information

CWE-327 Use of a Broken or Risky Cryptographic Algorithm

CWE-352 Cross-Site Request Forgery (CSRF)

CWE-602 Client-Side Enforcement of Server-Side Security

CWE-759 Use of a One-Way Hash without a Salt

CAPEC-16 Dictionary-based Password Attack

CAPEC-22 Exploiting Trust in Client (aka Make the Client Invisible)

CAPEC-66 SQL Injection

CAPEC-94 Man in the Middle Attack

CAPEC-108 Command Line Execution through SQL Injection

References

1. Almorsy, M., Grundy, J., Ibrahim, A.S.: Automated software architecture security risk analysis using formalized signatures. In: 35th International Conference on Software Engineering (ICSE), pp. 100–109 (2013)

2. Basin, D., Clavel, M., Doser, J., Egea, M.: Automated analysis of security-design models. Inf. Softw. Technol. **51**, 815–831 (2009)

3. Berger, B., Sohr, K., Koschke, R.: Extracting and analyzing the implemented security architecture of business applications. In: 2013 17th European Conference on Software Maintenance and Reengineering (CSMR), pp. 285–294 (2013)

4. Bunke, M., Sohr, K.: An architecture-centric approach to detecting security patterns in software. In: Erlingsson, Ú., Wieringa, R., Zannone, N. (eds.) ESSoS 2011. LNCS, vol. 6542, pp. 156–166. Springer, Heidelberg (2011)

5. Chess, B., West, J.: Secure Programming with Static Analysis. Addison-Wesley, Reading (2007)

6. Clavel, M., da Silva, V., Braga, C., Egea, M.: Model-driven security in practice: an industrial experience. In: Schieferdecker, I., Hartman, A. (eds.) ECMDA-FA 2008. LNCS, vol. 5095, pp. 326–337. Springer, Heidelberg (2008)

7. Criteria, C.: Common Criteria for Information Technology Security Evaluation-Part 1: Introduction and general model (2009). http://www.commoncriteriaportal.org/files/ccfiles/CCPART1V3.1R3.pdf
8. Dhillon, D.: Developer-driven threat modeling: lessons learned in the trenches. IEEE Secur. Priv. **9**(4), 41–47 (2011)
9. Hernan, S., Lambert, S., Ostwald, T., Shostack, A.: Uncover Security Design Flaws Using the STRIDE Approach. MSDN Magazine, November 2006. http://msdn.microsoft.com/en-us/magazine/cc163519.aspx
10. Holzschuher, F., Peinl, R.: Performance of graph query languages: comparison of cypher, gremlin and native access in neo4j. In: Proceedings of the Joint EDBT/ICDT 2013 Workshops, EDBT 2013, NY, USA, pp. 195–204. ACM, New York (2013) http://doi.acm.org/10.1145/2457317.2457351
11. Jung, C., Rudolph, M., Schwarz, R.: Security evaluation of service-oriented systems with an extensible knowledge base. In: 2011 Sixth International Conference on Availability, Reliability and Security (ARES), pp. 698–703 (2011)
12. Jürjens, J., Shabalin, P.: Automated verification of UMLsec models forsecurity requirements. In: Baar, T., Strohmeier, A., Moreira, A., Moreira, S.J. (eds.) UML 2004 - The Unified ModelingLanguage: Modeling Languages and Applications. LNCS, vol. 3273. Springer, Heidelberg (2004)
13. Kuhlmann, M., Sohr, K., Gogolla, M.: Comprehensive two-level analysis of static and dynamic rbac constraints with uml and ocl. In: Proceedings of the 2011 Fifth International Conference on Secure Software Integration and Reliability Improvement, pp. 108–117. IEEE Computer Society, Washington, DC (2011)
14. Manadhata, P.K., Wing, J.M.: An attack surface metric. IEEE Trans. Softw. Eng. **37**(3), 371–386 (2011)
15. Mantel, H.: Preserving information flow properties under refinement. In: IEEE Symposium on Security and Privacy, p. 78 (2001).http://computer.org/proceedings/s%26p/1046/10460078abs.htm
16. McGraw, G.: Software Security: Building Security In. Addison-Wesley, Reading (2006)
17. Microsoft: Microsoft Security Development Lifecycle (SDL) - Version 5.0. https://www.microsoft.com/en-s/download/details.aspx?displaylang=en&id=12285 (2010)
18. Mitre: CWE/SANS Top 25 Most Dangerous Software Errors (2015). Accessed: January 15, 2015 http://cwe.mitre.org/top25
19. Peine, H., Jawurek, M., Mandel, S.: Security goal indicator trees: a model of software features that supports efficient security inspection. In: 11th IEEE High Assurance Systems Engineering Symposium, HASE 2008, pp. 9–18 (2008)
20. Sabelfeld, A., Myers, A.C.: Language-based information-flow security. IEEE J. Sel. Areas Commun. **21**(1), 5–19 (2003)
21. Schaad, A., Borozdin, M.: Tam2: Automated threat analysis. In: Proceedings of the 27th Annual ACM Symposium on Applied Computing, pp. 1103–1108 (2012)
22. Schrettner, L., Fülöp, L.J., Ferenc, R., Gyimóthy, T.: Visualization of software architecture graphs of java systems: managing propagated low level dependencies. In: Proceedings of the 8th International Conference on the Principles and Practice of Programming in Java, PPPJ 2010, pp. 148–157. ACM, New York (2010). http://doi.acm.org/10.1145/1852761.1852783
23. Schumacher, M.: Security Engineering with Patterns - Origins, Theoretical Models, and New Applications. LNCS, vol. 2754. Springer, Heidelberg (2003)
24. Swiderski, F., Snyder, W.: Threat Modeling. Microsoft Press, Redmond (2004)

On the Static Analysis of Hybrid Mobile Apps
A Report on the State of Apache Cordova Nation

Achim D. Brucker[1]([⊠]) and Michael Herzberg[2]

[1] Department of Computer Science, The University of Sheffield, Sheffield, UK
`a.brucker@sheffield.ac.uk`
[2] SAP SE, Vincenz-Priessnitz-Strasse 1, 76131 Karlsruhe, Germany
`michael.herzberg@sap.com`

Abstract. Developing mobile applications is a challenging business:
developers need to support multiple platforms and, at the same time,
need to cope with limited resources, as the revenue generated by an average app is rather small. This results in an increasing use of cross-platform
development frameworks that allow developing an app once and offering
it on multiple mobile platforms such as Android, iOS, or Windows.

Apache Cordova is a popular framework for developing multi-
platform apps. Cordova combines HTML5 and JavaScript with native
application code. Combining web and native technologies creates new
security challenges as, e. g., an XSS attacker becomes more powerful.

In this paper, we present a novel approach for statically analysing
the foreign language calls. We evaluate our approach by analysing the
top Cordova apps from Google Play. Moreover, we report on the current
state of the overall quality and security of Cordova apps.

Keywords: Static program analysis · Static application security
testing · Android · Cordova · Hybrid mobile apps

1 Introduction

Developing mobile applications is a challenging business: developers need to
support multiple platforms, but also have to cope with limited resources, as the
revenue generated by an average app is rather small. In principle, there are three
different approaches: (1) native apps, (2) mobile web apps, or (3) hybrid apps.
Native apps are built using platform specific technologies (e. g., Swift for iOS or
Java for Android). They have the advantage that they can use all platform specific features. *Mobile web apps* are on the other end of the spectrum: they are web
apps developed using standard web technologies (i. e., HTML5 and JavaScript)
and, thus, run on every device with a modern web browser. As a downside, they
are only very shallowly, if at all, integrated into the mobile platform and can

---

A.D. Brucker–Parts of this research were done while the author was a Security
Testing Strategist and Research Expert at SAP SE in Germany.

J. Caballero et al. (Eds.): ESSoS 2016, LNCS 9639, pp. 72–88, 2016.
DOI: 10.1007/978-3-319-30806-7_5

only access device features that are supported by HTML5. *Hybrid apps* combine the advantages of native and mobile apps; they allow developing most of the application using platform independent technologies, where small platform specific plugins enable the developer to access all device features that a native application can access.

Due to the increased market pressure for supporting multiple mobile platform as well as the increased demand to save development costs, more and more mobile apps are developed as hybrid apps. Thus, hybrid development frameworks such as PhoneGap (http://phonegap.com/), Trigger.io (https://trigger.io/), or Apache Cordova (https://cordova.apache.org/) are becoming more and more popular. This is not only true for small independent studios developing mobile apps, also large enterprise software vendors such as SAP are recommending the hybrid approach as the default development model to their developers. SAP offers its own extension of Apache Cordova, called SAP Kapsel, that is used both by SAP as well as its customers for developing mobile enterprise apps.

From a security development perspective, hybrid apps pose several challenges. We need to be aware that, e.g., a XSS attacker becomes much more powerful as he might be able to break out of the JavaScript environment and inject code that is executed in the context of the native part of the app—resulting in a much larger attack surface. The combination of web technologies and native mobile code is not yet supported by state of the art automated security testing tools in general and static application security testing (SAST) tools in particular. SAST tools are the back-bone of a holistic security testing strategy [3] and are widely used in the software industry [5,6].

We address this problem by developing a static code analysis approach that supports hybrid mobile apps developed using Apache Cordova. In more detail, our contributions are twofold: (1) we present a novel technique providing the basis for detecting data-flows in hybrid mobile apps, and (2) we report on our lessons learned from applying our approach to a large number of top Cordova apps from Google Play.

2   Apache Cordova and Its Security Model

In this section, we briefly introduce Apache Cordova and provide a general overview of the particular security challenges of Cordova apps.

2.1   Apache Cordova Architecture and Programming Model

Cordova is a framework for developing mobile apps using HTML5 and JavaScript while still allowing full access to the device features.

Architecture. Figure 1 shows the architecture of an Android Cordova app. The main part, i.e., the application logic and the user interface, are written in HTML5, CSS, and JavaScript. This part is executed in an extended WebView that provides, besides the HTML5 API, also a dedicated Cordova JavaScript

API. The latter allows, via the Cordova Native API, to access various Cordova Plugins. The Cordova Plugins are written in the platform's programming language (e.g., in Java for Android). Cordova ships with many default plugins; additional plugins are offered by third party providers or can implemented by the application developer.

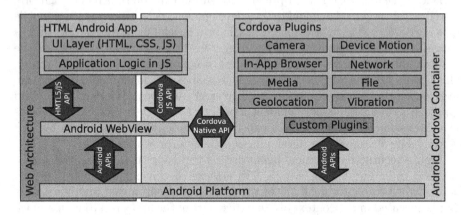

Fig. 1. The Android Cordova Architecture

Our approach also works for extensions of Cordova such as PhoneGap by Adobe or SAP Kapsel by SAP that mostly provide additional plugins.

An Example Cordova Plugin. Let us assume we want to implement a Cordova plugin that allows for searching the contacts database. Listing 1.1 (Listing 1.2) shows an excerpt of the JavaScript (Java) of the plugin implementation.

Listing 1.1 shows a JavaScript function `showPhoneNumber` that can be used to implement the business logic of a Cordova app. The `exec` method (Line 5–6) is the core of the foreign language interface of Cordova. It takes five arguments: 1. a callback that is invoked in case of a successful termination of the native call, 2. a callback that is invoked in case of a erroneous termination of the native call, 3. a string that identifies the name of Java class that implements the native function, 4. a string that identifies the action that should be executed by the native function, and 5. a list containing the arguments of the native function.

```
  function showPhoneNumber(name) {
2   var successCallback = function(contact) {
      alert("Phone number:" + contacts.phone);
4   }
    exec(successCallback, null, "ContactsPlugin", "find",
6       [{"name" : name}]);
  }
```

Listing 1.1. Contacts Plugin Example: JavaScript

```
 1   class ContactsPlugin extends CordovaPlugin {
     boolean execute(String action, CordovaArgs args,
 3                        CallbackContext callbackContext) {
       if ("find".equals(action)) {
 5       String name = args.get(0).name;
         find(name, callbackContext);
 7       } else if ("create".equals(action)) ...
     }
 9   void find(String name, CallbackContext callbackContext) {
     Contact contact = query("SELECT ... where name=" + name);
11   callbackContext.success(contact);
     }
13  }
```

Listing 1.2. Contacts Plugin Example: Java

The Cordova framework delegates this call to the **execute** method of the Java class **ContactsPlugin** (Listing 1.2, Line 2), which delegates the call, based on the action, to the **find** method (Line 9). The **find** method uses a SQL query to find the contact information and passes it to the success callback (Line 11). The information is then passed back to the corresponding JavaScript method (Listing 1.1, Line 2).

2.2   Security Considerations for Cordova Apps

On the one hand, Cordova apps are HTML5 applications, i.e., they share all typical features (e.g., JavaScript code that is downloaded at runtime) and security risks (e.g., XSS) of web applications (see, e.g., [19,23] for an overview of these risks). On the other hand, Cordova apps share the features (e.g., full device access) and security risk (e.g., SQL injections, privacy leaks) of native apps (see, e.g., [17,27] for an overview of these risks).

To limit the typical web application threats, WebViews are re-using the well-known security mechanism from web browsers such as the same-origin policy [10]. Moreover, WebViews are separated from the regular web browsers on Android, e.g., WebViews have their own cache and cookie store. Still, there are subtle differences that make implementing secure Cordova apps even for experienced web application developers a challenge [9,10].

A plugin is a mechanism for drilling holes into the sandbox of a WebView, making the traditional web attacker much more powerful as, e.g., an XSS attack might grant access to arbitrary device features. The root cause for such vulnerabilities can be located in Cordova itself (e.g., CVE-2013-4710 or CVE-2014-1882) or in programming and configuration mistakes by the app developer.

There have been several works introducing more fine-grained access control mechanism for the cross-language interface in hybrid mobile apps, particularly Cordova, such as NoFrak [10], MobileIFC [22], and others [12,21]. They all identified the breach of the sandbox security and that Cordova fails to restrict access to plugins by untrusted JavaScript code as the major security and privacy concern. To remedy this breach, they propose modifications to the hybrid framework which mitigate attacks by introducing fine-grained access control and modifications to Android's permission model. Apache Cordova is certainly in need of such additions. This way, existing hybrid applications could be secured without modification, reducing the potential implications of vulnerabilities such as XSS. This running time protection paired with tools helping the app developers to secure their apps in the first place, such as presented in this paper, is certainly a good combination to ensure a secure experience when using hybrid apps.

3   Static Analysis for Finding Cross-Language Flows

In this section, we present our approach for building a uniform call graph for Cordova apps with connected Java and JavaScript parts. This call graph is the basis for a cross-language data-flow analysis which enables an end-to-end static program analysis of Cordova apps.

3.1   Modelling Cordova

The usage patterns of cross-language calls depends heavily on the underlying framework, e. g., Cordova. Thus, to implement a cross-language analysis, one can either model the underlying framework or analyse the application including the cross-language framework itself. In our work, we decided to model the Cordova framework due to two reasons: 1. Modelling Cordova avoids the need for re-analysing the Cordova source code for each app and 2. data-flows within the framework code are not of interest to the app developer.

Since the official documentation regarding plugins is rather sparse, many observations are based on the officially provided plugins.

The usual cross-language control flow in a Cordova app follows a JavaScript-to-Java-to-JavaScript scheme: Starting in the JavaScript part, a call to exec transfers the flow to the Java side, where the requested native action is executed. When finished, the Java part calls one of the two callbacks that were passed to the exec call, after which the flow transfers back to the JavaScript part.

We model Cordova implicitly by four Cordova specific heuristics. The purpose of the first two is finding the JavaScript callbacks passed to the exec call; they are the targets of the Java-to-JavaScript call chain link. The third heuristic is concerned with finding the Java callers of this link. The fourth one filters out cross-language calls which have been reported by the first three heuristics, but are very unlikely to be correct.

The JavaScript-to-Java calls are easier to detect and thus not addressed by the heuristics, because the exec calls are rather static and carry enough information in their service and action parameters[1] to deduct the Java call target.

---

[1] For more information on the usage of these two parameters, see https://cordova. apache.org/docs/en/latest/guide/hybrid/plugins/.

Mocking the Cross-Language Call Interface. Cordova's `exec` method is the heart of its cross-language interface, thus a precise modelling of calls to this method is key. The actual implementation of this method, `androidExec` in `cordova.js`, is not useful for detecting cross-language calls statically for at least two reasons: 1. The heavy use of dynamic language features by Cordova is challenging; e. g., the callback functions passed to `exec` are being stored by `androidExec` in a global dictionary and are only used much later. Thus, it is very hard to determine statically when the callback functions are called. As a result, these calls will not get modelled by typical building algorithms, which is fatal as they are the targets of the calls from Java-to-JavaScript. 2. The algorithms for building JavaScript call graphs are often context-insensitive. As all cross-language calls from JavaScript-to-Java are done via the one `exec` method offered by Cordova, this becomes a problem. We want to be able to relate the passed callback functions to the other parameters, which provide important information about the part on the Java side which will later call these callbacks. Therefore, context-sensitivity for the calls to `exec` is vital.

Solution. Both issues are addressed by our heuristic *ReplaceCordovaExec*, which automatically pre-processes the JavaScript source code. The core idea is to search for all `cordova.exec` and `exec` calls and replace each of them with a call to a freshly created method with a unique name that calls the callbacks.

```
function showPhoneNumber(name) {
  var succCb = function(contact) {
    alert("Number:"+contacts.phone);
  }



  exec(succCb, null,
      "ContactsPlugin",
      "find", [{"name" : name}]);
}
```

Listing 1.3. Before: Example of mocking the cross-language call interface

```
function showPhoneNumber(name) {
  var succCb = function(contact) {
    alert("Number:"+contacts.phone);
  }
  function stub1(succ, fail, service,
      action, args) {
    succ(null);
    fail(null);
  }
  stub1(succCb, null,
      "ContactsPlugin",
      "find", [{"name" : name}]);
}
```

Listing 1.4. After: Example of mocking the cross-language call interface

Recall the JavaScript part of our example, Listing 1.1. Listing 1.3 shows it again in a shorter version, and Listing 1.4 shows the modifications made by *ReplaceCordovaExec*. A new method `stub1`, which replaces the `exec` call, is introduced that makes the calls to the success and fail callbacks explicit.

The renaming of the *ReplaceCordovaExec* takes into account that the result of invoking `require("cordova/exec")` can be assigned to an arbitrary variable. This call is not shown in the example, but often used to obtain the `exec` method. Overall, *ReplaceCordovaExec* introduces local context-sensitivity into our analysis approach.

Emulating the Module Loading Mechanism. Cordova provides its own JavaScript module mechanism, i. e., it provides two functions for structuring JavaScript code: `define` and `require`. When a Cordova app is assembled, the plugins' JavaScript code is converted into modules, and bigger plugins use those modules, too, to separate their code.

There are basically two major challenges when searching for uses of the plugins: 1. determining which object gets returned by a call to `require`, and 2. helping the call graph builder understand what is behind the global plugin variables under which Cordova makes the plugins available.

Solution. Both issues are addressed by our heuristic *ConvertModules*, which automatically pre-processes the JavaScript source code. The object that gets returned by the require call is determined by whatever gets assigned to the `module.exports` field inside the factory function. Thus, we replace the require and `module.exports` references with a global unique variable, derived from the unique module id. Now, any call graph builder will be able to track this new global object and connect the corresponding method calls. For the plugin modules, one additional transformation needs to be applied: For all global variables (there may be more than one), which are specified in the plugin's configuration file, a statement is added to the plugin definition which assigns the variable that is created by the first transformation to the queried global variable. Normally, these variables get defined at runtime when Cordova loads the plugins, but this transformation now hard-codes these definitions into the module.

```
define("com.contacts",
  function(require, exports, module){
    exports.find =
      function(succCb, name) {
        exec(succCb, null,
          "ContactsPlugin", "find",
          [{"name" : name}]);
      };
  });
...
var succCb = function(contact) {
  alert("Number:"+contacts.phone);
}
plugins.contacts.find(succCb,
                "Peter");
```

Listing 1.5. Before: Example of emulating the module loading mechanism

```
define("com.contacts",
  function(require, exports, module){
    plugins.contacts.find =
      function(succCb, name) {
        exec(succCb, null,
          "ContactsPlugin", "find",
          [{"name" : name}]);
      };
  });
...
var succCb = function(contact) {
  alert("Number:"+contacts.phone);
}
plugins.contacts.find(succCb,
                "Peter");
```

Listing 1.6. After: Example of emulating the module loading mechanism

Recall our phone number example: Listing 1.5 shows an exemplary definition of the contacts plug-in with an export declaration as it would look like after being imported by Cordova. We transform this export declaration using the plugin's global variable (see Listing 1.6). As a result, the relation between this global variable and the actual plugin method becomes statically apparent.

Data-Flow Heuristic Based on Action String. While the first two heuristics enable finding the *targets* of the calls from Java-to-JavaScript related to each `exec` call, finding the *callers* poses its own challenges: when execution is transferred to the Java side, the passed callback functions can be called via a `CallbackContext` Java object. This object offers three methods which get mapped to the two callback calls: `success`, `error`, and `sendPluginResult`. Given such a call somewhere on the Java side, how does one determine the possible JavaScript targets?

All `exec` calls of a plugin are mapped to a single Java `execute` method. Thus, it is not clear how calls to methods of `CallbackContext` object map to the JavaScript callbacks. During runtime, Cordova decides based on the `feature` string which class's `execute` to call, and passes the supplied `action` string to the plugin's `execute` implementation. Commonly, each `exec` call has only one possible value for each of the two parameters, so it is possible to limit the number of Java-to-JavaScript connections by utilising these context information.

Another challenge is the frequent use of the command pattern in the `execute` method, e. g., when dealing with threads. As the calls on the `CallbackContext` object are then actually done somewhere deep in the thread library, call graph builders will not attribute this call even to the `execute` method, which is a problem since the context information supplied by the JavaScript `exec` call is needed.

Solution. As `callbackContext` calls are only of interest when an `exec` call is encountered in the JavaScript code, the parameters passed to `exec` can be used as context information when looking for the `callbackContext` calls. First the Java class and its `execute` method that corresponds to the `feature` string need to be found. As Cordova keeps a mapping from those string values to Java classes, the class is looked up there.

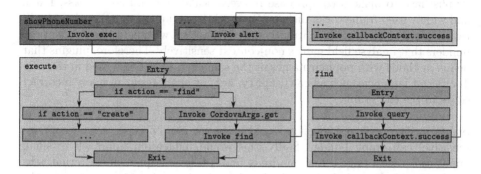

Fig. 2. The control flow graph and call graph of the example in Listings 1.1 and 1.2, including two cross-language edges

To determine which `callbackContext` calls are reachable from the beginning of the `execute` method, a two-fold reachability analysis is conducted for each call site. Figure 2 illustrates the involved control flow graph and call graphs.

1. First, using the Java call graph, we compute all possible call chains without cycles from the `execute` method to the method which contains the call to the particular `callbackContext`. If the `execute` method is not a predecessor in this call graph, the `callbackContext` call is considered not reachable. If it is reachable though, all those invoke instructions in the `execute` method are determined through which the `callbackContext` call is eventually reachable.
2. Using these invoke instructions as well as the `action` parameter and the control flow graph of the `execute` method, a more precise reachability analysis is conducted. For each invoke instruction, all possible paths through the control flow graph from the entry of the `execute` method to the invoke are determined. For all found paths, the `action` parameter is taken into account; as many plugins implement an `execute` method using many `if-else` clauses based on `action`, the paths are checked for statements similar to `"get".equals(action)`. If the `action` strings do not match, the path can be discarded as impossible, as it can never be taken during runtime.

If there are any paths left after the two-fold reachability analysis, the reachable `callbackContext` calls need to be classified as either being a `success` or `fail` callback call. This is done by deciding whether the method called on the `callbackContext` is either `success`, `error`, or `sendPluginResult` (and here, which status codes are possibly passed). Eventually, the corresponding `success` and `fail` connections can be reported as calls from Java-to-JavaScript.

Filtering Frameworks. The static construction of *precise* call graphs for JavaScript programs is challenging [7]. Approaches for building JavaScript call graphs have to make a compromise between scalability and correctness. Large and widely used frameworks such as jQuery (https://jquery.com/) or AngularJS (https://angularjs.org) can currently only be handled with field-based call graph builders that analyse field names non-context sensitively. Therefore, plugins that define methods with popular names or the same names as those used in the core JavaScript language such as `call`, `apply`, `get`, or `open`, result in many incorrect edges in the call graph.

Solution. The preferred solution would be to use more precise (e. g., a context-sensitive) call graph builder. Sadly, this would reduce our approach to small applications with only a few hundreds lines of JavaScript code. Alternatively, we could exclude such frameworks from our analysis. As this would make the analysis of apps based on frameworks that change the way the JavaScript code is written, e. g., frameworks promoting an asynchronous programming style, impossible, this approach is also not feasible.

Thus, we filter the problematic functions after the call graph is constructed based on further information such as the file names. This approach, on the one hand, allows balancing correctness and scalability of the static analysis. On the other hand, the configuration of the filter need to be adapted to fit new frameworks that might emerge.

3.2   Implementation

We implemented our approach, in particular a unified call graph builder for Cordova apps, using the WALA framework (http://wala.sf.net).[2] Our prototype allows to process Android binaries (i. e., APK files) directly. Using WALA's Java front-end, the analysis of Java source of Android apps can be supported easily as well. For parsing the Dalvik binary code and the JavaScript, we rely on the front-ends provided by WALA.

First, we apply the *ReplaceCordovaExec* and *ConvertModules* heuristic to the JavaScript parts of the application. Then we use WALA for building the call graphs for the JavaScript and Java parts of a Cordova app. After building the Java and JavaScript call graphs independently, we traverse both call graphs for connecting the cross-language calls. The result is a unified call graphs that allows implementing further static analysis methods that can uniformly traverse the Java and JavaScript parts of a Cordova app.

4   The State of Cordova App Security (and Quality)

In this section, we evaluate our approach for building uniform call graphs for Cordova apps as well as report on our findings based on analysing Cordova apps from the top Android app category of the Google Play Store, three Cordova apps from SAP, and one artificial app specifically written for this work. Our evaluation is two-fold in order to assess the scalability and quality of our analysis.

4.1   Popularity of Cordova and Benchmark Selection

We took the Top 1000 apps (as ranked by Google in spring 2015) from Google Play and checked if these apps contain a `config.xml` file that belongs to the Cordova framework. Using this criterion, we could identify 50 Cordova apps. Thus, according to our analysis, only 5 % of the Top 1000 apps are using Cordova.

As SAP usually distributes its applications directly to its customers, we did not expect SAP apps within the Top 1000 apps category. To include SAP apps and their specific characteristics in our analysis, we have selected three mobile enterprise apps from SAP that are based on SAP Kapsel and SAP's OpenUI5 JavaScript framework (for details, see http://openui5.org/).

Finally, we implemented one test app, called Damn Vulnerable Hybrid Mobile App (DVHMA), that intentionally contains vulnerabilities and different coding styles to serve as a controlled test bed for our analysis.[3]

---

[2] Our prototype is available at https://github.com/DASPA/DASCA.

[3] The DVHMA app is available at https://github.com/ZertApps/DVHMA.

4.2 Scalability

To evaluate the runtime behaviour and, thus, the scalability of our approach, we analysed all 54 apps of our test set. Our prototype is able to analyse 52 out of the 54: two apps from the Top 1000 are obfuscated in such a way that the WALA front-ends are not able to analyse them at all.

Our analysis can build the unified call graph for 50 % of the apps in under 30 min and for all but one within 12 h. The memory consumption was in all but one cases under 8 GB. The benchmarks have been run on a virtual machine running Ubuntu 14.04 using six cores of an Intel Xeon CPU E7-4830v2 @ 2.20 GHz and 12 GB of RAM. Due to space reasons, we omit the detailed results. Thus, our prototype is able to analyse typical Android apps on modern modern workstations and notebooks.

In general, the runtime for building the language specific call graphs is mainly influenced by the complexity in terms of the number of function calls as call depth and only to a minor extend by the code size. This is true for both the Java as well as the JavaScript part. For building the unified call graphs, the number of cross-language calls is, given the pre-computed call graphs for each language, the main influence for building the unified call graphs.

4.3 Quality

To assess the quality of our analysis, we selected eight apps (six from the top apps as ranked by the Google Play Store, one from SAP, and our artificial test app). We did a thorough manual code review either on the original source code (for the app from SAP and our test app) or on the result of de-compiling the binary (for the six apps from Google Play). Our manual code review focused on finding all cross-language calls.

As a manual code review is a time consuming task, we limited the analysis to eight apps that we consider a good representation of the overall population of Cordova apps: Table 1 shows that the most commonly used plugins from the six apps from Google Play are the same ones as from the 50 apps. In addition, we have chosen a typical SAP app as well as our test apps that captures our expertise based on a shallow analysis of a larger number of Cordova apps. We consider the distribution of plugins as most relevant for our work, as cross-language calls are most often located in plugins. Thus, this analysis allows us to assess the quality of the unified call graphs with respect to capturing cross-language calls.

The following four sections will compare the *manually* found cross-language calls with the ones reported from the prototype. We will focus on the calls from Java-to-JavaScript. The calls from JavaScript-to-Java are relatively easy to find, thanks to the structure of Cordova's function interface. Therefore, the prototype found all these calls.

Two values are especially important when evaluating the quality [1]:

$$R = \frac{TP}{TP + FN} \quad \text{(recall)} \qquad\qquad P = \frac{TP}{TP + FP} \quad \text{(precision)}$$

Table 1. The ten most used plugins from each test set

(a) Plugins from the 50 apps   (b) Plugins from the six manually analysed apps

Plugin	#		Plugin	#
device	26		device	5
inappbrowser	25		inappbrowser	5
dialogs	20		dialogs	2
splashscreen	18		splashscreen	2
network-information	14		console	2
file	14		network-information	1
console	12		file	1
camera	11		camera	1
statusbar	11		statusbar	1
PushPlugin	11		PushPlugin	1

where TP is the number of correctly found cross-language calls, FP the number of falsely reported ones, and FN the number of missed calls.

Informally, *recall* is defined as the number of correctly found calls divided by the number of calls which should have been found and *precision* is defined as the number of correctly found calls divided by the number of calls reported.

ReplaceCordovaExec. This heuristic is necessary to identify any Java-to-JavaScript calls at all. Without it, the callback functions on the JavaScript side will not get modelled, which is bad since they are the targets of those calls from the Java side. As can be seen in Table 2a, the precision with just *ReplaceCordovaExec* is already very good. However, as is represented by the bad recall, there are also many incorrect calls being reported. But before we will present the results of *FilterJavaCallSites* and *FilterJSFrameworks*, which will lead to less errors, we will present the results for another heuristic aimed at increasing the number of found calls.

ConvertModules. The main purpose of this heuristic is to model the module mechanism and thus allow finding more calls from Java-to-JavaScript. However, this effect is only observed on one of the eight apps: our artificially created one. The explanation is simple; this heuristic enables tracking callback functions through the Cordova plugin mechanism, from the application code to the actual call to `exec`. Surprisingly, our app was the only one of those eight to create and pass callbacks from application code.

The errors for two apps are significantly reduced. This is because assigning the functions to `module.exports` is not ambiguous anymore and does not result in the field-based call graph builder vastly overestimating method invocations.

FilterJavaCallSites. Adding this heuristic, two effects can be observed in Table 2c: The number of errors is greatly reduced, but at the cost of a few

Table 2. The quality of the found cross-language calls from Java-to-JavaScript

(a) *ReplaceCordovaExec*

App	Hits	Misses	Errors	Recall	Prec.
app_{01}	4	0	400	1%	100%
app_{02}	3	0	8	28%	100%
app_{03}	30	0	5804	1%	100%
app_{04}	1	0	2315	1%	100%
app_{05}	3	0	47	6%	100%
app_{06}	246	0	1567	14%	100%
sap_{01}	3	0	32	9%	100%
DVHMA	5	5	8	39%	50%

(b) *ReplaceCordovaExec* and *ConvertModules*

App	Hits	Misses	Errors	Recall	Prec.
app_{01}	4	0	394	2%	100%
app_{02}	3	0	8	28%	100%
app_{03}	30	0	4574	1%	100%
app_{04}	1	0	1157	1%	100%
app_{05}	3	0	47	6%	100%
app_{06}	246	0	1552	14%	100%
sap_{01}	3	0	32	9%	100%
DVHMA	10	0	9	53%	100%

(c) *ReplaceCordovaExec*, *ConvertModules*, and *FilterJavaCallSites*

App	Hits	Misses	Errors	Recall	Prec.
app_{01}	3	1	397	1%	75%
app_{02}	2	1	0	100%	67%
app_{03}	28	2	2829	1%	94%
app_{04}	1	0	0	100%	100%
app_{05}	2	1	12	15%	67%
app_{06}	239	7	444	35%	98%
sap_{01}	2	1	0	100%	67%
DVHMA	10	0	0	100%	100%

(d) Using all heuristics

App	Hits	Misses	Errors	Recall	Prec.
app_{01}	3	1	6	34%	75%
app_{02}	2	1	0	100%	67%
app_{03}	28	2	2323	2%	94%
app_{04}	1	0	0	100%	100%
app_{05}	2	1	4	34%	67%
app_{06}	239	7	443	36%	98%
sap_{01}	2	1	0	100%	67%
DVHMA	10	0	0	100%	100%

cross-language calls missed. The misses come from the fact that this heuristic relies on being able to trace the `callbackContext` call back to the `execute` call. Some plugins, however, store their `CallbackContext` object for later use, e.g., when a listener for changes of the network state triggers. In these cases, other possibilities than simply discarding these call sites are also imaginable: Instead of reporting the callback functions from no `exec` call as targets, the callbacks from all `exec` calls could be reported, resulting possibly in a vast over approximation.

Most of the errors which are still reported are related to the file plugin. Here, the developers used a utility method which translates a lot of different exception types into different `callbackContext` calls. However, not all actions are able to throw all of them. This distinction is not made by this heuristic and would require a more sophisticated reachability analysis.

FilterJSFrameworks. Cordova apps contain significant amounts of framework code. As expected, this heuristic increases the recall by a great amount as can be seen in Table 2d, because cross-language calls related to these frameworks are filtered. However, as the detection of framework code is currently only based on the file name, apps who repackage all JavaScript code into one big file will not see any improvements. Also, not all errors are related to JavaScript frameworks, so some errors coming from incorrectly found calls within the apps themselves will not get filtered.

4.4 Noticeable Findings About the Apps

How Developers Use the Cordova Framework. The way the Cordova framework is used differs wildly among the 50 apps. Many apps do, in fact, use Cordova as intended: The app is written in JavaScript, the Java part is unmodified and simply loads the entry-point HTML file which is set in the Cordova configuration file. Some apps, however, significantly change the Java part. The most extreme apps do not ship any HTML or JavaScript code in the APK and simply specify one hard coded URL in Java to be loaded, which is often just the mobile version of their website, hosted in a remote location.

Some apps chose a middle ground: They may first load Activities like regular Android apps, and may embed HTML and JavaScript code only into some parts of the app, where Cordova Plugins may be used to communicate back and forth. Such irregular Cordova apps are the exception and are significantly harder to statically analyse, as they change the way Cordova is integrate into the app.

How Developers Use the Cordova Plugins. Many plugins take callback functions and pass them through to their `exec` call. Especially for plugins which do not simply yield a result which can be passed to the success callback, e. g., when the plugin is just supposed to execute a command, there are often no fail callbacks being provided, either. Some of these actions could indeed fail, which would not get propagated through to the app code itself, though, because no fail callback has been passed. This seems to indicate a lack of proper error handling for many apps, and is one of the reasons why the *ConvertModules* heuristic did not find any additional calls in the apps.

How Cordova Plugins Are Written. Plugins generally have the character of libraries, where the JavaScript part does rarely more than encapsulate the `exec` calls. There are also no other mechanism used to conduct cross-language calls. The official Cordova plugins adhere to these guidelines. Our work is intended for this kind of plugins.

Anyone can write Cordova plugins, and not all developers adhere to these guidelines. One found plugin, apparently written just for this specific app, does not contain any JavaScript code; instead, the `exec` calls are done right in the app code itself. Other plugins represent the other extreme and implement quite a bit of the plugin logic on the JavaScript side, which could have been as well written in Java. Again some other plugins do not even use `exec` to communicate with their Java side, but use methods which are also used internally in the Cordova framework. The reason for these unnecessary uses of workarounds remains unclear.

One plugin found in those Cordova apps is special in a different way: Combining Java and JavaScript was apparently not enough, as the APK contained some native libraries accessed via JNI to do some basic arithmetic calculations. As JSON strings get passed from the JavaScript part via Java to the C part, the attack surface gets even larger.

5 Related Work

There is a large body of work that uses static program analysis for finding security vulnerabilities in JavaScript-based web applications [11,16,24,26] as well as dealing with the privacy concerns of Android apps [4,13,18,20].

While cross-language calls in the form of foreign language interface such as the Java Native Interface (JNI) are not new, there are surprisingly few works that address the problem of static program analysis across such interface. Among those few there is SafeJNI [25], which statically ensures that unsafe native code cannot bypass Java's type-safety. Another example is the work of Li and Tan [15], who developed a static analysis framework to find bugs related to the different use of exceptions in Java and native code.

The most closely related work is HybriDroid. The development of HybriDroid seems to have started by Lee et al. [14] roughly at the same time as we started our work. With HybriDroid, we share the overall goal: detecting security vulnerabilities as well as leakage of private information in hybrid mobile applications on Android. In contrast to our work, HybriDroid analyses not the cross language interface of Cordova, but the low-level interface provided by Android and does not yet support Cordova. Thus, HybriDroid works rather independently from the framework (e. g., Cordova) used for developing a hybrid app and therefore reports also cross-language calls that our approach might miss, e. g., in case a Cordova developer does use the low-level functions in addition to the mechanism offered by Cordova. In exchange, our approach allows for better explanations of found issues to Cordova developers. Moreover, we expect a better scalability of our approach. Still, as both approach are very young, it is too early for a detailed comparison of the actual implementations.

The next most closely related works are FlowDroid [2] and SCanDroid [8]. Both are tools supporting the Android life-cycle model and are able to build call graphs for *native* Android apps as well as perform a static data-flow analysis for finding security vulnerabilities as well as privacy violations. For our work, SCanDroid is of particular interest, as it is based on Wala which makes it very attractive to extend its data flow analysis to support our unified call graphs. Extending the data flow analysis of SCanDroid would require developing support for the JavaScript part of our unified call graph as well as the cross language calls. In addition, the Android life-cycle events that are specific to the JavaScript part need to be added.

6   Conclusion and Future Work

We presented a novel approach for constructing a uniform call graph for hybrid mobile apps using the Cordova framework. Our evaluations show that the generated calls graphs are, with respect to the cross language calls, very accurate. Their quality, though, depends on the used call graph builder for JavaScript.

As future work, we plan to develop a data-flow analysis (e.g., extending SCanDroid [8]) on top of the uniform call graphs that will allow for detecting programming related vulnerabilities in Cordova apps such as SQL injections and to enforce policies such as "only local JavaScript code shall be allowed to access the address book" statically, i.e., at development time.

Still, the presented approach is already applicable to real Cordova applications. When the apps from the test set have been manually examined, it quickly became apparent that any tool helping with properly programming Cordova apps is useful. One app even used a custom Cordova plugin which contains libraries written in C++ that were used by the Java code, so detecting cross-language calls does not stop at just Java and JavaScript and can certainly be extended.

Acknowledgements. We would like to thank Jens Heider and Stephan Huber from Fraunhofer SIT who provided us with the initial list of Cordova apps for our evaluation. This research was partially supported by the Federal Ministry for Education and Research (BMBF) in the context of the project ZertApps (http://www.zertapps.de/).

References

1. Anderson, P.: Measuring the value of static-analysis tool deployments. IEEE Secur. Priv. **10**(3), 40–47 (2012)
2. Arzt, S., Rasthofer, S., Fritz, C., Bodden, E., Bartel, A., Klein, J., Le Traon, Y., Octeau, D., McDaniel, P.: Flowdroid: Precise context, flow, field, object-sensitive and lifecycle-aware taint analysis for Android apps. In: PLDI 2014, pp. 259–269. ACM (2014)
3. Bachmann, R., Brucker, A.D.: Developing secure software: A holistic approach to security testing. Datenschutz und Datensicherheit (DuD) **38**(4), 257–261 (2014)
4. Batyuk, L., Herpich, M., Camtepe, S.A., Raddatz, K., Schmidt, A.D., Albayrak, S.: Using static analysis for automatic assessment and mitigation of unwanted and malicious activities within android applications. In: Malicious and Unwanted Software (MALWARE), pp. 66–72. IEEE (2011)
5. Bessey, A., Block, K., Chelf, B., Chou, A., Fulton, B., Hallem, S., Henri-Gros, C., Kamsky, A., McPeak, S., Engler, D.: A few billion lines of code later: using static analysis to find bugs in the real world. Commun. ACM **53**, 66–75 (2010)
6. Brucker, A.D., Sodan, U.: Deploying static application security testing on a large scale. In: Katzenbeisser, S., Lotz, V., Weippl, E. (eds.) GI Sicherheit 2014, Lecture Notes in Informatics, vol. 228, pp. 91–101. GI (2014)
7. Feldthaus, A., Schafer, M., Sridharan, M., Dolby, J., Tip, F.: Efficient construction of approximate call graphs for JavaScript IDE services. In: 2013 35th International Conference on Software Engineering (ICSE), pp. 752–761. IEEE (2013)

8. Fuchs, A.P., Chaudhuri, A., Foster, J.S.: SCanDroid: automated security certification of android applications. Technical report CS-TR-4991, Department of Computer Science, University of Maryland, College Park (2009)
9. Georgiev, M., Iyengar, S., Jana, S., Anubhai, R., Boneh, D., Shmatikov, V.: The most dangerous code in the world: validating SSL certificates in non-browser software. In: CSS, pp. 38–49. ACM (2012)
10. Georgiev, M., Jana, S., Shmatikov, V.: Breaking and fixing origin-based access control in hybrid web/mobile application frameworks. In: NDSS 2014. The Internet Society (2014)
11. Guha, A., Krishnamurthi, S., Jim, T.: Using static analysis for AJAX intrusion detection. In: World Wide Web, pp. 561–570. ACM (2009)
12. Jin, X., Wang, L., Luo, T., Du, W.: Fine-grained access control for HTML5-basedmobile applications in Android. In: ISC (2013)
13. Kim, J., Yoon, Y., Yi, K., Shin, J., Center, S.: Scandal: static analyzer for detecting privacy leaks in android applications. MoST (2012)
14. Lee, S., Dolby, J., Ryu, S.: Hybridroid: Analysis framework for Android hybrid applications (2015)
15. Li, S., Tan, G.: Finding bugs in exceptional situations of JNI programs. In: CCS, pp. 442–452. ACM (2009)
16. Madsen, M., Livshits, B., Fanning, M.: Practical static analysis of javascript applications in the presence of frameworks and libraries. In: Foundations of Software Engineering, pp. 499–509. ACM (2013)
17. McGraw, G.: Software Security: Building Security In. Addison-Wesley, Boston (2006)
18. Mohr, M., Graf, J., Hecker, M.: Jodroid: Adding android support to a static information flow control tool. In: Conference on Programming Languages (2015)
19. Rubin, A.D., Geer Jr., D.E.: A survey of web security. Computer 31(9), 34–41 (1998)
20. Shabtai, A., Fledel, Y., Elovici, Y.: Automated static code analysis for classifying android applications using machine learning. In: CIS, pp. 329–333. IEEE (2010)
21. Shehab, M., AlJarrah, A.: Reducing attack surface on Cordova-based hybrid mobile apps. In: Workshop on Mobile Development Lifecycle, pp. 1–8. ACM (2014)
22. Singh, K.: Practical context-aware permission control for hybrid mobile applications. In: Stolfo, S.J., Stavrou, A., Wright, C.V. (eds.) RAID 2013. LNCS, vol. 8145, pp. 307–327. Springer, Heidelberg (2013)
23. Stuttard, D., Pinto, M.: The Web Application Hacker's Handbook: Discovering and Exploiting Security Flaws. Wiley, New York (2011)
24. Taly, A., Erlingsson, Ú., Mitchell, J.C., Miller, M.S., Nagra, J.: Automated analysis of security-critical JavaScript apis. In: SP, pp. 363–378. IEEE (2011)
25. Tan, G., Appel, A.W., Chakradhar, S., Raghunathan, A., Ravi, S., Wang, D.: Safe Java native interface. In: Secure Software Engineering, pp. 97–106 (2006)
26. Tripp, O., Pistoia, M., Fink, S.J., Sridharan, M., Weisman, O.: Taj: effective taint analysis of web applications. ACM Sigplan Not. 44(6), 87–97 (2009)
27. Tsipenyuk, K., Chess, B., McGraw, G.: Seven pernicious kingdoms: a taxonomy of software security errors. IEEE Secur. Priv. 3(6), 81–84 (2005)

Semantics-Based Repackaging Detection
for Mobile Apps

Quanlong Guan[1]([✉]), Heqing Huang[2], Weiqi Luo[1], and Sencun Zhu[2]

[1] Jinan University, Guangzhou, China
{gql,luoweiqi}@jnu.edu.cn
[2] Department of Computer Science and Engineering,
The Pennsylvania State University, University Park, PA 16802, USA
{hhuang,szhu}@cse.psu.edu

Abstract. While Android app stores keep growing in size and in number, app repackaging has become a major threat to the health of the mobile ecosystem. Different from many syntax-based repackaging detection techniques, in this work we propose a semantic-based approach, RepDetector, which is more robust against code obfuscation attacks. To capture an app's semantics, our approach extracts input-output states of core functions in the app and then compare function and app similarity. We implement a prototype of RepDetector, and evaluate it against various obfuscation technologies. The results show that our approach can detect repackaged apps effectively. It is also at least a hundred times faster than Androguard.

1 Introduction

In recent years, the mobile application world has been expanding dramatically. As of Oct. 2015, Google Play has over 1.5 millions of apps available for downloading. Since high popularity leads to more downloads, many popular Android apps have been copied or repackaged in recent years. Attackers can easily repackage an app under their own names or embed advertisements to earn pecuniary profits. They can also modify a popular app by inserting malicious payloads into the original app and leverage its popularity to accelerate malware propagation. Moreover, because of the popularity of the Android platform, many unofficial app markets exist. Most of them do not perform careful sanity check of uploaded apps. Thus, app repackaging in the Android platform has become very serious, which is increasingly hurting the app ecosystem. App developers are discouraged for loss of revenue, and app users may be deceived from installing malware.

To detect app repackaging, recently various approaches have been explored. Some of them detect repackaged or cloned apps based on code features, such as DroidMOSS [30], Juxtapp [13], DNADroid [8], AnDarwin [12]. For example, DroidMOSS [30] extracts app fingerprints through fuzzy hashing; Juxtapp [13] uses feature hashing for code similarity analysis. While such approaches are capable of recognizing code that is syntactically similar, they are not effective under semantic-preserving obfuscation, where repackaged apps are functionally

© Springer International Publishing Switzerland 2016
J. Caballero et al. (Eds.): ESSoS 2016, LNCS 9639, pp. 89–105, 2016.
DOI: 10.1007/978-3-319-30806-7_6

the same but syntactically different. The syntactic differences include instruction reordering, interleaved methods, opaque branch insertions or the substitution of semantically equivalent control structures. Moreover, new evasion solutions or obfuscation methods have also been explored recently [15, 27]. Hence, false negative rates of these approaches could be very high under such attacks.

In this work, we take a semantic-based approach to detecting repackaged apps. In our approach, named RepDetector, the input-output relationship of a function is captured to express its semantics. As long as a repackaged app preserves the critical semantics of the original app, according to our approach, their similarity would be high. In summary, this paper makes the following technical contributions:

- We propose a semantic-based approach to detecting repackaged apps. It can tolerate certain noise insertion or sophisticated obfuscation, and hence is obfuscation-resilient.
- We capture the input-output states of functions with state flow graphs to describe the semantic behaviors of an app. Then we introduce an effective algorithm to compare the similarity of functions from different apps by an SMT solver and detect semantic repackaging with Mahalanobis distance.
- We implement a prototype of RepDetector, and evaluate its detection accuracy and efficiency with both known repackaged apps, obfuscated apps, and apps from Google Play. The result shows that RepDetector is capable of scanning real-world Android repackaging apps with obfuscation resiliency. It is also over a hundred times faster than a well-known detection tool Androguard [4].

2   Overview

Problem Statement: Mobile app repackaging is an approach used by an adversary to change an existing app while keeping its main functionality. After altering an app's code, data, ad library, or structure, the adversary re-publishes the new app to the app store for profits or malware propagation. The cost of repackaging an Android app is low. An app's bytecode can be disassembled into an intermediary representation, which is easy for human to read. After that, the code may be quickly understood and extra code may be inserted. After performing the bytecode manipulation on the intermediary representation, the modified version can be directly assembled back to a functional Android application with tools like APKtool or Smali/Baksmali. Although obfuscation tools like Proguard [18] and Dexguard [3] may be employed to confuse the adversary, obfuscation is not sufficient to prevent repackaging. An experienced adversary may obtain the data/control flow graphs and guess the meaning of functions in the app. Moreover, the adversary may modify the app's code through various obfuscation techniques while keeping its function equivalent. It will be harder to detect this type of repackaging behavior.

Figure 1 shows a running example. We consider the Dalvik bytecode of two methods(*function_add(II) and function_Grid(III)*) from a legitimate app(minion-fun-1.2.apk) and a repackaged app(card-sharks.apk), respectively.

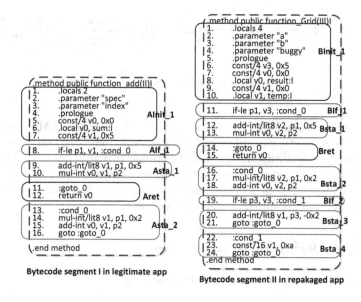

Fig. 1. Dalvik bytecode of an app and its obfuscated version

function_add(II) (*Segment I*, for short) has been divided into several blocks, labeled as {*Ainit_1, Alf_1, Asta_1, Aret, Asta_2*}. Likewise, the *function_Grid (III)* (*Segment II*, for short) is also split into several blocks, labeled as {*Binit_1, Blf_1, Bsta_1, Bret, Bsta_2, Blf_2, Bsta_3, Bsta_4*}. The code between *Segment I* in legitimate app and *Segment II* is quite different. Some noisy variables, such as {*buggy,temp*}, and redundant instructions have been injected in *Segment II*. These instructions contain opaque branch and junk code, labeled as {*Blf_2, Bsta_3 and Bsta_4*}. In fact, the functional behaviors of both methods in Fig. 1 are extremely similar: both perform the similar calculation and return the same result in the end, although *Segment II* has obfuscated the syntax structure of *Segment I*.

The similarity between Android programs can be reflected in different aspects, including syntax-level birthmark, GUI features, and resource features. Syntax features [13,21], such as fingerprints or feature hashing of mobile code, have been applied for repackaging detection. However, such schemes will not work well under semantic-preserving transformation/obfuscation.

Architecture: To detect repackaging under code obfuscation, we propose RepDetector in this work. The architecture of RepDetector is described in Fig. 2. The inputs are the Android apps probably from different Android markets. RepDetector consists of four major modules: core class and function extraction, function output states construction, function similarity measurement and app similarity measurement. Core functions along with important classes will be extracted according to bytecode and the manifest file of an app. We then construct a state flow graph for these core functions and compute the output

Fig. 2. The architecture of RepDetector

states of each function. With a Satisfiability Modulo Theories (SMT) solver, we then check the semantic equivalence of two core functions based on their output states. Finally, we quantify the similarity between two apps (each consisting of some core functions) using Mahalanobis distance. Our method uses symbolic execution, but it is simplified to compute by merging the flow states. The details of our approach are presented in the next section.

3   System Design of RepDetector

3.1   Core Classes and Functions Extraction

Each APK file contains files like *classes.dex, AndroidManifest.xml,* and sub-directories like *Res, META-INF.* The *classes.dex* file is a Dalvik executable generated from the compiled classes. Tools like *Dadexer* [1] *and Dex2jar* [2] can be used to decompile it into classes directly from the APK or JAR file. We note that not all classes are relevant to the core functionality of an app, while repackaging keeps the app's core functionality. Hence, for repackaging detection, we will only consider the functionality-relevant classes. The manifest file(*AndroidManifest.xml*) presents essential information about an app to the Android system. The functionality-relevant classes include principal components of an app: activities($<$ *activity* $>$), services ($<$ *service* $>$), broadcast receivers ($<$ *receiver* $>$), and content providers ($<$ *provider* $>$). We directly retrieve a list of such classes by parsing the *AndroidManifest.xml* file. and then construct the Class Invocation Graph (CIG) for the list of classes.

In a CIG, each node represents a class, and a directed edge between two nodes represents the existence of a function (i.e., method) invocation relationship between them. Our CIG takes into account class relationship and function invocations by static analysis. Moreover, we calculate the weight for each class in the CIG based on several attributes (e.g., fan-in and fan-out of a class node in CIG). By setting a threshold on weight, we can filter out the classes whose weights are below the threshold. The remaining classes are called *core classes.* Following a similar procedure, we can identify the *core functions* in these core classes. Another condition for core functions is that they must be created and defined by the developer (or a potential re-packager), not defined by Android libraries or third-party libraries.

Specifically, we create a whitelist, which includes Android SDK and third-party libraries. When analyzing an individual core method, we will check whether it invokes other methods. If an invoked method is from Android SDK or a third-party library in the whitelist, we will summarize its features such as class type and variable types. With such features, we can determine whether two invoked methods are equivalent or not in a later stage. On the other hand, when an invoked method is also defined by the developer (or a potential repackager), we will follow into the invoked method. This process is repeated until an invoked method contains no user defined method. The output states of a callee method are then jointed into the states of other instructions in the caller method. Our work does not consider function invocations in native code though. It was reported that only a small fraction (5 %) of apps in the Android market contain native codes [30].

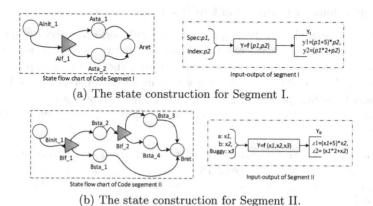

(a) The state construction for Segment I.

(b) The state construction for Segment II.

Fig. 3. The state flow chart and input-output state for Segment I and Segment II.

3.2 Output Semantics Construction

Our next step is to analyze the structure of these core functions through State Flow Chart (SFC). An SFC is constructed from control flow graph (CFG) of every core function, and the nodes in the SFC are either basic blocks in the CFG or relevant conditional instructions (e.g., if-branching instruction). Thus, the SFC is a directed graph that clearly displays how the program states flow among *instructions*. Using the code segments in Fig. 1 as an example, the SFCs of these two segments are extracted and shown in Fig. 3. In the left hand of Fig. 3(a), Segment-I includes initial block, one constraint, two statements, and one return. There are five nodes in total: {$Ainit\_1$, Alf_1, $Asta\_1$, $Aret$, $Asta\_2$}. Segment-II has two constraints and eight nodes, as showed in the left hand of Fig. 3(b): {$Binit\_1$, Blf_1, $Bsta\_1$, $Bret$, $Bsta\_2$, Blf_2, $Bsta\_3$, $Bsta\_4$}.

Definition 1 *(Function's Output State). Let set $P = [p_1, ..., p_n]$ be input register parameters (totally n) of a core function, and set $Y = [y_1, ..., y_m]$ be all possible output states (totally m) of the function. $Y = f(P)$ is a set of symbolic formulas on P generated from the state flow chart of this function.*

From the SFC, we generate the output states by symbolic execution. For example, as shown on the right side of Fig. 3, Segment I has two register parameters: *spec:* p_1 and *index:* p_2; thus, the input parameter set $P_I = [p_1, p_2]$. After several blocks are executed, the output state set Y_I contains two elements y_1, y_2: $y_1 = (p_1 + 5) * p_2$ and $y_2 = (p_1 * 2 + p_2)$. Since Segment II was injected one junk parameter *buggy:* x_3, compared to Segment I, its input parameter set is $P_{II} = [x_1, x_2, x_3]$. Furthermore, some noisy code such as opaque branch was inserted in Segment II. Nevertheless, its output state set Y_{II} contains similar symbols as in Segment I.

3.3   Equivalence Measurement of Two Functions

To determine how similar two apps are, we will perform pairwise similarity measurement between their core functions. Given two core functions from a pair of apps, we measure how semantically similar they are by comparing their input/output states. As it is unknown which input parameters of one function correspond to which parameters of the other function, we hence try different permutations of input parameters to check the equivalence. A form definition is given below.

Definition 2 *(Pairwise Equivalence of Input Register Variables). Given two variable sets:* $P_I = [p_I^1, ..., p_I^n]$ *and* $P_{II} = [p_{II}^1, ..., p_{II}^k]$, $n \leq k$. *let* $\lambda(P_{II})$ *be a permutation of the variables in* P_{II}. *A pairwise equivalence for* P_I *and* P_{II} *is defined as:*

$$\bigwedge_{i=1}^{n} [p_I^i = \lambda_i(P_{II})] \longrightarrow P_I = \lambda(P_{II}).$$

where p_I^i *and* $\lambda_i(P_{II})$ *are the i-th variables in* P_I *and* $\lambda(P_{II})$, *respectively.*

For each output state in one core function, we check whether there exists an equivalent output state in the other core function with certain combination of inputs.

Definition 3 *(Output Equivalence of Two Functions). Let* P_I *and* P_{II} *be the input sets,* Y_I *and* Y_{II} *be the sets of output states, respectively; if we have:* $\forall y_1 \in Y_I, \exists y_2 \in Y_{II}, P_I = \lambda(P_{II}), y_1 = f_I(P_I), y_2 = f_{II}(P_{II})$., *then we check:* $P_I = \lambda(P_{II}) \longrightarrow f_I(P_I) = f_{II}(P_{II})$ *where* $f_I(P_I)$ *and* $f_{II}(P_{II})$ *are the symbolic formulas of* y_1 *and* y_2, *respectively.*

Example. From Fig. 3, we generate all state results representing the input-output relationship from two function segments.

$$y_1 = (p_1 + 5) * p_2; \qquad z_1 = (x_1 + 5) * x_2$$
$$y_2 = p1 * 2 + p_2; \qquad z_2 = x1 * 2 + x_2;$$

We then check the input parameters' equivalence by permutation through Definition 2. Comparing in pairwise their equivalence by Definition 3, we find that

$$(p_1 = x_1) \wedge (p_2 = x_2) \longrightarrow (y_1 = z_1)$$
$$(p_1 = x_1) \wedge (p_2 = x_2) \longrightarrow (y_2 = z_2)$$

The detailed comparison procedure is presented in Algorithm 1. The inputs are the core functions F_I and F_{II} from two different apps. The algorithm mainly involves three steps. First, it analyzes the input variables P_I of function F_I and P_{II} of F_{II}, and then constructs the output state sets Y_I and Y_{II} of two functions by the method **Outstate**(F_I) and **Outstate**(F_{II}). After that, it compares in pairwise each permutation of P_{II} with P_I. For each comparison, it uses Satisfiability Modulo Theories (SMT) to check their equivalence . SMT is a decision procedure that can handle various types of arithmetics and other decidable theories, formulas (e.g.,$(a+b < 5) \wedge (a-c > 3 \vee d < 2)$) with propositional variables. It can decide the satisfiability of formulas containing uninterpreted function symbols with equality, linear real and integer arithmetic, bit-vectors, etc. Because Dalvik bytecode is register-based, in this work, our tool extracts the input variables as abstract register variables instead of their actual types. Later when we apply an SMT solver (CVC4 [17]) to check whether the input variables of two functions are equivalent or not, we simplify the procedure by treating the input parameter types as integer and boolean. Theoretically, this simplification could introduce false positives, because different input types may be treated as equivalent, In practice, however, two core functions (i.e., their symbolic formulas) are matched only when their logic is identical, or very similar in the case of false match due to our simplification. For functions with relatively complex logic, false function matching could be unlikely. On the other hand, this simplification has two benefits. First, the input to the SMT solver becomes simpler, making it more efficient in computation. Second, it adds a type of obfuscation resilience when an attacker attempts to evade detection by simply changing variable types (e.g., integer to double or to float). Lastly, we note that some function code contains *intents*, a special type of objects in Android. CVC4 cannot directly recognize and handle intents, so we get around the limitation by using array structure instead. When an app's code uses *intents* to pass data between activities, e.g., through "intent.putExtra()" operations, we simulate such operations of intents by the put operations of arrays.

Second, while the input variables' equivalence is satisfied, the two output states y_i and y_j are checked similarly by **Checksat()** to see whether they are semantically equivalent or not. Specifically, for all permutations $\lambda(P_{II})$ and $(y_i, y_j) \in Y_I \times Y_{II}$, it asks the SMT whether the following formula (the negation of Definition 3) is *unsatisfiable*:

$$P_I = \lambda(P_{II}) \wedge y_i \neq y_j.$$

If so, the comparison result of Checksat() $T_{ij} = 0$. Otherwise, we say the two output states y_i and y_j are equivalent, and the comparison result $T_{ij} = 1$. Then we normalize the accumulative results of comparison by cardinality $m = |Y_I|$ and $n = |Y_{II}|$, respectively. In our running example (Fig. 3), both output variables of two functions are equivalent, so the accumulative comparison result is 2. The size of both output state set is 2, so their normalized similarity score is 1. Now assume Y_I has the same two output states, but Y_{II} has one more output state

---

Algorithm 1. Semantic Equivalence Measurement of Two Core Functions

---

Input: Two Core functions from different apps: F_I, F_{II}
Output: *Similarity* $sf, sf^\flat$
1: $sf = 0, \ sf^\flat = 0$
2: *Temp Matrix* T
3: Set P_I : *the input variables of* F_I,
4: Set P_{II} : *the input variables of* F_{II},
5: Set $Y_I \leftarrow$ **Outstate**(F_I), $Y_{II} \leftarrow$ **Outstate**(F_{II})
6: **while** $\{\lambda(P_{II})\}$ **do**
7:    **if** (**Checksat**$(P_I, \lambda(P_{II})) = 0$) **then**
8:       $\lambda(P_{II}) \leftarrow$ *next permutation of combination list for* P_{II};
9:       *Continue*;
10:   **else**
11:      **for each** $y_i \ \in \ Y_I \ (i = 1 \ to \ m)$ **do**
12:         **for each** $y_j \ \in Y_{II} \ (j = 1 \ to \ n)$ **do**
13:            *Assert* $p_I = \lambda(P_{II})$
14:            $T_{ij} \leftarrow$ **Checksat**(y_i, y_j)
15:         **end for**
16:      **end for**
17:      $\lambda(P_{II}) \leftarrow$ *next permutation of combination list for* P_{II};
18:      $sf \leftarrow max(sf, \frac{\sum_{i=1}^{m} T_{ij}}{m}), \ sf^\flat \leftarrow max(sf^\flat, \frac{\sum_{j=1}^{n} T_{ij}}{n})$;
19:      *Break*;
20:   **end if**
21: **end while**
22: **return** $sf; \ sf^\flat$

(i.e., $n = 3$). Their accumulative comparison result is still 2, but the normalized similarity score is different, depending on the roles of comparison. If Y_I is compared against Y_{II}, the denominator for normalization is $|Y_I| = 2$, so the final score is $sf = 1$. If Y_{II} is compared against Y_I, the denominator is $|Y_{II}| = 3$, so the final score is $sf^\flat = 2/3$. If two functions are the same or very similar, both sf and $sf^\flat$ should be close to 1. On the other hand, for two different functions, both scores should be close to 0. Finally, the algorithm returns the two normalized similarity scores sf and $sf^\flat$.

3.4   Similarity Comparison Between Apps

Finally, RepDetector measures the similarity scores of all core functions between two apps. Given app A and app B, which have k and l core functions, respectively, we perform pairwise comparison of core functions and obtain two similarity scores sf and $sf^\flat$ for each pair based on Algorithm 1. Let SF_i (or $SF_i^\flat$) be the vector consisting of many sfs (or $sf^\flat$s) when comparing the i-th core function of app A with each of l core functions of app B, we define two similarity matrices MF and $MF^\natural$ as follows:

$$MF = \begin{bmatrix} SF_1 \\ SF_2 \\ \cdots \\ SF_k \end{bmatrix}, \quad MF^\natural = \begin{bmatrix} SF_1^\flat \\ SF_2^\flat \\ \cdots \\ SF_l^\flat \end{bmatrix}$$

MF is the matrix for comparing app A against app B, and $MF^\natural$ is the matrix for comparing B against A. If the two apps are similar, these two matrices would contain similar values. To measure the difference between these two matrices, we use the Mahalanobis distance [10], $d(A, B)$, as our metric. It is a dissimilarity measure between app A and app B of the same distribution with the covariance matrix Σ. For the two apps, their Mahalanobis distance can be calculated as follows:

$$d(A, B) = \sqrt{(MF - MF^\natural)^T \Sigma^{-1} (MF - MF^\natural)}. \tag{1}$$

Here, the inverse of matrix Σ^{-1} is the inverse covariance matrix of Σ.

$$\Sigma = \mathrm{E}\left[MF \cdot MF^\natural\right] - \mathrm{E}\left[MF\right]\mathrm{E}\left[MF^\natural\right]) \tag{2}$$

where $\mathrm{E}\left[MF \cdot MF^\natural\right], \mathrm{E}\left[MF\right]$ and $\mathrm{E}\left[MF^\natural\right]$ are the expected value or mean of the vectors $MF \cdot MF^\natural$, MF and $MF^\natural$, respectively. The smaller the Mahalanobis distance, the higher similarity of two apps.

4 Performance Evaluation

In this section, we present the evaluation of our tool RepDetector. The implementation of RepDetector consists of 6380 lines of Java code. Our experiments were conducted on a 20-machine cluster. Each machine is equipped with a Core i7 3.2 GHz CPU and 16 GB memory and its operating system is Ubuntu Linux 10.04.

4.1 Study I: Detection Accuracy with Known Samples

The first objective of our evaluation is to evaluate the detection accuracy of RepDetector, in comparison with Androguard [4] and ViewDroid [27]. We will compare it with a few more detection algorithms later on. We show the number of false positives (FP), number of false negatives (FN), and accuracy ACC [11]. ACC is defined as follows.

$$ACC = 1 - \frac{FP + FN}{\Sigma\,(Total\,Population)} \tag{3}$$

We select 1,000 repackaged app samples from a previous dataset [5]). These apps cover different ranges of file sizes and have been verified as repackaged apps. Specifically, our samples contain 183 groups of apps from various categories, such as game, social, and books. Each group consists of the original app and its repackaged version(s). The maximum number of repackaged apps in one group

Table 1. Comparison of detection accuracy among Androguard,ViewDroid and RepDetector

μ	FP			FN			ACC		
	A	V	R	A	V	R	A	V	R
0.7	47	31	9	12	13	0	94.1 %	95.6 %	99.1 %
0.75	32	31	6	23	18	0	94.5 %	95.1 %	99.4 %
0.8	27	23	0	46	25	8	92.7 %	95.2 %	99.2 %
0.85	16	21	0	107	36	19	87.7 %	94.3 %	98.1 %

μ: Similarity   Threshold.   A: Androguard.
V: ViewDroid R: RepDetector

is 81. We compare detection accuracy by counting FP (when apps in different groups are reported as repackaging), FN (when apps in the same group are not reported as repackaging) and computing ACC.

Table 1 shows the evaluation result. In the first column, we set four different similarity thresholds μ, from 0.7 to 0.85. When μ increases, the number of FP instances decreases for all algorithms, but both Androguard and ViewDroid has more FP instances than RepDector. Androguard treats all the bytecode of an app equally, no matter where it comes from; hence, some common libraries could introduce high similarities between different apps. In the case of ViewDroid, some apps in the test dataset are very simple in GUI, so their view graphs extracted by ViewDroid are small with few edges, leading to false positives. RepDetector only examines the core functions and classes, but excludes the third-party libs or the Android framework's bytecode. Thus its FP instances are much less than Androguard. The false negatives of Androguard clearly increases with threshold μ, whereas ViewDroid and RepDector achieve better performance. The FN instances of ViewDroid are mostly caused by some "add-on" functions, which introduce differences in view graphs but with some common functionality code. The last three columns show the detection accuracy (ACC) of these algorithms. The accuracy of RepDetector exceeds 98 %. ViewDroid has stable accuracy at around 95 %. Androguard's accuracy is below 95 % and its lowest accuracy in our experiment setting is 87.7 %, when it reports incorrect results with 123 instances.

4.2   Study II: Efficiency

Figure 4 shows the time for pairwise comparison of apps in our testing app set. Specifically, Fig. 4(a) and (b) depict the average execution time for pairwise comparison by Androguard and RepDetector, respectively. Over 95 % of comparisons in Androguard take $5 \sim 15$ s, whereas in RepDetector 98 % pairwise comparisons require less than 0.12 s. From the table in Fig. 4(c), one can further see that the min and max execution times of Androguard are 0.66 s and 17.7 s, respectively. For RepDetector, the min execution time is 0.009 seconds and the max time is 0.14 s. All these numbers indicate that RepDetector is at least one hundred times faster than Androguard. Androguard has low efficiency because

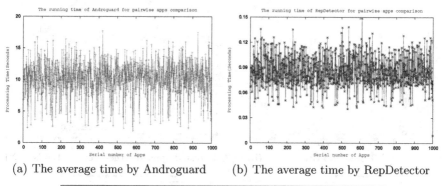

(a) The average time by Androguard   (b) The average time by RepDetector

Repackaging Detection Tool	The min Time of app's detection	The Max time of app's detection
Androguard	0.66 s	17.7 s
RepDetector	0.009 s	0.14 s

(c) The min and max time for pairwise comparison of apps by
Androguard and RepDetector

Fig. 4. The average running time for pairwise comparison of apps by Androguard and
RepDetector

it spends too much time in compressing apps' bytecode by the NCD algorithm
and generating features for the apps. Differently, RepDetector only needs to
handle core functions and classes, which are only a fraction (sometimes a small
fraction) of all functions and classes used by an app. Its handling cost is determined by the efficiency of SMT and the complexity of core functions, not the
app size or the total amount of code. We have also compared RepDector with
ViewDroid. The average execution time of ViewDroid for testing each pair is
about 11 s, much higher than RepDetector. This is because ViewDroid involves
relatively expensive subgraph isomorphism detection.

To further explain the high efficiency of RepDetector, we perform a statistical
study with our dataset. Figure 5(a) shows the sizes of these 1,000 tested apps.
99 % of them are smaller than 50MB and 80 % are below 20MB. Only two apps
are larger than 100MB. Figure 5(b) shows the numbers of lines of code (LOC)
in the bytecode of these apps. While 90 % apps have LOC between 10^4 and $10^{5.5}$
($\lg N$ =4 to 5.5), the LOC in all core methods (selected by setting a weight
threshold as 5) are only between $10^{2.5}$ and $10^{3.5}$. This shows a reduction on
LOC needed for analysis by around a hundred times, which is the main reason
for RepDetector's higher efficiency than Androguard. Figure 5(c) further shows
the number of core classes in each app. About 80 % of apps, the number of their
core classes is below 30. And 50 % of apps have less than 10 core classes. Finally,
Fig. 5(d) shows that in 90 % tested apps, the ratio between number of core classes
and number of whitelisted libraries is below 0.2.

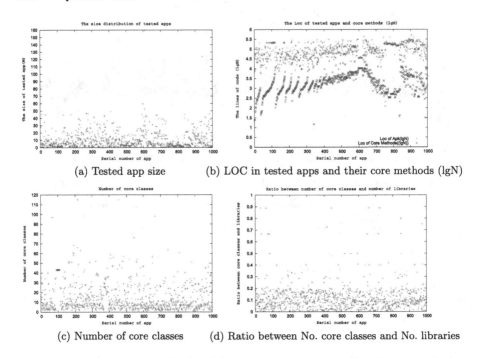

(a) Tested app size   (b) LOC in tested apps and their core methods (lgN)

(c) Number of core classes   (d) Ratio between No. core classes and No. libraries

Fig. 5. The statistics of tested apps

4.3   Study III: Obfuscation Resilience

We now evaluate the robustness of RepDetector against code obfuscation. To obfuscate the original apps for repackaging, we may leverage obfuscation tools like ProGuard [18] and DexGuard [3]), which are able to shrink, optimize, and obfuscate Java source code. However, they cannot perform complicated obfuscation such as data or control flow obfuscation; hence, we resort to another tool [15], which is a dedicated framework for evaluating the obfuscation resilience of Android repackaging detection algorithms. The tool includes 37 types of obfuscation from SandMark [7], including instruction reorder, layout obfuscation, control structure obfuscation and data flow obfuscation, etc. It provides both single obfuscation (each type applies one type of obfuscation) and serialized multiple obfuscation (multiple obfuscation are applied one after another). There are textual or syntactical differences when apps are repackaged by single or serialized multiple obfuscation. For the evaluation, we compare RepDetector with Androguard and three other techniques, including DroidMOSS [30], AST-Distance [21] and AST-Coverage [21]. Except Androguard, the other three tools are not publicly available, so we implemented a version for each of them based on their used algorithms.

We randomly choose 200 apps from our crawled app set and evaluate the true positives and false positives of these algorithms by setting different app similarity detection thresholds. We then use the ROC curve to compare them.

Figure 6 shows the detection results under single or serialized multiple obfuscation. The x-axis is the false positive rate (FPR) and the y-axis is the true positive rate(FPR),

Figure 6(a) shows the ROC curves under single obfuscation. RepDetector performs the best among these algorithms. The detection accuracy of AST-Distance and AST-Coverage drops significantly, while DroidMOSS and Androguard have acceptable performance. As a concrete example, when the false positive rate is 0.32, the true positive rates are: 0.97 (RepDetector), 0.94(Androguard), 0.87 (DroidMoss), 0.72 (AST-Coverage) and 0.63 (AST-Distance), respectively. The ROC curve of AST-Distance is near the diagonal section because AST-Distance can hardly handle obfuscated code.

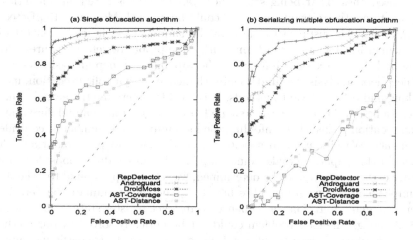

Fig. 6. ROC of algorithm detection accuracy under code obfuscation

We then compare RepDetector with other four detection algorithms under serialized multiple obfuscation. For example, an app may be obfuscated by noise code injection, register rename and followed by opaque branch injection. However, not every app can go through multiple obfuscators successfully. For comparison, we have eliminated the failure cases. Figure 6(b) shows the detection results when apps are obfuscated with duplicate registers, node spliter, buggy code, method madness. We can see that all algorithms degrade their accuracy, compared to the single obfuscation case. This is reasonable because each additional obfuscation changes the code more from its original and makes it harder to detect code similarity. On the other hand, RepDetector still outperforms the other algorithms significantly. Moreover, RepDetector can handle semantics-based obfuscation: computation obfuscation, program reconstruction, data transformation and instruction reordering, because it relies on semantic analysis of each app instead of syntax characteristics. Androguard may partially detect semantic obfuscation such as instruction reordering, but it cannot detect complicated obfuscation. DroidMOSS is less effective than Androguard and RepDetector because it is mainly based on syntax fingerprints and has hence little effect on code obfuscation. AST-Coverage and AST-Distance have almost no resilience to complicated obfuscation.

5   Discussion

In this section, we discuss some limitations of our work.

Adversaries may split an app's function into multiple smaller functions. In addition, the adversaries may merge several unrelated functions into a big one using redundant code. Crussell et al. [9] mentioned that those kinds of subversions are difficult to detect for most similarity detection methods, including PDG-based ones. That being said, since RepDetector is based on semantics of core functions, the relevant semantics could be merged. Moreover, RepDetector uses Mahalanobis distance to measure the similarity, so it may even tolerate some semantics-changing transforms with an appropriate detection threshold.

Second, RepDetector is limited by the capability of loop analysis. As far as we know, loop analysis is a fundamental challenge for symbolic execution, model checking and other relevant methods. It is hard to determine the actual number of iterations in a loop and even a single loop may generate many different symbolic execution paths when unfolding the loop into a large number of equivalent statements. We also need to specify pre-conditions and post-conditions altogether. In this paper, we make some simplification in handling loops. When the number of iterations cannot be determined, we set an upper bound threshold (in our implementation we set the threshold to 5) and then terminate the iterations. After that many iterations, a sequence of abstract register states will be generated. Although the loop problem could not be perfectly solved by our method, the generated sequence can still catch in some degree the semantic meaning of the loop body. Since our ultimate goal is to detect code similarity, not to understand the exact meaning of a function, we believe our current treatment of this loop problem will not cause big errors.

6   Related Work

Android Application Repackaging Detection. A number of techniques for mobile app repackaging detection have been proposed previously. DroidMOSS [30] computes a series of fuzzy hashes of each method in an app and combines them together. It then compares the fuzzy hashes of two APKs to detect app similarity This approach cannot capture the semantic information of the Dalvik bytecode. PiggyApp [29] uses program dependency graph (PDG) as the core feature to detect piggybacked apps and employ the nearest neighbor searching algorithm to improve scalability. The PDG based approach has also been proposed in DNADroid [8], where the PDG for every method in the Dex file is computed, and a graph isomorphism algorithm is used to calculate the similarity between the computed PDGs. AnDarwin [9] improves DNADroid's detection efficiency by extracting features using LSH and comparing similar apps with Min-Hash algorithm. PDGs comparison usually suffers from the inefficiency in subgraph isomorphism detection. Also, as shown by Huang et al. [15], it is straightforward for plagiarists to insert redundant code (with data dependency) to obfuscate the PDGs.

In Black-Hat 2011, Androguard [4] was proposed with several techniques on Android app reverse engineering and repackaging detection, It generates the Normalized Compression Distance (NCD [14]) of the signatures extracted from the victim and suspicious apps to determine the similarity. Signatures are created based on features extracted from a generalized representation of the Dalvik bytecode – the opcode sequences. Although this approach is able to capture the high-level semantic information of the code, including control flow graph (CFG), a simple CFG flattening obfuscation can defeat its effectiveness on repackaging detection. Huang et al. [15] proposed a framework for evaluating repackaging detection algorithms of Android apps by generating several obfuscators. Our techniques is robust against most of the obfuscators that preserve the semantics of the app.

Recently, Shao et al. [23] showed that it is possible to detect repackaging clones by app resource features. Zhang et al. [27] have proposed a detection technique called ViewDroid, which constructs core features based on the app UI. ViewDroid is robust against most of the code-level obfuscation. However, for apps with a small set of UI components, it might easily produce some false positives. SmartDroid [28] also uses user interfaces to find user interactions that will trigger sensitive APIs dynamically. Chen et al. [6] proposes a code clone detection technique, which performs the geometry-characteristic-based encoding of control flow graph. While the technique is computationally efficient, it cannot deal with app repackaging using code obfuscation techniques. We note that these techniques (including ours) target at solving different challenges on repackaging detection, they each have some unique strengths. Therefore, they are complementary to each other in nature.

Software Plagiarism Detection. MOSS [22] applies local fingerprinting to detect source code plagiarism. Lim et al. [19] leveraged stack pattern based birthmark, which requires the source code and are vulnerable to some types of code obfuscation. Myles et al. [20] analyzed executables statically and used k-gram to perform similarity measurement. However, it is not robust to instruction reordering and junk instruction insertion. There are also dynamic software birthmarks based software plagiarism detection methods, including core values based birthmark [16, 26] and system call based birthmark [24, 25]. These dynamic methods are not efficient enough for large-scale repackaging detection in Android markets.

7   Conclusions

App repackaging is a way to change an original app into one with the same or similar functionality. It has become a popular tool for malware propagation and for piracy. It is very important for repackaging detection algorithms to be resilient to code obfuscation. While many approaches have been proposed, scalability and obfuscation resilience remains a challenge. In this paper, we propose a tool called RepDetector, which is designed to find repackaging apps by semantic similarity. Our evaluation has demonstrated its effectiveness and efficiency. In our future work, we plan to conduct a much larger scale evaluation with apps from different markets.

Acknowledgments. We thank the anonymous reviewers for their valuable comments and Dr. Nick Nikiforakis for shepherding our paper. The work of Guan and Luo was supported by the Science and Technology Planning Project of Guangdong Province, China (2014A040401027, 2012A080102007, 2015A030401043). The work of Huang and Zhu was partially supported by NSF CCF-1320605.

References

1. Dedexer. http://dedexer.sourceforge.net/
2. Dex2jar. https://code.google.com/p/dex2jar/
3. Dexguard. http://www.saikoa.com/dexguard
4. Desnos, A.Z.: Androidguard. https://code.google.com/p/androguard/
5. Chen, K., Liu, P., Zhang, Y.: Achieving accuracy and scalability simultaneously in detecting application clones on android markets. In: Proc. of ICSE (2014)
6. Chen, K., Wang, P., Lee, Y., Wang, X., Zhang, N., Huang, H., Zou, W., Liu, P.: Finding unknown malice in 10 s: mass vetting for new threats at the google-play scale. In: Proceedings of the 24th USENIX Conference on Security Symposium, pp. 659–674. USENIX Association (2015)
7. Collberg, C.S., Myles, G., Huntwork, A.: Sandmark-a tool for software protection research. IEEE Secur. Priv. **1**(4), 40–49 (2003)
8. Crussell, J., Gibler, C., Chen, H.: Attack of the clones: detecting cloned applications on android markets. In: Foresti, S., Yung, M., Martinelli, F. (eds.) ESORICS 2012. LNCS, vol. 7459, pp. 37–54. Springer, Heidelberg (2012)
9. Crussell, J., Gibler, C., Chen, H.: Scalable semantics-based detection of similar android applications. Technical report (2012). ucdavis.edu
10. De Maesschalck, R., Jouan-Rimbaud, D., Massart, D.L.: The mahalanobis distance. Chemom. Intell. Lab. Syst. **50**(1), 1–18 (2000)
11. Fawcett, T.: An introduction to ROC analysis. Pattern Recogn. Lett. **27**(8), 861–874 (2006)
12. Gibler, C., Stevens, R., Crussell, J., Chen, H., Zang, H., Choi, H.: Characterizing android application plagiarism and its impact on developers. In: Proceedings of MobiSys (2013)
13. Hanna, S., Huang, L., Wu, E., Li, S., Chen, C., Song, D.: Juxtapp: A scalable system for detecting code reuse among android applications. In: Proceedings of DIMVA (2013)
14. Hemel, A., Kalleberg, K.T., Vermaas, R., Dolstra, E.: Finding software license violations through binary code clone detection. In: Proceedings of MSR. ACM (2011)
15. Huang, H., Zhu, S., Liu, P., Wu, D.: A framework for evaluating mobile app repackaging detection algorithms. In: Huth, M., Asokan, N., Čapkun, S., Flechais, I., Coles-Kemp, L. (eds.) TRUST 2013. LNCS, vol. 7904, pp. 169–186. Springer, Heidelberg (2013)

16. Jhi, Y.C., Wang, X., Jia, X., Zhu, S., Liu, P., Wu, D.: Value-based program characterization and its application to software plagiarism detection. In: Proceedings of the 33rd International Conference on Software Engineering, pp. 756–765. ACM (2011)
17. King, T., Barrett, C., Tinelli, C.: Leveraging linear and mixed integer programming for SMT. In: Formal Methods in Computer-Aided Design, FMCAD 2014, pp. 139–146. IEEE (2014)
18. Lafortune, E.: Proguard. http://proguard.sourceforge.net/
19. Lim, H., Park, H., Choi, S., Han, T.: Detecting theft of Java applications via a static birthmark based on weighted stack patterns. IEICE - Trans. Inf. Syst. **E91-D**(9), 2323–2332 (2008)
20. Myles, G., Collberg, C.S.: K-gram based software birthmarks. In: SAC (2005)
21. Potharaju, R., Newell, A., Nita-Rotaru, C., Zhang, X.: Plagiarizing smartphone applications: attack strategies and defense techniques. In: Proceedings of ESoSS (2012)
22. Schleimer, S., Wilkerson, D.S., Aiken, A.: Winnowing: local algorithms for document fingerprinting. In: Proceedings of ACM SIGMOD International Conference on Management of Data (2003)
23. Shao, Y., Luo, X., Qian, C., Zhu, P., Zhang, L.: Towards a scalable resource-driven approach for detecting repackaged android applications. In: Proceedings of ACSAC. ACM (2014)
24. Wang, X., Jhi, Y., Zhu, S., Liu, P.: Behavior based software theft detection. In: Proceedings of 16th ACM Conference on Computer and Communications Security (CCS) (2009)
25. Wang, X., Jhi, Y.C., Zhu, S., Liu, P.: Detecting software theft via system call based birthmarks. In: Computer Security Applications Conference, ACSAC 2009. Annual, pp. 149–158. IEEE (2009)
26. Zhang, F., Jhi, Y., Wu, D., Liu, P., Zhu, S.: A first step towards algorithm plagiarism detection. In: Proceedings of the 2012 International Symposium on Software Testing and Analysis. ACM (2012)
27. Zhang, F., Huang, H., Zhu, S., Wu, D., Liu, P.: Viewdroid: Towards obfuscation-resilient mobile application repackaging detection. In: Proceedings of ACM WiSec, pp. 25–36. ACM, New York, NY, USA (2014)
28. Zheng, C., Zhu, S., Dai, S., Gu, G., Gong, X., Han, X., Zou, W.: SmartDroid: an automatic system for revealing UI-based trigger conditions in Android applications. In: Proceedings of the second ACM workshop on Security and privacy in smartphones and mobile devices, pp. 93–104. ACM (2012)
29. Zhou, W., Zhou, Y., Grace, M., Jiang, X., Zou, S.: Fast, scalable detection of piggybacked mobile applications. In: Proceedings of ACM CODASpPY (2013)
30. Zhou, W., Zhou, Y., Jiang, X., Ning, P.: Detecting repackaged smartphone applications in third-party android marketplaces. In: Proceedings of ACM CODASpPY (2012)

Accelerometer-Based Device Fingerprinting for Multi-factor Mobile Authentication

Tom Van Goethem$^{(\boxtimes)}$, Wout Scheepers, Davy Preuveneers, and Wouter Joosen

iMinds-DistriNet-KU Leuven, Leuven, Belgium
wout.scheepers@student.kuleuven.be,
{tom.vangoethem,davy.preuveneers,wouter.joosen}@cs.kuleuven.be

Abstract. Due to the numerous data breaches, often resulting in the disclosure of a substantial amount of user passwords, the classic authentication scheme where just a password is required to log in, has become inadequate. As a result, many popular web services now employ risk-based authentication systems where various bits of information are requested in order to determine the authenticity of the authentication request. In this risk assessment process, values consisting of geo-location, IP address and browser-fingerprint information, are typically used to detect anomalies in comparison with the user's regular behavior.

In this paper, we focus on risk-based authentication mechanisms in the setting of mobile devices, which are known to fall short of providing reliable device-related information that can be used in the risk analysis process. More specifically, we present a web-based and low-effort system that leverages accelerometer data generated by a mobile device for the purpose of device re-identification. Furthermore, we evaluate the performance of these techniques and assess the viability of embedding such a system as part of existing risk-based authentication processes.

1 Introduction

In September 2014, an attacker managed to access a privileged account on Bugzilla, a bugtracker software used by Mozilla, by simply using the credentials of the user that were leaked by a data breach on an unrelated website. As a result, the adversary was able to access security-sensitive information on flaws reported in the Firefox browser. By leveraging the obtained information, the attacker eventually tried to exploit unwitting visitors of a news website.

This is just one of the many stories where an attacker managed to gain access to a user's account, and along with the tendency of users to either re-use passwords between different websites or choose weak passwords [1], serves as a good example that passwords are no longer fitting for strong authentication. As a countermeasure to these threats, multi-factor authentication systems have recently gained in popularity. These multi-factor authentication schemes require a user to provide multiple elements that can be used to prove his identity. A popular choice is two-factor authentication, where a randomly generated passcode is sent by SMS to the user's mobile phone. By requiring such a token, it becomes

© Springer International Publishing Switzerland 2016
J. Caballero et al. (Eds.): ESSoS 2016, LNCS 9639, pp. 106–121, 2016.
DOI: 10.1007/978-3-319-30806-7_7

very challenging for an adversary to log in to a user's account, as this would require him to both know the password, as well as intercept the SMS message.

Although the additional factor substantially improves security, there exist several drawbacks, mainly with regards to deployment and costs, that prevent tokens over SMS from being widely adopted as a second factor in authentication mechanisms. Multi-factor authentication, which can be seen either as an alternative solution, or as a complement to the two-factor authentication system, attempts to tackle these drawbacks. In multi-factor authentication systems, various pieces of information on the authenticating agent are gathered during the authentication process. This information, which can consist of the user's IP address, behavioral and contextual information, or a fingerprint of the browser he is using to authenticate, is then compared against the user's typical behavior. In case some elements from this information deviate from what is expected, the user is either denied access, or is required to use a stronger authentication method for verification, e.g. by using the aforementioned two-factor authentication.

The strength of this type of multi-factor authentication system is strongly dependent on the trustworthiness of the acquired information. While it has been shown that fingerprint information gathered by a modern browser on a desktop computer typically yields high entropy [2], the fingerprints obtained from mobile devices carry a lot of similarity [3,4], making them an unsuitable candidate in a multi-factor authentication system. In this paper, we focus on improving the reliability of multi-factor authentication in the web, and more specifically, on mobile devices. We show how current authentication systems can be augmented with a sensor-based fingerprint, in order to evaluate whether the authenticated user is using a trusted device. By computing an adaptive similarity score, the identity provider can determine the risk that the request is illicit, and take the appropriate actions. We exemplify this type of authentication system by developing a web-based system that leverages accelerometer data in combination with controlling the mobile phone's vibration motor. Finally, we evaluate a proof-of-concept implementation on the usability and feasibility of deploying such a system in real-world scenarios.

Our main contributions are:

- An accelerometer-based device fingerprinting mechanism with adaptive similarity scores as a suitable candidate for multi-factor mobile authentication
- The integration of such a web-based multi-factor authentication solution in a contemporary identity and access management (IAM) system
- A feasibility assessment of such a system in real-world scenarios

The rest of this paper is structured as follows: in Sect. 2, we sketch a brief background on fingerprinting and authentication. In Sect. 3, we motivate our approach and describe the implementation of a proof-of-concept application. In Sect. 4, we evaluate the application. In Sect. 5, we briefly reflect on related work, and finally, we conclude our work in Sect. 6.

2   Background

Fingerprinting browsers and devices has a variety of applications, both nefarious as well as beneficial. The majority of use cases require the unique identification of a specific user, whether it is used to prevent fraud, or to track a user over different sessions. This tracking is done by gathering information about the browser and system that is being used. Examples include identification of the browser version [2], canvas fingerprinting [5], and enumerating fonts and plugins that are installed on the browser or system [6]. In most desktop environments, these methods, or a combination thereof, provide a fingerprint that can uniquely identify a user. Consequently, it is not surprising that fingerprinting is a suitable technique for multi-factor authentication systems.

However, for mobile devices, which often share the same hardware and do not allow as many customizations of the system, device fingerprints of devices that share the same brand often result in exactly the same identifier. Moreover, Spooren et al. found that the majority of fingerprinted properties are predictable [3], preventing these techniques from being used as part of authentication processes. In an attempt to overcome these shortcomings, researchers have evaluated alternative approaches that can be leveraged to collect identifying information.

Although mobile devices of the same type most likely share the same hardware, certain sensors may suffer from microscopic imperfections caused during the manufacturing process. Because the variations on the sensor data are most likely unique for each device, they become an interesting target for fingerprinting [7]. For example, Lukas et al. found that digital cameras expose a certain pattern noise, which can then be used for identification [8]. Similarly, Das et al. show that imperfections in device microphones and speakers induce anomalies in the sound that is produced and recorded [9]. Again, the variations among devices can be used for fingerprinting purposes.

Another interesting approach, is to analyze the data returned by a mobile device's accelerometer. In their research, Dey et al. describe how imperfections induced to accelerometer chips can be leveraged to create a unique fingerprint [10]. For the purpose of authentication in the context of the web, accelerometer data is particularly interesting because this data is exposed by the majority of mobile browsers, and does not require the user's consent.

3   Approach and Implementation

The sensor used for fingerprinting in this work is the accelerometer. An accelerometer measures the acceleration force that is applied to the device along the three physical axes. This acceleration is expressed in meter per second squared (m/s^2). The reasons for using this sensor is that (1) nowadays every smartphone has an accelerometer, (2) accelerometer data is accessible through JavaScript in a mobile browser and (3) there is recent work about using accelerometer data for sensor fingerprinting [10].

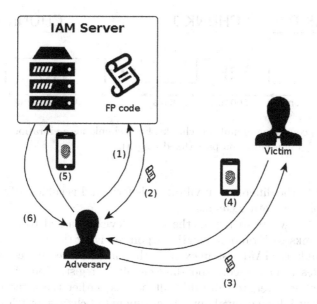

Fig. 1. Fingerprint phishing misuse scenario, enabling spoofing of the user agent.

3.1 Enduring the User Agent and Fingerprint Spoofing Threat

In contrast to prior work, where native APIs were used to obtain accelerometer data for tracking purposes, our research focuses on leveraging the accelerometer data, as exposed through browser APIs, for the purpose of improving the reliability of a multi-factor authentication system. Since the collection of fingerprinting data is done purely on the side of the client, an additional concern to be taken into consideration in our work is the mitigation against spoofing attacks where an adversary poses as the legitimate user. Consider the misuse scenario in Fig. 1. Imagine an adversary who knows the credentials of a victim. He reads the fingerprinting code and sets up a phishing site to steal a user's device fingerprint. When prompted to log in, the adversary provides the stolen credentials along with the fingerprint and is authenticated to the system. It is clear that this scenario should not be possible, and thus a more clever approach is needed.

3.2 Mitigation Against Spoofing Attacks

To counter the spoofing threat, our mitigation provides following mitigation. Upon registration, several *traces* are collected, each consisting of multiple *chunks*. A chunk contains a vibration part, i.e. a short period of time during which the device's vibration motor is enabled using the `navigator.vibrate()` API, and a non-vibration part. The length (in ms) of the vibration parts will increase for each chunk within a trace, as depicted in Fig. 2. Upon registration, a trace is collected from the user's device and for each chunk, features are extracted. The assumption made here is that each chunk will have a different vibration

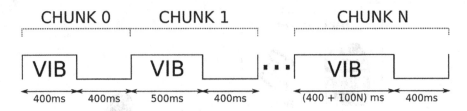

Fig. 2. A trace consisting of multiple chunks. Each chunk has a vibration part (increasing length) and a non-vibration part (fixed length).

behavior due to the difference in vibration length, and is sufficiently robust and distinguishable from other chunks.

When a user wants to log in, the user's device is asked to provide sensor data from chunks with random lengths. Upon receiving the sensor data for each requested chunk, the IAM system extracts the same features and compares them to the features from the corresponding chunk at registration. This approach makes a spoofing attempt more difficult because either the attacker needs to obtain sufficient information about a large amount of chunks, or trick the victim in providing data for the requested chunks. We consider the former attack to be unlikely, as this would require the attacker to collect accelerometer data during a considerable amount of time, something that would easily alert the user of the wrongdoing. Although the second attack, where the adversary just collects information on the requested chunks, may be viable under certain circumstances, it should be noted that the proposed mechanism is not a stand-alone authentication system. This means that an adversary not only needs to "forge" the accelerometer-based fingerprint, he also needs to know the user's credentials as well as uncover the expected values for the other aspects that are required by the multi-factor authentication system.

3.3   Sensor Data Collection and Fingerprint Extraction

For the collection of accelerometer data, we developed an HTML web application using the jQuery Mobile framework[1], collecting chunks as illustrated in Fig. 2. The chunks are numbered from 0 until N. The first chunk has a vibration part of 400 ms. For each subsequent chunk, the vibration part increases with 100 ms. The length of the non-vibration part remains fixed at 400 ms. This length was defined experimentally; 400 ms without vibration is long enough for the device to return to a motionless state. When this non-vibration length is less, it is hard to distinguish between chunks because of noise the accelerometer still registers due to the momentum of the device. This results in chunk 0 having a length of 800 ms, chunk 1 having a length of 900 ms and so on. Chunk N has length $(400 + 100N)$ ms. In this paper, we only consider accelerometer data that was collected when the device was placed on a hard surface, e.g. a table top, because

---

[1] https://jquerymobile.com/.

we found it to result in more robust data, and is still viable in the context of authentication. This also makes potential attack scenarios more difficult, since this requires an adversary to either trick the user in placing his phone on a hard surface until a sufficient amount of accelerometer data has been obtained, or estimate the expected accelerometer values from the values that originate from vibrating on an unknown surface. While not completely unfeasible, this approach significantly constraints the viability and plausibility of attack scenarios.

The extraction of fingerprintable features is done in Matlab. After loading the data, the timestamps are normalized. Next, for each datapoint, the Root Sum Squared (RSS) is taken of the values on the three axes. According to the chunk-stamps taken at collection, the trace is split into the corresponding chunks. For each chunk, 8 time-domain features are extracted. These features are the *mean, standard deviation, average deviation, skewness, kurtosis, RMS amplitude, minimum* and *maximum*. These extracted features will be evaluated on their robustness and distinguishability among devices.

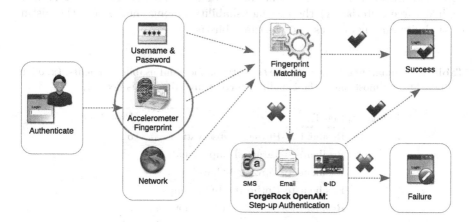

Fig. 3. Integrating accelerometer-based device fingerprinting in contemporary identity and access management systems.

3.4   Integration in Identity and Access Management Systems

We have integrated our solution in OpenAM 12, a contemporary identity and access management system. OpenAM offers device fingerprinting and matching capabilities using client-side and server-side JavaScript technology. As shown in our previous work [3], the built-in fingerprinting code is not well suited for mobile devices. In this work, we adapted the JavaScript code to call our service to process accelerometer traces and chunks. The additional benefit of this integration is that OpenAM and our solution can be independently scaled out.

4   Evaluation

4.1   Qualitative Evaluation

The prototype is evaluated thoroughly based on the framework proposed by Stajano et al. [11]. This framework provides an evaluation methodology and benchmark for web authentication proposals. For evaluation purposes it makes use of a taxonomy of 19 security, privacy and usability benefits. Table 1 shows a summary of all evaluated benefits. From the 19 benefits, the prototype offers 10 benefits completely. There are 6 benefits that are almost offered by the prototype. One benefit (*scalable-for-users*) could not be evaluated, and is left as future work.

4.2   Quantitative Evaluation

Feature Analysis. For the design of the feature matching algorithm, the extracted features from the chunks are evaluated. All features are evaluated against three criteria: (1) the distinguishability among chunks, (2) the distinguishability among devices and (3) the robustness.

Table 1. Summarizing table with the security, privacy and usability benefits. (x: offers the benefit; o: almost offers the benefit; ? further investigation is needed)

Category	Benefit	Prototype
Security	Resilient-to-Physical Observation	o
	Resilient-to-Targeted-Impersonation	o
	Resilient-to-Throttled-Guessing	x
	Resilient-to-Unthrottled-Guessing	o
	Resilient-to-Internal-Observation	x
	Resilient-to-Leaks-from-Other-Verifiers	x
	Resilient-to-Phishing	o
	Resilient-to-Theft	o
	No-Trusted-Third-Party	x
	Requiring-Explicit-Consent	x
Privacy	Unlinkable	x
Usability	Memorywise-Effortless	
	Scalable-for-Users	?
	Nothing-to-Carry	o
	Physically-Effortless	
	Easy-to-Learn	x
	Efficient-to-Use	x
	Infrequent-Errors	x
	Easy-Recovery-from-Loss	x

As we investigate whether the device's sensor fingerprint can be used for authentication, the features trivially have to be distinguishable among devices. If this would not be the case, a fingerprint from any random device could be used to log in. The features also need to be robust, i.e. they cannot deviate too much from each other when new data is collected from the same device. In case this would not hold, the features would have little meaning, as their values deviate too much for every feature extraction of raw data.

We analyzed the criteria above and depicted them in Fig. 4. This figure depicts 8 accelerometer-based features (mean, standard deviation, average deviation, skewness, Kurtosis, root mean square, minimum and maximum) on 3 different mobile devices. The length of each box plot represents the robustness of a given feature on a particular device. This gives an idea of how consistently the device produces the extracted features for different measurements. As the robustness of the features is device dependent, this value can be used as a risk measure for a device-specific adaptive scoring function. The distinguishability corresponds to the extent to which the box plots overlap. The more they overlap, the more the values lie in the same range and are hereby harder to distinguish from each other.

The most important conclusion of this analysis is that the short chunks contain more entropy than the long chunks. This is mainly due to the exponential behavior of the mean, skewness and RMS amplitude features. The minimum feature is bad for distinguishing among chunks and devices and hereby useless.

Matching Algorithm. Now that the behavior of the features is known, it is possible to design and implement a matching algorithm that checks whether a login trace will be accepted or not.

Upon registration, a series of registration traces is collected. This number of traces will be defined empirically in the next section. Using the data from the multiple registration trace, we calculate an interval, which consists of a lower- and upper-bound percentile of the observed values. During the login process, a single trace, consisting of a randomized subset of chunks, is requested. For each chunk in this login trace, all features are compared to those that were extracted during the registration process. A certain feature, for a certain chunk is marked as accepted when it falls within the boundaries of the registered values. Subsequently, a score is computed for each chunk, which is based on the accepted feature-values. Because the robustness of each feature is different, we attributed each feature a certain weight, based on their distinguishability. The final score of a specific chunk is determined as the sum of the weights of all accepted features. When this sum exceeds a certain threshold, the chunk is marked as accepted.

For a successful authentication attempt, a ratio, which can be user-defined, of all probed chunks should be classified as accepted. This ratio is directly related to the difficulty of passing a login attempt: if this ratio is set to a high value, and all chunks must match the values from the registration process, it becomes more likely that an legitimate login trace will fail due to momentary measurement inaccuracies. In our evaluation, we found that a ratio of $\frac{3}{4}$ provides a balanced end-result.

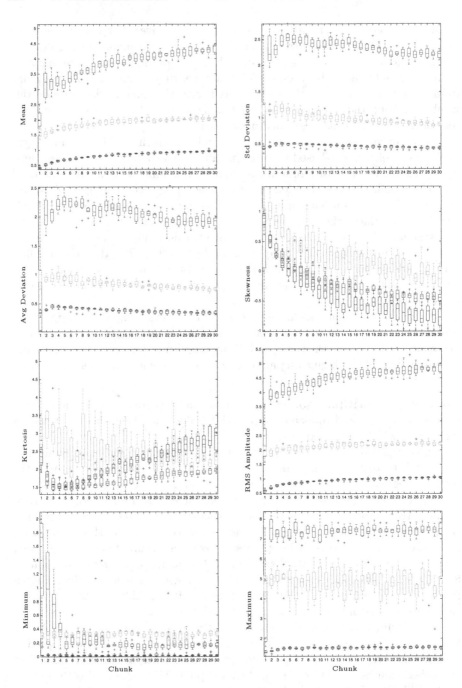

Fig. 4. Distinguishability and robustness of 8 accelerometer-based features of different mobiles devices (red: Motorola Moto G, green: Huawei P8 Lite, blue: Google Nexus 4). The length of each box plot is a measure for the device-specific robustness of the feature (Color figure online).

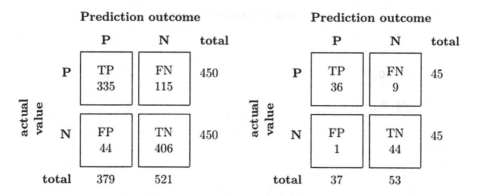

Fig. 5. Confusion matrices for all 15 devices. Left: chunks, right: traces.

4.3   Experiment Setup

For conducting experiments, 10 registration traces consisting of 10 chunks were gathered together with 3 login traces. The dataset consists of data from 15 devices: 6 Google Nexus 5's, 3 Google Nexus 4's, 3 OnePlus One's, a Samsung Galaxy S4 mini, a Samsung Galaxy S5 and an LG G3.

Performance Metrics. For measuring classification performance, each login trace of the users device is checked by the matching algorithm. From this, the amount of *true positives* TP_i and *false negatives* FN_i is calculated for each chunk i. A true positive occurs when a chunk collected by the registered device is accepted, i.e. a chunk that needed to be classified as accepted is accepted. A false negative occurs when a chunk collected by the registered device is rejected, i.e. a chunk that needed to be classified as accepted is rejected.

To investigate how the algorithm behaves for login traces from different devices, three login traces are selected randomly. These traces will be referred to as *alien traces*. The results of the alien traces are used to calculate the *false positives* FP_i and *true negatives* TN_i for each chunk i. A false positive occurs when a chunk collected from an alien trace is classified as accepted, i.e. a chunk that needed to be classified as rejected is accepted. A true negative occurs when a chunk collected from an alien trace is classified as rejected, i.e. a chunk that needed to be classified as rejected is rejected.

To measure the performance of the matching algorithm (which acts as a binary classifier), the *true positive rate* (TPR) and *false positive rate* (FPR) are defined as follows (for each chunk i):

$$TPR_i = \frac{(TP_i)}{(TP_i + FN_i)}, FPR_i = \frac{(FP_i)}{(FP_i + TN_i)}$$

These metrics can be plotted as an ROC-curve, where the FPR is shown on the X-axis and the TPR on the Y-axis. It is obvious that in an authentication system, the focus should be on minimizing the false positive rate. A trace

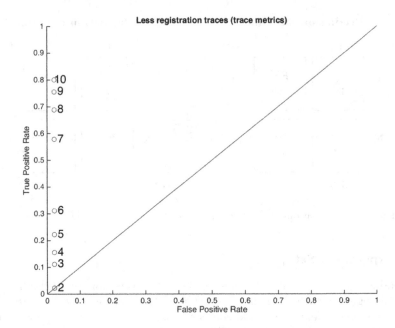

Fig. 6. ROC with the number of registration traces varying from 10 to 2. The threshold factor $t = \frac{1}{2}$.

of an alien device should not be marked as accepted by the matching algorithm. Otherwise logging in would be possible from any device, making accelerometer-based fingerprinting useless for authentication.

Results. We investigate the performance of the matching algorithm on 10 registration traces and 10 probed chunks. The results are shown in a confusion matrix in Fig. 5. As there are 15 devices that test 6 traces, the results show the classification of 900 chunks and 90 traces. For the chunk classification, the TPR and FPR are 0.7444 and 0.0978 respectively. Hereby, the classification is considered good, as the false positive rate is low. When the trace classification's TPR and FPR are calculated, they yield 0.8000 and 0.0222 respectively. The true positive rate is even higher and the false positive rate lower, indicating a good trace classification.

The false positives of the chunks can be considered insignificant, as they almost all are filtered by the trace acceptance (which depends on the ratio parameter). Only one alien trace was classified as accepted. As the user still needs to fill in a username and password, the classification algorithm is definitely sufficient.

For each device, the number of registration traces is varied from 10 to 2. The lower bound is 2 because this is the minimum amount of registration traces needed to calculate the intervals used for matching. The intuition behind lowering the amount of registration traces is that the calculated intervals will be less accurate. This inaccuracy will probably result in more chunk rejections. There will be more FPs and FNs as the number of registration traces is reduced.

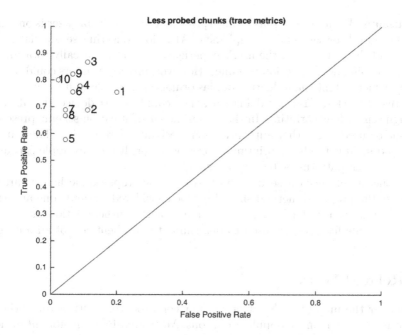

Fig. 7. ROC with number of probed chunks varying from 10 to 1.

The next experiment investigates what happens when the amount of registration traces is reduced. The amount of registration traces are varied from 10 to 2. The results in Fig. 6 show the classification performance of the traces. As can be seen, the true positive rate declines when we reduce the amount of registration traces. This result confirms the intuition that the intervals used for matching become less accurate when the number of registration traces decreases. Notably, the false positive rate remains more or less constant. This means that a reduction in registration traces only induces more false negatives, and no false positives. In the results described above, the threshold factor t was set to $\frac{1}{2}$. It is clear that there is a trade-off between usability and security. The more registration traces are collected, the better matching and hereby the security. But taking many registration traces impact the usability by imposing a longer registration time.

Figure 7 shows what happens when the number of probed chunks decreases. All classifications yield approximately the same results. The intervals stay as accurate as before because they stay the same. Because every chunk has a good classification performance, every possible subset of chunks could be used for logging in. This behavior is desirable, as it should be possible to collect any subset for enforcing the spoofing mitigation approach. When a small subset of chunks is probed, the probability of fingerprint phishing is lower because an attacker his a harder time collecting the data of every possible chunk-probe.

Discussion. We note that sensor fingerprinting through the browser is only possible because browser vendors implement APIs for collecting sensor data. The purpose of this is to enrich the user's experience, like automatically rotating the display or using device motion in games. However, browser vendors could investigate to which extent the induction of bias or noise is possible to mitigate privacy concerns of tracking [12]. By doing this, they could distort device identification through sensor fingerprinting. In that case, the authentication scheme presented can not be used, and other authentication mechanism are needed. These methods could include the fingerprinting of user behavior, like for example capturing a user's search patterns or battery consumption.

Furthermore, since the security properties of the proposed mechanism are not absolute, the proposed method should not be considered a standalone authentication mechanism, but rather part of a multi-factor authentication mechanism where multiple features are used to determine the authenticity of an authentication request.

5   Related Work

To counter the imminent threat of account compromise, multi-factor authentication is one of the most popular solutions. Motivated by their ubiquitousness, the mobile devices of users are often leveraged in a multi-factor authentication setup. For instance, mobile devices can be trusted to compute a One Time Password [13], or in authentication schemes, where they are used to scan a QR-code [14], communicate over Near Field Communication (NFC) technology [15], or emit and receive inaudible soundwaves [16]. Unfortunately, the majority of these solutions can only be used when the user attempts to authenticate on his desktop computer or laptop. Contrastingly, in this paper, we focus on improving the security of users who attempt to authenticate using just their mobile device [17].

For this purpose, we leverage sensor data generated by the accelerometer chip in the mobile device, which was found to be unique due to manufacturing imprecisions [7,10]. Prior research has evaluated using the accelerometer in the context of authentication. In contrast to fingerprinting the imprecisions of the accelerometer chip, Wang et al. leveraged the accelerometer data to analyze a user's gestures, and perform authentication on the basis of the uniqueness of the gestures [18]. Similarly, Mayrhofer and Gellersen proposed a device-to-device authentication mechanism where two devices are shaken together, and thus share very similar accelerometer data, in order to generate authenticated, secret keys [19].

In the context of analyzing sensor data to uniquely identify devices, researchers have evaluated various sensors for imperfections. Examples include pattern noise exposed the camera [8,20,21], imperfections of a device's acoustic components [9,22], or using the gyroscope to recognize speech [23]. We can further generalize our solution by fusing behavioral information about how people interact with the world, exploiting user specific traits about app usage, location, stylometry, keystroke dynamics, phone calls, etc. [24–27].

6 Conclusion

In this work, our objective was to extend an authentication system where browser fingerprinting is used as an additional authentication factor. The scope of the system is limited to authentication with mobile devices, where security and privacy threats imposed by fingerprinting are taken into account. The contributions of this paper are (1) the extension of a state-of-practice authentication system architecture for fingerprinting, (2) a prototype implementation of this architecture using accelerometer sensor data and (3) a critical assessment of this prototype.

The main conclusion is that sensor fingerprinting can be used for authentication. However, the quantitative evaluation has shown that there is a trade-off between usability and security, depending on the amount of traces used for registration. The limitations of this work are the absence of a large scale sensor fingerprint assessment, the focus on only one sensor for fingerprinting and the use of a limited set of features for fingerprint extraction.

Future work could conduct a more extensive scalability assessment, with regard to the uniqueness of sensor fingerprints and the performance of the system on a large scale. As only the accelerometer sensor is used, it would also be interesting to investigate to which extent other sensors such as speakers, microphones and cameras can be fingerprinted for authentication purposes.

Acknowledgment. This research is partially funded by the Research Fund KU Leuven, and by the MediaTrust and TRU-BLISS projects funded by iMinds.

References

1. Florencio, D., Herley, C.: A large-scale study of web password habits. In: Proceedings Of The 16th International Conference on World Wide Web, pp. 657–666. ACM (2007)
2. Eckersley, P.: How unique is your web browser? In: Atallah, M.J., Hopper, N.J. (eds.) PETS 2010. LNCS, vol. 6205, pp. 1–18. Springer, Heidelberg (2010)
3. Spooren, J., Preuveneers, D., Joosen, W.: Mobile device fingerprinting considered harmful for risk-based authentication. In: Proceedings of the Eighth European Workshop on System Security, pp. 6. ACM (2015)
4. Hupperich, T., Maiorca, D., Kührer, M., Holz, T., Giacinto, G.: On the robustness of mobile device fingerprinting: can mobile users escape modern web-tracking mechanisms? In: Proceedings of the 31st Annual Computer Security Applications Conference, pp. 191–200. ACM (2015)
5. Mowery, K., Shacham, H.: Pixel perfect: Fingerprinting canvas in html5. Proceedings of W2SP (2012)
6. Acar, G., Juarez, M., Nikiforakis, N., Diaz, C., Gürses, S., Piessens, F., Preneel, B.: Fpdetective: Dusting the web for fingerprinters. In: Proceedings of the 2013 ACM SIGSAC Conference on Computer & Communications Security, pp. 1129–1140. ACM (2013)
7. Bojinov, H., Michalevsky, Y., Nakibly, G., Boneh, D.: Mobile device identification via sensor fingerprinting. arXiv preprint (2014). arxiv:1408.1416

8. Lukas, J., Fridrich, J., Goljan, M.: Digital camera identification from sensor pattern noise. IEEE Trans. Inf. Forensics Secur. **1**(2), 205–214 (2006)
9. Das, A., Borisov, N., Caesar, M.: Do you hear what i hear?: fingerprinting smart devices through embedded acoustic components. In: Proceedings of the 2014 ACM SIGSAC Conference on Computer and Communications Security, pp. 441–452. ACM (2014)
10. Dey, S., Roy, N., Xu, W., Choudhury, R.R., Nelakuditi, S.: Accelprint: imperfections of accelerometers make smartphones trackable. In: Proceedings of the Network and Distributed System Security Symposium (NDSS) (2014)
11. Bonneau, J., Herley, C., van Oorschot, P., Stajano, F.: The quest to replace passwords: a framework for comparative evaluation of web authentication schemes. In: 2012 IEEE Symposium on Security and Privacy (SP), pp. 553–567, May 2012
12. Das, A., Borisov, N., Caesar, M.: Exploring ways to mitigate sensor-based smartphone fingerprinting. CoRR abs/1503.01874 (2015)
13. Aloul, F., Zahidi, S., El-Hajj, W.: Two factor authentication using mobile phones. In: IEEE/ACS International Conference on Computer Systems and Applications, AICCSA 2009, pp. 641–644. IEEE (2009)
14. Dodson, B., Sengupta, D., Boneh, D., Lam, M.S.: Secure, consumer-friendly web authentication and payments with a phone. In: Gris, M., Yang, G. (eds.) MobiCASE 2010. LNICST, vol. 76, pp. 17–38. Springer, Heidelberg (2012)
15. Alpár, G., Batina, L., Verdult, R.: Using NFC phones for proving credentials. In: Schmitt, J.B. (ed.) Measurement, Modelling, and Evaluation of Computing Systems and Dependability and Fault Tolerance. LNCS, vol. 7201, pp. 317–330. Springer, Heidelberg (2012)
16. Google: Slicklogin
17. Preuveneers, D., Joosen, W.: Smartauth: dynamic context fingerprinting for continuous user authentication. In: Proceedings of the 30th Annual ACM Symposium on Applied Computing, SAC 2015, pp. 2185–2191. ACM, New York (2015)
18. Wang, H., Lymberopoulos, D., Liu, J.: Sensor-based user authentication. In: Abdelzaher, T., Pereira, N., Tovar, E. (eds.) EWSN 2015. LNCS, vol. 8965, pp. 168–185. Springer, Heidelberg (2015)
19. Mayrhofer, R., Gellersen, H.-W.: Shake well before use: authentication based on accelerometer data. In: LaMarca, A., Langheinrich, M., Truong, K.N. (eds.) Pervasive 2007. LNCS, vol. 4480, pp. 144–161. Springer, Heidelberg (2007)
20. Chen, M., Fridrich, J., Goljan, M., Lukáš, J.: Determining image origin and integrity using sensor noise. IEEE Trans. Inf. Forensics Secur. **3**(1), 74–90 (2008)
21. Bertini, F., Sharma, R., Iannì, A., Montesi, D.: Profile resolution across multilayer networks through smartphone camera fingerprint. In: Proceedings of the 19th International Database Engineering & Applications Symposium, pp. 23–32 (2015)
22. Chen, D., Mao, X., Qin, Z., Wang, W., Li, X.-Y., Qin, Z.: Wireless device authentication using acoustic hardware fingerprints. In: Wang, Y., Xiong, H., Argamon, S., Li, X.Y., Li, J.Z. (eds.) BigCom 2015. LNCS, vol. 9196, pp. 193–204. Springer, Heidelberg (2015)
23. Michalevsky, Y., Boneh, D., Nakibly, G.: Gyrophone: recognizing speech from gyroscope signals. In: Proc. 23rd USENIX Security Symposium (SEC 2014). USENIX Association (2014)
24. Fridman, L., Weber, S., Greenstadt, R., Kam, M.: Active authentication on mobile devices via stylometry, application usage, web browsing, and GPS location. CoRR abs/1503.08479 (2015)

25. Antal, M., Szabo, L.Z., Laszlo, I.: Keystroke dynamics on android platform. Procedia Technol. **19**, 820–826 (2015). 8th International Conference Interdisciplinarity in Engineering, INTER-ENG 2014, Tirgu Mures, Romania, 9–10 October 2014
26. Li, F., Clarke, N.L., Papadaki, M., Dowland, P.: Active authentication for mobile devices utilising behaviour profiling. Int. J. Inf. Sec. **13**(3), 229–244 (2014)
27. Shi, E., Niu, Y., Jakobsson, M., Chow, R.: Implicit authentication through learning user behavior. In: Burmester, M., Tsudik, G., Magliveras, S., Ilić, I. (eds.) ISC 2010. LNCS, vol. 6531, pp. 99–113. Springer, Heidelberg (2011)

POODLEs, More POODLEs, FREAK Attacks Too: How Server Administrators Responded to Three Serious Web Vulnerabilities

Benjamin Fogel[✉], Shane Farmer, Hamza Alkofahi, Anthony Skjellum, and Munawar Hafiz

Auburn University, Auburn, AL 36849, USA
bnf0001@auburn.edu, munawar.hafiz@gmail.com

Abstract. We present an empirical study on the patching characteristics of the top 100,000 web sites in response to three recent vulnerabilities: the POODLE vulnerability, the POODLE TLS vulnerability, and the FREAK vulnerability. The goal was to identify how the web responds to newly discovered vulnerabilities and the remotely observable characteristics of websites that contribute to the response pattern over time. Using open source tools, we found that there is a slow patch adoption rate in general; for example, about one in four servers hosting Alexa top 100,000 sites we sampled remained vulnerable to the POODLE attack even after five months. It was assuring that servers handling sensitive data were more aggressive in patching the vulnerabilities. However, servers that had more open ports were more likely to be vulnerable. The results are valuable for practitioners to understand the state of security engineering practices and what can be done to improve.

1 Introduction

Security is a game involving two parties: attackers and secure system developers. An attacker discovers a new vulnerability in a software, attackers launch fresh attacks that exploit that vulnerability, secure system developers fix the vulnerability and generate patches, and attackers move on to find the next vulnerability. But, there is another major actor involved: the users of the software. Users are supposed to accept patches and update their software. They introduce the human element to the process, the Achilles Heel of the secure software engineering process. Too often, users fail to update their software even when a patch is available for a long time. This is where the battle for viable security is lost.

For web applications, it is chiefly the responsibility of the server administrators to keep their servers secure, giving the web application developers a layer of security. A web vulnerability that targets the server infrastructure targets all servers that host *any* applications. This is why a server vulnerability (e.g., a vulnerability in the SSL protocol implementation in the server) attracts more attention and requires prompter response than an application vulnerability (e.g., an SQL injection vulnerability). But, do server administrators respond to these vulnerabilities in a timely fashion?

© Springer International Publishing Switzerland 2016
J. Caballero et al. (Eds.): ESSoS 2016, LNCS 9639, pp. 122–137, 2016.
DOI: 10.1007/978-3-319-30806-7_8

A few studies have reported a common pattern: server administrators are generally slow to respond to reported vulnerabilities [27,35]. Durumeric and colleagues [9] studied how server administrators updated their systems to prevent the Heartbleed vulnerability. They found that server administrators actually promptly fixed the vulnerability in the first couple of weeks after the vulnerability was reported, but the response rate flattened after that period. Other than these studies, some sources periodically track servers in order to collect information [31]. But all of these works provide an aggregate view of the response rate of the server administrators against a specific vulnerability. They do not attempt to explore the results considering other factors, e.g., the demographics of the servers that are vulnerable, how the servers are configured, etc.

This paper describes how the server administrators of the top 100,000 Internet sites responded to the infamous POODLE, POODLE TLS, and the more recent FREAK vulnerabilities. We also analyzed the vulnerable servers considering the category of web applications hosted on the servers and the configuration of the servers (how many TCP or UDP protocol ports are open in the servers). The data collection and analysis process is entirely automated and is immediately usable.

We found that many servers remained vulnerable specifically to POODLE even after five months past widespread notification of the vulnerability. Particularly, 23 % of the servers hosting the top 100,000 Alexa sites we sampled still remained vulnerable to POODLE. Response to FREAK vulnerability was also slow in general, but the administrators proactively fixed the vulnerability even before it was disclosed. Sites that handle sensitive information responded quickly to fix the reported vulnerabilities. Also, sites that have more open ports—perhaps connoting weakness in server administration—were more likely to remain vulnerable.

This paper makes the following contribution:

– It describes a study in which we scanned the top Internet sites and identified their security status against three recently reported vulnerabilities of SSL protocol (Sect. 3). The process is fully automated.
– It reports how promptly the server administrators responded to the vulnerabilities (Sect. 4). It analyzes the data to identify several factors that may be indicative of the problems faced by the server administrators. We describe the approaches in practice and suggest what works, what does not, and what needs to change (Sect. 5) for better compliance.

The results and implications of this work are valuable for secure software engineering practitioners who seek to fix the bottleneck in patch delivery, target the server administrators who are more likely to be slow to update, and to resolve the issues in the update process. Additional information and raw data are on the project webpage: http://sites.google.com/site/WebSSLStudy.

2   Background

Because we focus on SSL security threats here, we begin with background on SSL, then discuss the three vulnerabilities.

2.1   SSL Protocol

In 1994, the Secure Sockets Layer (SSL) was invented by Netscape Communications as a response to Internet security concerns [15]. In the years following its introduction, SSL underwent several modifications in order to improve security. There were two version released after SSL: SSL 2.0 and SSL 3.0. In 1999, SSL 3.0 underwent another improvement then was renamed by the Internet Engineering Task Force (IETF) as the Transport Layer Security (TLS) Protocol [8].

Most HTTP connections are secured with either SSL or TLS. Either of these cryptographic protocols used in conjunction with the HTTP protocol creates the HTTPS protocol. Although each version improved on the prior versions, the newer protocols never supplanted the older; fallback remained valid. Many browsers and servers still use the weaker SSL 2.0 and SSL 3.0 versions; this backward compatibility was the root cause of the reported vulnerabilities.

2.2   POODLE Vulnerability

POODLE (Padding Oracle On Downgraded Legacy Encryption) attack was reported by Adam Langley of the Google security team on October 14, 2014 [17]. The key is to trick a server to downgrade to a weak version of SSL (SSL 3.0). The SSL/TLS version is typically the highest version supported by both the client and the server. However, in an attempt to maintain a continuous connection, clients may downgrade to lower versions. By downgrading to SSL 3.0, weak cipher suites can be used to expose encrypted data. Langley found a single byte can be decrypted on average in 256 requests [17]. An attacker can launch a man-in-the-middle attack and decrypt messages between the client and the server after a downgrade.

The Google security team proposed a patch before the time of disclosure that a flag (TLS_FALLBACK_SCSV) be added to SSL/TLS implementations on both clients and servers [22]. The flag disallows downgrading from TLS versions to SSL 3.0 and lower, thus preventing the attack. However, a client that connects with SSL 3.0 continues to work. An alternate approach is to disable SSL 3.0 altogether, but this will block clients that exclusively encrypt with SSL 3.0 or lower.

2.3   POODLE TLS Vulnerability

On December 8, 2014, another vulnerability similar to the original POODLE attack was released. This vulnerability exploits similar padding flaws without the need for downgrading. Since TLS is an upgrade of SSL 3.0, vulnerabilities found in SSL 3.0 founds its way into some TLS implementations [18].

POODLE TLS arises from an implementation flaw; it can affect all versions of the SSL/TLS protocol. In order to prevent it, vulnerable SSL/TLS implementations need to be reimplemented. Typically, a server administrator will not perform this implementation. Most sites must wait for software vendor updates to secure against this vulnerability.

2.4 FREAK Vulnerability

FREAK (Factoring RSA Export Keys) attack was reported by a research team led by Karthikeyan Bhargavan on March 3, 2015 [4]. The attack exploits an implementation error that allows a man-in-the-middle attack. An attacker can downgrade a non-export cipher suite to an RSA export cipher suite. An export cipher suite is a weak cipher suite (less than 512 bit key) that is used to retain compatibility between US and non-US sites, since the US government restricted keys only 512 bits or smaller to be exported. Although the restriction has been lifted, the weak export cipher suites are still supported to ensure compatibility. The attack is highly exploitable—a key can be extracted in only 8 hours using $100 on an Amazon EC2 instance [4].

Server administrators need to remove all RSA export cipher suites from their accepted cipher suite collection to stay protected. Clients can protect themselves by upgrading to a browser that does not support any RSA export cipher suites.

3   Study Design: Server Scan

We created a client that probed a server and collected various parameters about its SSL implementation. The parameters denote if the server is vulnerable to POODLE, POODLE TLS, and FREAK attacks. We also collected information about port configurations.

Scanning Methodology. In order to collect data about whether a server can downgrade to SSL 3.0 (POODLE vulnerability), we created a client script using a vulnerable version of OpenSSL toolkit. We used the OpenSSL client program, s_client, available in OpenSSL version 1.0.1j [30]. This version provides the TLS_FALLBACK_SCSV flag option which is required to gather data about the POODLE attack. The client script also collected information about SSL/TLS versions supported by the sites.

To collect information about the weakness in TLS implementation (POODLE TLS vulnerability), we used the TLS Prober tool [25] developed by Opera Software ASA. It was one of the first open source tools to provide information about whether a server was vulnerable to POODLE TLS.

In order to collect information about whether the servers used export ciphers (FREAK vulnerability), we needed to collect information about the supported cipher suites. We used the ciperscan tool [32] that utilizes a custom version of OpenSSL (the one used for POODLE scan) to provide cipher suite information. It produces various information including cipher suite name, cipher suite priority, and cipher suite supported protocols.

Table 1. Study timeline and time taken for each scan

Scan date	Time needed (Hours)	Sites scanned
Nov 13, 2014	104	37,494
Nov 22, 2014	147	44,502
Dec 12, 2014	264	58,123
Dec 28, 2014	245	53,749
Jan 18, 2015	244	57,843
Feb 12, 2015	270	43,193
Mar 1, 2015	306	57,738
Mar 21, 2015	304	57,670

We used the Nmap tool [20] to collect data about ports in the servers. Nmap is able to identify the state of the port, possible services running on a port, and a possible reason of a port's state. Recent work has explored faster network scanners, e.g., ZMap [11], but we opted for Nmap since it is the most well known and widely used tool.

Threats to the validity of the data collection process must be considered. Since standard port connection requests were issued, any mechanism to provide incorrect feedback to these requests would be reflected in our data. In general, firewalls intercept connection requests; since firewalls discard unwanted incoming requests, servers protected by firewalls show as not available in our data.

A more serious source of noise in the data is load-balancing or request redirection. It is commonplace for requests incoming to a known address to be redirected to one of a *set* of servers; this is not accounted for in our testing, and the results from this set of servers would be reported as a single machine. This is a potential source of error since we could sample different machines of such groups over time while categorizing them as a single machine. It is our opinion that this is not a serious methodological flaw as servers used in load-balancing operations such as these tend to be relatively symmetric.

All the scans were split between two servers. The primary server ran Ubuntu on an eight core, 3.5 GHz processor with 32 GB of RAM. The secondary server ran Debian on an eight core, 3.5 GHz processor with 8 GB of RAM. Most web scanning was performed on the primary server, with the secondary server assisting the scanning workload.

Study Timeline. We started collecting information about the POODLE vulnerability on Nov 13, 2014—twenty nine days after the vulnerability was reported. Table 1 shows the scan dates and the time required for each scan. The scans took between 4 to 13 days to complete. During this process, we collected information about 61,600 unique servers. It is noteworthy that we were unable to collect information about *all* the servers in *all* the scans due to a result of servers being offline, otherwise unavailable, or because of changes in connection policies. We explain how we dealt with missing information in Sect. 4 in RQ1.

We started collecting information about POODLE TLS vulnerability on Dec 8, 2014—four days after the vulnerability was reported. Because of the additional information, the scan time is longer than the first two scans ($\approx$10 days).

We collected the cipher suite information from the first scan in November; hence, the information about whether a server is vulnerable to the FREAK attack was already available. No additional data was needed.

We scanned for 68 common ports using Nmap starting on March 1, 2015. An additional scan on March 21, 2015 found that these ports remained unchanged, and this study assumes port status to be static during the analysis. Table 1 shows the extra overhead for these scans ($\approx$13 days).

4   Results

We collected information by scanning the servers and by contacting with the server administrators. Specifically, we wanted to explore four research questions.

RQ1. How promptly and effectively did the server administrators react and respond to the vulnerabilities?

RQ2. Did server administrators favor one kind of patch over another when fixing the reported vulnerabilities?

RQ3. Can we identify varying levels of security within certain server demographics?

RQ4. Is there a correlation between the way the servers are administered and their security status?

The questions focus on how the server administrators responded to POODLE, POODLE TLS, and FREAK vulnerabilities.

RQ1. How promptly and effectively did the server administrators react and respond to the vulnerabilities?

> **Key Result:** A lot of servers remain vulnerable to POODLE attack, with 23 % of the servers hosting top 100,000 Alexa sites we sampled. Server administrators reacted better after the Heartbleed vulnerability was reported.

Table 1 lists the dates we performed the eight scans and the varying number of sites contacted per scan. During the first scan, we faced some network connectivity problems. However, the remaining scans also could not collect the information from a fixed number of sites. Our data had to be adjusted to include sites absent from certain scans.

For each vulnerability, we considered the eight scan results as a sequence $\mathbb{S}$, $\mathbb{S} = s_1, s_2, ..., s_8$, in which $s_i = V$, if the parameters indicate that a server hosting a site is vulnerable, $s_i = NV$, if a server is not vulnerable, and $s_i = ND$, if no scan data is available.

We made two corrections to handle missing data. First, if a server's scan data is unavailable for the initial scan, but is found to be vulnerable during

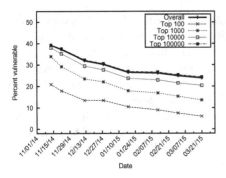

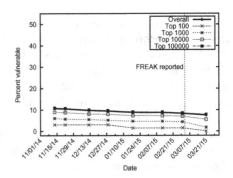

Fig. 1. POODLE vuln. in top sites   **Fig. 2.** FREAK vuln. in top sites

a latter scan, the server is considered to be vulnerable in the previous scans. So, for any sub sequence, $\mathbb{S}_{sub} = s_0, s_1..., s_i$, in which $s_i = V$ and $s_k = ND$ where $0 \leq k \leq i - 1$, we converted $s_k = V$ for all k. But, for the same case with $s_i = NV$ and $s_k = ND$ where $0 \leq k \leq i - 1$, we did not consider the server's data further. Second, if scan data is missing in intermediate scans, we considered the scan sequence only if the missing data is surrounded by unchanged status. In this case, we converted the missing status to be the same as the surrounding status. So, consider any sub sequence, $\mathbb{S}_{sub} = s_i, s_{i+1}, ..., s_{j-1}, s_j$ where $i < j-1$. If $s_k = ND$ where $i + 1 \leq k \leq j - 1$ and $s_i = s_j$, we converted, $s_k = s_i$ where $i + 1 \leq k \leq j - 1$. Otherwise, we discarded the scan sequence for a server from our analysis, since it would be impossible to predict which scan first identified the status change.

We collected information about 61,600 unique servers during our scan. We excluded 2,009 servers because we could not correct for missing data. The remaining 59,591 servers were considered for results. Out of these, 67 were among the servers hosting the top 100 Alexa sites, 575 were among the top 1,000 sites, 5,590 were among the top 10,000 sites, and 53,359 were among the top 100,000 sites. The ranking categories were exclusive—a site in the top 100 sites is not counted again in the top 1,000 sites category.

We expected that most of the sites will be either non-vulnerable or vulnerable in all the scans, or initially deemed as vulnerable but fixed in a latter scan. However, there were a few exceptions. There were 54 sites that patched against the POODLE attack, but reverted back to a vulnerable status later. For example, one of these sites had patched against the POODLE attack by using the TLS_FALLBACK_SCSV flag until our Dec 12, 2014 scan where they had removed it. The patch could have been removed due to compatibility issues.

Figure 1 shows the servers vulnerable to POODLE at different scans. It shows that a lot of servers remain vulnerable even after 5 months have passed, 23.85 % on average. Unsurprisingly, highly ranked sites had a lower vulnerability percentage when compared to lower ranked sites. Only 6 % of the top 100 sites remain vulnerable. As the top 100 sites are expected to have the best security,

Table 2. POODLE TLS vulnerability in all servers

Scan date	Vulnerable servers	% changed
Dec 12, 2014	541	—
Dec 28, 2014	541	0.00 %
Jan 18, 2015	503	7.02 % ⇓
Feb 12, 2015	418	16.90 % ⇓
Mar 1, 2015	360	13.88 % ⇓
Mar 21, 2015	360	0.00 %

this result is expected. Correspondingly, the top 1,000 had a higher vulnerability percentage than the top 100 with 13 % remaining vulnerable.

Very few sites were vulnerable to POODLE TLS ($< 1\%$). Table 2 shows the actual number of servers vulnerable during a particular scan. Interestingly, SSL Pulse [31] reported that about 10.1 % of the Alexa top one million sites were vulnerable to POODLE TLS on December 7, a day before the vulnerability was reported. It may happen that a lot of servers were fixed in the first four days, although it is unlikely since the vulnerability is difficult to fix. Also, they reported that 7.3 % of the top one million sites remained vulnerable a month later. This suggests that the servers hosting less popular sites (after top 100,000) are more vulnerable. We did not perform any other analysis on POODLE TLS because of the small number of vulnerable servers in our dataset.

Figure 2 shows the sites that are vulnerable to the FREAK attack. Sites have been removing RSA export cipher suites at a steady rate during the entire study as shown by the declining slope even before the vulnerability was reported. There could be two different reasons. Sites would patch early if there was a pre-disclosure notification detailing the vulnerability. The miTLS team had notified many sites of the FREAK vulnerability prior to their disclosure [4]. On the other hand, export cipher suites have been known to be less secure. In order to increase security, sites could have preemptively removed all export cipher suites to avoid a future vulnerability. However, we did see a slightly sharp decline for all rank categories except the top 100,000 immediately after disclosure. Predictably, higher ranked websites were less vulnerable.

Since we did not collect information about POODLE vulnerability in the first 29 days, we may have missed an exponential drop in vulnerable servers after the disclosure. There are no data points regarding this from other sources. Had there been a sharp decline, the patching response could be compared to the previously-reported vulnerable response—an exponential decline in vulnerable servers, followed by a steady, flat period [9,27,35]. The response to the FREAK attack showed a similar vulnerability response pattern after the vulnerability was reported; but the interesting aspect is the gradual decline long before the vulnerability was reported.

Heartbleed, POODLE, and FREAK were the three most important vulner-abilities reported on SSL/TLS protocol during the last one year. Servers were

Table 3. Type of patch used to fix POODLE

Scan date	Total sites Patched	By disabling SSL 3.0	By adding Flag
Nov 22, 2014	1043	651	392
Dec 12, 2014	3740	2083	1657
Dec 28, 2014	1083	516	567
Jan 18, 2015	2449	1117	1332
Feb 12, 2015	198	91	107
Mar 1, 2015	1160	587	573
Mar 21, 2015	600	480	120

aggressively fixed in the wake of Heartbleed [9]: within a month, only 3.1 % of the top 1 million sites were vulnerable. However, even after five months, 13.96 % of the top 100,000 sites remain vulnerability to POODLE. One explanation may be that administrators think that POODLE is harder to exploit [5]. Also, Heartbleed was easier to fix than POODLE, because administrators cannot just discard SSL 3.0.

RQ2. Did server administrators favor one kind of patch over another when fixing the reported vulnerabilities?

Key Result: For the POODLE attack, server administrators slightly favored disabling SSL 3.0 over using the TLS_FALLBACK_SCSV flag.

There are two patching techniques for fixing POODLE vulnerability: by disabling SSL 3.0 or by adding the proposed TLS_FALLBACK_SCSV flag. Table 3 shows the total number of sites patched during a particular scan and the patching technique followed by the server administrators. The results show that server administrators who patched against the POODLE attack slightly favored disabling SSL 3.0. There were 4748 sites choosing to add the TLS_FALLBACK_SCSV flag compared to 5524 sites choosing to disable SSL 3.0. This could be because server administrators prefer the simplicity and effectiveness of disabling SSL 3.0.

Server administrators' slow response to fix POODLE vulnerability compared to the prompt response to fix Heartbleed suggested that they may be concerned about compatibility (RQ1). However, the administrators who actually updated against POODLE chose to take the simpler route and risk being incompatible to older clients that only use SSL 3.0 protocol or lower. Administrators may favor removing a vulnerable technology instead of patching due to perceived weakness.

The FREAK vulnerability is patched by removing RSA export cipher suites. Server administrators have continually removed RSA export cipher suites throughout the study (RQ1). There are many non-export cipher suites that

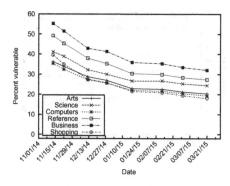
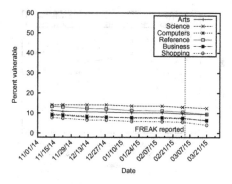

Fig. 3. POODLE in different sites   **Fig. 4.** FREAK in different sites

are commonly in use today. Server administrators could remove export cipher suites without encountering compatibility problems. Since the vulnerability was fixed preemptively even before the disclosure, perhaps server administrators would remove security weaknesses if compatibility can be maintained.

RQ3: Can we identify varying levels of security within certain server demographics?

> **Key Result:** Sites dealing with sensitive information showed a better response to prevent the reported POODLE and FREAK vulnerabilities.

Figures 3 and 4 show the categories of the sites hosted on the servers and the percentage of servers that remained vulnerable. The site categories were collected from Alexa analytics. Sites categorized as shopping or business were the least vulnerable to POODLE attack. Since shopping and business sites often deal with financial information, security should be very important. Sites that likely dealt with non-sensitive information, such as arts or society, were more vulnerable to POODLE attack.

For FREAK vulnerability (Fig. 4), the percentage of vulnerable servers remained consistent for all categories until the disclosure date. After the disclosure, servers hosting some categories of sites showed a quicker response (higher patching rate). These were shopping, business, and regional–again as expected.

RQ4: Is there a correlation between the way the servers are administered and their security status?

> **Key Result:** Servers with more open ports consistently had higher vulnerability rates than servers with fewer open ports.

One reason behind the slow response of server administrators may come from the fact that many servers are loosely administered. We collected information

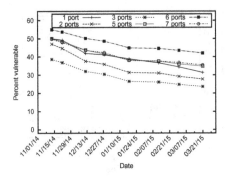   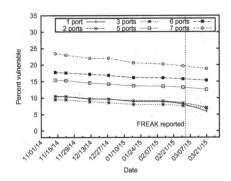

Fig. 5. POODLE vuln. and port status   **Fig. 6.** FREAK vuln. and port status

about open ports in the servers under study. We wanted to explore if a server with many open ports is more likely to remain vulnerable to POODLE and FREAK attacks. This is only a heuristic that may (or may not) represent a loosely managed server. A server can keep many ports open if the applications hosted on the server require them. However, we collected information about ports that are typically blocked; therefore, leaving these open may denote loose administration.

We found the median number of open ports on all servers to be 14 with a standard deviation of 6.08. This shows that many of the top 100,000 sites have multiple ports open. The top 100 sites contained no sites having more than four ports open. More so, only 15 of the top 100 sites had more than two ports open. This suggests that top servers—arguably better administered ones—do have fewer ports open.

Figures 5 and 6 show the vulnerable servers categorized by the number of ports open. The different scans show similar slow responses in each category. But servers with fewer ports open were less vulnerable than servers with more ports open. Particularly, sites that had less than 5 open ports were less vulnerable to POODLE and FREAK than sites containing 5 or more open ports. This is evident in Fig. 6 (FREAK attack), but the difference is not much in Fig. 5 (POODLE attack). Interestingly, the patching rates were the same; no disparity was found there. Note that the percentage of vulnerable sites does not linearly follow the number of open ports.

A server port can be open, or blocked, or filtered. We collected status information about 68 different ports (Sect. 3). For this illustration, we considered the states of five well known ports—ports 21 (FTP), 22 (SSH), 23 (telnet), 25 (SMTP), 8080 (HTTP alternate)—for 58,495 servers; we could not collect the port information for the remaining 1,096 servers due to a default timeout in our script of the Nmap program. The test in this case was whether a site was vulnerable to POODLE. We applied a decision tree learning algorithm (C4.5) with 10-folds cross validation. A pruned decision tree is shown in Fig. 7. The precision for this tree is 0.526. From the tree, we can infer:

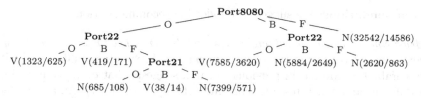

O=open, B=blocked, F=filtered, V=vulnerable, N=not vulnerable

Fig. 7. Decision tree showing closed/filtered ports are more likely to be traits of a not-vulnerable server

- In general, sites with blocked or filtered ports were less vulnerable to POO-DLE.
- Port 23 (telnet) and Port 25 (SMTP) had no impact in this tree; these were blocked by most of the servers anyway.
- Sites that had ports filtered had a lower vulnerability rate than sites with blocked ports. This trend was the same when we considered the status of all 68 ports. Since a filtered port may suggest a site using a firewall, sites with protection equivalent to a firewall showed less vulnerabilities than sites that did not employ filtering.

5   Discussion

Servers remained highly vulnerable to POODLE months after disclosure (RQ1). Server administrators were more reactive after the Heartbleed vulnerability, perhaps because it was the first vulnerability in the line (in recent times), it was easier to exploit, and it was easier to fix without disrupting server configuration. However, when some feature is known to be bad—e.g., obsolete export ciphers—server administrators preemptively take actions as shown in their response to FREAK attack.

We identified that server administrators prefer disabling SSL 3.0 over patching SSL 3.0 with the TLS_FALLBACK_SCSV flag (RQ 2). Given that disabling SSL 3.0 is the simpler solution, and protects against future SSL 3.0 vulnerabilities, we find that server administrators prefer an effective yet simple solution to security, even disabling a feature over patching one and continuing to support (RQ2).

We found that servers that host applications handling sensitive data are more responsive, perhaps because of the pressure from web developers (RQ3).

Lastly, we were able to classify servers as loosely or strictly managed and identify trends within the two classifications. (RQ4). Using our heuristic, we found servers that were loosely managed to be more at risk to any given vulnerability. Oppositely, servers that are strictly managed had a much lower risk to vulnerabilities and observed better patching rates.

Recommendations: We suggest the following recommendations.

- *Measuring response to vulnerabilities and Internet health should be a regular activity.* Comparing the trends among the three vulnerabilities and with vulnerabilities covered in previous research suggests that every vulnerability is different and generates different kinds of responses (RQ1). An automated, periodic study can even categorize vulnerabilities based on previously-studied response patterns. Such a study should concentrate on analyzing the data from multiple perspectives (RQ3 and RQ4).
- *There should be a mechanism to actively 'nudge' server administrators into adopting patches.* Our work hinted at the lack of awareness of administrators about vulnerabilities (RQ1). This is supported by works on actively notifying administrators that have reported considerable success [9,26].
- *Targeted campaigns should be launched to fix vulnerabilities.* Our analysis suggests that some groups are more likely to be vulnerable (RQ3). For example, servers that host non-sensitive data could be a target of an Internet-wide campaign because they are more likely to remain vulnerable. The more high profile servers are more likely to update anyway.
- *Patch providers should design patches that are less disruptive.* The goal is to explore for a zero-downtime patch or a hot update that also does not have incompatibility issues (Follows from RQ1).

6   Threats to Validity

There are several threats to validity of our study; here we describe them following the four classic tests and discuss how they have been mitigated.

External Validity. There may be a concern about generalizability of our results. The top 100,000 sites may not be representative of the practices of the entire web. However, the servers hosting the top sites receive a larger portion of traffic and should be better administered. So, their patching rate can be considered as an upper bound for the rate in the entire web.

Internal Validity. Internal validity is mainly a concern for explanatory studies. Since ours is not an explanatory study, it does not have a threat to internal validity from the interpretation aspect.

Construct Validity. We described several issues about scanning servers and interpreting missing data and how we handled them (Sects. 3 and 4).

Reliability. We describe an automated approach to periodically collect information. Although our approach is not repeatable for the same vulnerabilities (cannot go back in time), the periodic scan data validates each other. There is also a chance of misinterpreting survey data. We used structural coding to avoid confusion [28]. Coding was done by the two authors who reached consensus.

7   Related Work

Many studies have focused on security of Internet protocols, e.g., non-compliance to HTTP [1], weakness of client authentication [12], weakness of OpenID implementations [29], weakness of adoption of new HTTPS features [16], etc.

The three web vulnerabilities in our study target weaknesses in the SSL protocol and its implementation. Murray [24] did a survey of early SSL servers and reported that—in 2000—about one-third SSL servers supported a weak version of the algorithm. Later, Lee and colleagues [19] reported that a lot of servers continued to use weak SSL 2.0, which would have made them vulnerable to the POODLE and POODLE TLS attacks. They also reported that many of these servers used old "export" cipher suites, which would have made them vulnerable to FREAK attack. Other than these, several studies have reported the weakness of SSL certificates [10,14], although this aspect is not directly linked with any of the three vulnerabilities we studied.

There are a few longitudinal studies on how server administrators fixed a newly reported vulnerability, e.g., a SSH CRC vulnerability that was reported in 2001 [26], a vulnerability in BIND [7], and the Code Red worm [23]. But these were either done in a limited scope, or did not concentrate on the response rate of server administrators. Rescorla [27] studied the response after a buffer overflow vulnerability in OpenSSL. He observed that server administrators updated their systems slowly in general, but the update rate was higher right after the vulnerability was reported and after a worm exploiting the vulnerability (the Slapper worm) was deployed. Similar pattern was reported by Yilek and colleagues [35] in their study of the aftermath of the Debian OpenSSL Bug. This response pattern was common in all these studies, but our study showed that the response rate can differ (RQ1).

A recent study followed the response of server administrators in response to the Heartbleed vulnerability. Durumeric and colleagues [9] studied top 1 million Alexa sites two days after the Heartbleed vulnerability was reported and continued the study for about two months (another one shot study monitored the servers for this vulnerability one day after the report [2]). Our study monitored the servers for a longer period, covered three different vulnerabilities, and reported different response trends. Durumeric and colleagues' study reported that only about 3 % servers remained vulnerable after two months. However, our study found that a higher percentage of the more important servers (hosting top 100,000 pages) remained vulnerable to POODLE even after several months.

In 2012, SSL Labs created a project named SSL Pulse [31] which monitors Alexa's top 1 million sites on a monthly basis and reports general statistics about SSL/TLS implementations. They do not focus on vulnerabilities (e.g., no data for POODLE), but on overall SSL health. In contrast, we analyzed specifically for vulnerabilities and analyzed the vulnerable servers from different perspectives, e.g., type of the web pages hosted on servers, port configurations of servers, etc.

There has been a few studies focusing on the needs of server administrators; these focus mostly on the tools that server administrators use [3,6], the practices they follow [13,34], and how the tools should be designed to fit their unique requirements [21,33]. Our study explores on server administrators' awareness of the security patches, but its main purpose is to get an idea of the main reasons why it is difficult for server administrators to update their systems regularly.

8   Conclusion

In this paper, we presented our approach to Internet measurement. Our study and methodology revealed that even in major attack scenarios, response to well documented threats at major Internet sites is neither instantaneous nor quick in many cases.

Our approach to Internet measurement provides a proactive way to study on-going vulnerability of sites and offers a means to explore which kinds of vulnerabilities receive immediate remediation vs. those that are allowed to remain active. We found that 23 % of the servers we sampled remain vulnerable to POODLE vulnerability even after five months have passed after disclosure, which shows that server administrators reacted better after the Heartbleed vulnerability was reported. Understanding these activities and intervening would give secure software engineering practitioners a chance to win the game of security.

Acknowledgements. This was funded by the Auburn Cyber Research Center. We thank Paul Adamczyk, Farhana Ashraf, Jeff Overbey, Awais Rashid, and the anonymous reviewers for their comments.

References

1. Adamczyk, P., Hafiz, M., Johnson, R.: Non-compliant and proud: a case study of HTTP compliance. Technical report, UIUC (2008)
2. Al-Bassam, M.: Top Alexa 10,000 Heartbleed scan (2014). https://github.com/musalbas/heartbleed-masstest
3. Barrett, R., Kandogan, E., Maglio, P.P., Haber, E.M., Takayama, L.A., Prabaker, M.: Field studies of computer system administrators: analysis of system management tools and practices. In: CSCW 2004. ACM (2004)
4. Beurdouche, B., Bhargavan, K., Delignat-Lavaud, A., Fournet, C., Kohlweiss, M., Pironti, A., Strub, P.-Y., Zinzindohoue, J.K.: SMACK: state machine attacks (2015). https://www.smacktls.com/
5. Blevins, B.: POODLE SSL vulnerability doesn't equal Heartbleed, but still bad (2014)
6. Botta, D., Werlinger, R., Gagné, A., Beznosov, K., Iverson, L., Fels, S., Fisher, B.: Towards understanding it security professionals and their tools. In: SOUPS 2007. ACM (2007)
7. Cheswick, W., Bellovin, S., Rubin, A.: Firewalls and Internet Security: Repelling the Wily Hacker, 2nd edn. Addison-Wesley Professional, Reading (2003)
8. Dierks, T., Allen, C.: The TLS protocol
9. Durumeric, Z., Kasten, J., Adrian, D., Halderman, J.A., Bailey, M., Li, F., Weaver, N., Amann, J., Beekman, J., Payer, M., Paxson, V.: The matter of heartbleed. In: IMC 2014. ACM (2014)

10. Durumeric, Z., Kasten, J., Bailey, M., Halderman, J.A.: Analysis of the https certificate ecosystem. In: IMC 2013. ACM (2013)
11. Durumeric, Z., Wustrow, E., Halderman, J.A.: ZMap: fast internet-wide scanning and its security applications. In: SEC 2013. USENIX Association (2013)
12. Fu, K., Sit, E., Smith, K., Feamster, N.: Dos and don'ts of client authentication on the web. In: SSYM 2001. USENIX Association (2001)
13. Haber, E.M., Kandogan, E., Maglio, P.: Collaboration in system administration. Queue 8(12), 10:10–10:20 (2010)
14. Holz, R., Braun, L., Kammenhuber, N., Carle, G.: The SSL landscape: a thorough analysis of the x.509 PKI using active and passive measurements. In: IMC 2011. ACM (2011)
15. IBM developerWorks. The Secure Sockets Layer and Transport Layer Security. http://www.ibm.com/developerworks/library/ws-ssl-security/
16. Kranch, M., Bonneau, J.: Upgrading HTTPS in mid-air: an empirical study of strict transport security and key pinning. In: NDSS 2015. IEEE (2015)
17. Langley, A.: POODLE attacks on sslv3, October 2014
18. Langley, A.: The POODLE bites again, December 2014
19. Lee, H., Malkin, T., Nahum, E.: Cryptographic strength of SSL/TLS servers: current and recent practices. In: IMC 2007. ACM (2007)
20. Lyon, G.: Download the free nmap security scanner for linux/mac/unix or windows (2015). https://nmap.org/download.html
21. Mahendiran, J., Hawkey, K.A., Zincir-Heywood, N.: Exploring the need for visualizations in system administration tools. In: CHI EA 2014. ACM (2014)
22. Moeller, B.: TLS Signaling Cipher Suite Value (SCSV) for preventing protocol downgrade attacks
23. Moore, D., Shannon, C., Claffy, K.: Code-Red: a case study on the spread and victims of an internet worm. In: IMW 2002. ACM (2002)
24. Murray, E.: SSL server security survey (2000)
25. Opera Software ASA. operasoftware/tlsprober (2014). https://github.com/operasoftware/tlsprober
26. Provos, N., Honeyman, P.: ScanSSH - scanning the internet for SSH servers. In: LISA 2001. USENIX Association (2001)
27. Rescorla, E.: Security holes... who cares? In: SSYM 2003. USENIX Association (2003)
28. Saldana, J.: The Coding Manual for Qualitative Researchers. Sage Publications Limited, Singapore (2009)
29. Sun, S.-T., Beznosov, K.: The devil is in the (implementation) details: an empirical analysis of oauth sso systems. In: CCS 2012. ACM (2012)
30. The OpenSSL Project. OpenSSL 1.0.1j (2014). https://www.openssl.org/source/
31. TIM Trustworthy Internet Movement. SSL Pulse: Survey of the SSL implementation of the most popular web sites (2012)
32. Vehent, J.: jvehent/cipherscan (2014). https://github.com/jvehent/cipherscan
33. Velasquez, N.F., Weisband, S., Durcikova, A.: Designing tools for system administrators: an empirical test of the integrated user satisfaction model. In: LISA 2008. USENIX Association (2008)
34. Werlinger, R., Hawkey, K., Botta, D., Beznosov, K.: Security practitioners in context: their activities and interactions with other stakeholders within organizations. Int. J. Hum. Comput. Stud. 67(7), 584–606 (2009)
35. Yilek, S., Rescorla, E., Shacham, H., Enright, B., Savage, S.: When private keys are public: results from the 2008 debian OpenSSL vulnerability. In: IMC 2009. ACM (2009)

HexPADS: A Platform to Detect "Stealth" Attacks

Mathias Payer[✉]

Purdue University, West Lafayette, USA
`mathias.payer@nebelwelt.net`

Abstract. Current systems are under constant attack from many different sources. Both local and remote attackers try to escalate their privileges to exfiltrate data or to gain arbitrary code execution. While inline defense mechanisms like DEP, ASLR, or stack canaries are important, they have a local, program centric view and miss some attacks. Intrusion Detection Systems (IDS) use runtime monitors to measure current state and behavior of the system to detect an attack orthogonal to active defenses.

Attacks change the execution behavior of a system. Our attack detection system HexPADS detects attacks through divergences from normal behavior using attack signatures. HexPADS collects information from the operating system on runtime performance metrics with measurements from hardware performance counters for individual processes. Cache behavior is a strong indicator of ongoing attacks like rowhammer, side channels, covert channels, or CAIN attacks. Collecting performance metrics across all running processes allows the correlation and detection of these attacks. In addition, HexPADS can mitigate the attacks or significantly reduce their effectiveness with negligible overhead to benign processes.

1 Introduction

Software is constantly under attack using a wide set of attack vectors. The attack surface increases as more devices go online. Connected devices expose running services but also request services from untrusted parties through potentially vulnerable client-side software like web browsers.

Current systems leverage a wide range of different attack detection and protection mechanisms, many of them in combination. Protection mechanisms like Address Space Layout Randomization (ASLR) [21], Data Execution Prevention (DEP) [27], stack canaries [12] protect against some memory corruption attacks. Host-based protection mechanisms mitigate exploitation attempts of unknown or unpatched vulnerabilities in software but terminate the application whenever an attack is detected. Patching removes the vulnerability and mitigates attacks. Unfortunately, patches are not readily available when a vulnerability is disclosed. Intrusion Detection Systems (IDS) and Intrusion Prevention Systems

---

The stamp on the top of this paper refers to an approval process conducted by the ESSoS artifact evaluation committee chaired by Alessandra Gorla and Jacques Klein.

© Springer International Publishing Switzerland 2016
J. Caballero et al. (Eds.): ESSoS 2016, LNCS 9639, pp. 138–154, 2016.
DOI: 10.1007/978-3-319-30806-7_9

(IPS) on the other hand detect an attack before, during, or after it happened. Commonly, intrusion detection systems measure a set of parameters and check if the fingerprint matches any of the known signatures (note that signatures can be Turing complete verifiers). Network-based IDS like Bro [22] match network packets against known signatures and alert if an attack is detected. Host-based IDS collect information about a system and match this information against a set of rules or attack signatures. An IDS is either misuse-based, matching observed behavior with a set of attack signatures or anomaly-based, detecting divergences.

Existing host based defense mechanisms focus on memory corruption and code reuse attacks but offer limited to no protection against information leaks, side channel, and covert channel attacks. Existing host-based IDS detect a set of individual attacks by matching fingerprints of individual attacks against the runtime collected statistics but are limited to the collected software metrics. Due to the limited software metrics provided by the operating system itself, memory-based attacks like side channels and covert channels cannot be observed directly and are therefore *stealthy* (to available metrics). Software under attack it behaves differently compared to a regular execution. Lightweight, low performance overhead program analysis tools like performance counters (both hardware-based and software-based) allow a detailed fingerprinting of the execution behavior of software. We leverage the information collected from a set of specific probes to detect attacks through their anomalies, matching execution behavior of processes against attack classes. Using additional runtime metrics from the performance counters allows us to uncover these otherwise undetected attacks.

We propose HexPADS, a host-based, Performance-counter-based Attack Detection System that measures performance characteristics of all processes and detects attacks by matching a set of signatures. HexPADS is especially apt at detecting long running Covert and Side Channel (CSC) attacks. Compared to per-attack signatures, HexPADS uses broader per attack-vector signatures, generalizing signatures to all attacks in an attack class whenever possible, e.g., protecting against all CSCs by detecting cache performance anomalies instead of detecting specific cache attacks. HexPADS collects statistics about running processes and measures common performance parameters using existing low-overhead, hardware-based performance counters. To our knowledge, HexPADS is the first IDS that leverages per-process performance counters to detect attacks. In our evaluation we show that our prototype implementation achieves negligible (non-measurable) overhead and in a set of case studies we show how HexPADS detects (and mitigates) rowhammer [25], CSCs [11,18,24,37,39], and CAIN [2] attacks. Side-channel based information leaks are used to extract data from running systems and processes or to corrupt memory in the case of rowhammer. Such memory CSC attacks can, e.g., be used to break AES cryptographic key generation, or to break ASLR in the cloud [2]. The main contributions are:

1. Design of HexPADS, a host-based attack detection system that detects stealth attacks through fine-grained process monitoring using performance counters and performance metrics exported by the kernel.

2. Evaluation of a prototype implementation of our attack detection system that detects cache attacks, DRAM attacks like rowhammer, and memory deduplication attacks like CAIN at negligible overhead.
3. A discussion of mitigation mechanisms that protect against cache, DRAM, and memory deduplication attacks.

2 Threat Model and Attacker Goals

We assume a powerful threat model where the attacker can execute user-level code on the system. An attacker can achieve these capabilities either through a legitimate service on the system that offers the computational capabilities or through the exploitation of a service. HexPADS configures performance counters for all processes. To ensure integrity of our monitor, we assume that HexPADS is running as a separate process at higher privileges than the attacker and that the attacker cannot access the monitor or disable performance monitoring.

The trusted computing base contains the underlying hardware, hypervisor, and operating system. An alternative, hypervisor-based implementation would remove the operating system from the trusted computing base. We assume that the attacker does not have raw memory access and that we can rely on the performance counter results. We trust the integrity of memory, assuming that we detect attacks like rowhammer *before* memory is corrupted.

The attacker's goals are to escalate privileges, to communicate with other processes, to leak information, or to execute code while remaining undetected. HexPADS continuously monitors the system and detects an ongoing attack. Attack detection is inherently restricted to the precision of the measured run-time characteristics and limited by the effectiveness of the monitor to distinguish between benign behavior and attacks.

3 Background

HexPADS leverages existing process metrics and performance counters to collect information about all running processes. Both process metrics and performance counters are available and supported on all major operating systems. Here we give, without loss of generality, an overview of process metrics and performance counters on Linux systems.

3.1 Process Metrics

Operating systems continuously collect basic information about all running processes. This information is exposed to user-space to administer processes and to diagnose problems with user-space utilities. Linux provides the /proc pseudo-filesystem as an interface to kernel data structures which are accessible from user-space. The files in the exported directory are mostly read-only and used for informative purposes but kernel settings can be changed by writing to these files as well. Most Linux distributions make the /proc directory accessible to user-space processes, exposing information about all running processes.

Each running process has its own directory under the root /proc directory named after the process' PID. The file stat contains a wide range of process metrics, including name of the executable, process state, the PID of the parent, the process group, the associated terminal, the amount of page faults, total execution time in both user and kernel space, priority, number of threads for this process, when the process was started, memory limits for regions like heap or stack, which processor the task runs on, and the scheduling policy[1]. HexPADS collects all stat information.

The recommended way of using this information is to scrape all numerical directories in the /proc directory, thereby iterating over all running threads and processes. Tools like ps, top, or killall all leverage the files in the /proc directory to fulfill their tasks.

3.2   Performance Counters

Hardware performance counters are available in all major CPU architectures. These performance counters are special-purpose registers that collect information about the executed instructions. The names of the counted events differ between platforms and the number of available registers (and thereby the amount of performance events that can be sampled at the same time) is platform specific with low-end architectures generally featuring less performance counting infrastructure. An advantage of using hardware performance counters is that the overhead to count specific events is negligible (as the hardware is responsible for all the heavy lifting). The individual counters and their configuration are managed by the Performance Monitoring Unit (PMU).

The Intel x86 platform offers detailed configurable performance counters since the Intel Pentium. The Intel Core i7 family supports base level and enhanced architectural performance monitoring with four general-purpose, configurable performance counters (i.e., four types of events can be counted on any core at any point in time) [4, chapter 18.2]. In addition to counting, the Intel architecture also supports precise event-based sampling. Instead of counting the occurrences of an event, the PMU also takes a snapshot of the processors state at the time of the event. On x86, such a snapshot consists of the instruction pointer, stack pointer, and all general purpose registers. AMD processors have similar counters and hardware capabilities.

On Linux, the PMU can be configured using the perf_event_open system call (which does not have a libc-based wrapper but needs to be called using inline assembly). Some user-space programs, e.g., perf provide a command-line interface to the PMU and allow the collection of detailed performance events for executing software. The Linux perf_event interface tries to unify performance counter access across architectures and processor families. Performance counters can be assigned system-wide or per-process with a wide range of conditions (e.g., the processor the task runs on). After setting up the PMU, the event can be configured using the ioctl system call. Samples can be read explicitly by polling through a read system call or implicitly by setting up a signal that

---

[1] Additional information and details are available on the proc manpage.

is delivered whenever the counter reaches a pre-defined value (or the buffer used to store the samples when sampling overflows).

The only additional overhead when using hardware-based performance counters comes from (i) configuring the PMU whenever a process is scheduled and (ii) updating the aggregates whenever the process is interrupted. Collecting counters might incur some overhead during execution but these effects are hidden by the microarchitecture. In addition, if an event is sampled (and not just counted) then there is also additional cache pressure when samples are written into the sample buffer. The overhead of running performance counters alongside the executed software is in the noise (less than 1 %).

4   HexPADS Design

The core principle of HexPADS is to search for general attack behavior and attack artifacts in all running processes. The underlying hypothesis is that software attacks significantly change the environment or the behavior of a process or processes. Both the attacking process (if run on the same machine) and the attacked process (usually a service) will exhibit behavior that can be mapped to an attack. If an attacker uses, e.g., a cache-based CSC to communicate or to leak information from a benign process then the cache miss rate will increase significantly. Such changes can be observed by regularly checking key parameters of all running processes. A challenge for a detection mechanism is to detect attacks with few false positives. If applications run in phases then phase transitions can lead to a significant change in the observed behavior as well. A detection mechanism must be able to distinguish between phase changes and attacks. Figure 1 gives an overview of the HexPADS system. HexPADS leverages information from the operating system to collect core process characteristics of all running processes and uses the CPU's PMU to collect detailed low-level performance events from the underlying hardware.

We design HexPADS as a generic process behavior collection mechanism with a plugin-based detection subsystem for different attacks. The core of HexPADS continuously measures a set of parameters for all running processes at negligible

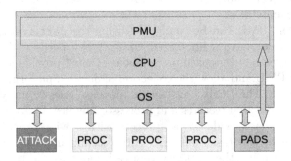

Fig. 1. Overview of the HexPADS system.

overhead. A flexible plugin interface extends the collection mechanism and allows detectors to analyze the behavior of processes. Each plugin detects a certain type of attack using past and current performance data of a process. HexPADS detects attacks by collecting and analyzing system information in five stages that are periodically repeated when the system is running:

1. To gather the necessary runtime information, HexPADS polls detailed process statistics of all running processes. This data is stored in a buffer across iterations to allow aggregate checks, e.g., page faults per iteration. This step takes care of registering new processes (including the setup of performance counters) and cleaning up dead processes.
2. Poll necessary performance counters for each running process. All performance counter results are stored in a buffer to allow aggregate checks and the counters are reset to 0.
3. Calculate performance statistics for each process to, e.g., allow checking if any measured parameter has changed rapidly.
4. Evaluate a set of attack signatures on the measured performance statistics for each process. If an attack signature matches the behavior of a process then a potential attack is detected.
5. If any potential attacks were reported, this step takes evasive or counter measures and reports the attack.

In its default configuration, HexPADS will collect the following performance counters: number of executed instructions, number of last level cache accesses, and number of last level cache misses. In addition to the performance counters, detectors can use the status information of each process as exported from the kernel, e.g., number of minor page faults, number of major page faults, and execution time are used in our signatures. In addition to this baseline, all other information available in the exported process' status can be used and additional counters can be configured. If the amount of desired performance events exceeds the available hardware registers, a time-based sampling scheme can multiplex the available registers (with some loss of precision). HexPADS uses a buffer to store the samples, all elements are initialized with the first measurement.

Attack detectors are functions that evaluate, based on the history of performance samples, if a process is either under attack or attacking another process. If an attack detector matches then it reports the potential attack and the PID to the attack reporting and mitigation module.

Distinguishing between attacking and attacked process is not always straight forward (e.g., a cache-based CSC attack will increase the cache misses in both the attacking and the attacked process). Countermeasures therefore cannot just kill the reported process and other mitigation strategies must be used. Any attack will be reported to the administrator who can decide on specific counter measures. In addition, HexPADS supports a set of automatic counter measures that can mitigate or slow down the attack. HexPADS, e.g., slows down the attacking process (reducing the bandwidth of CSC attacks), stops the attacking process until an administrator can evaluate the situation, or enforces specific scheduling decisions (e.g., pinning processes to disjoint processors[2]). Other mitigation strategies are possible as well, depending on the attack vector.

5 Implementation

Following a least privileges principle, HexPADS runs as a user-space daemon and collects information of all running processes. If multiple virtual machines share a single CPU then a HexPADS daemon must run on each VM. Results can then be collected by a central daemon and are evaluated across all running processes on all VMs. Our prototype implementation currently supports monitoring on a single system without distributing the results.

Our prototype follows the design outlined in Sect. 4 and implements the described analysis loop: it (i) crawls all running processes, updates status information, and initializes performance counters for new processes, (ii) polls the performance counters of all processes, (iii) calculates performance statistics, (iv) evaluates if an attack is in progress, and (v) deploys potential countermeasures against the affected processes. The ringbuffer for the measurements stores the last 60 samples and the scan interval is set to once each second.

The prototype is open-source[3] and the implementation uses less than 2,000 lines of C code. The prototype implementation includes the base framework, detectors for rowhammer, cache CSC attacks, and CAIN attacks and the slow-down and stop the process counter measures. The slow-down counter measure reduces the priority of the identified process and optionally pauses the process to some extend. The stop counter measure stops the process through the SIG_STOP signal. We discuss individual detectors in Sect. 6 as a set of three case studies.

6 Evaluation

Evaluating the performance overhead of HexPADS on a modern system shows that the increased protection results in negligible (non-measurable) performance overhead. Using a set of case studies, we show how HexPADS can detect different attacks: rowhammer attacks, cache CSCs, and CAIN attacks. We have run HexPADS with these detectors on both desktops and servers with regular workload for several days without false positives.

6.1 Performance Overhead

The perceived overhead for HexPADS is negligible and makes up for less than 1 % of CPU time on a single core on a modern system. To measure impact on other running processes we measured the performance overhead using the SPEC CPU2006 and PARSEC 3.0 benchmarks. We ran our experiment on an Ubuntu 14.04 system with an Intel Core i7-3770 CPU at 3.40 GHz with 4 cores

---

[2] Scheduling processes on disjoint cores is not enough as the last level cache is shared.

[3] The source code of HexPADS is available at http://github.com/HexHive/HexPADS.

Table 1. Performance results for HexPADS on SPEC CPU2006 and PARSEC. Native and HexPADS numbers are in seconds, overhead is in percent.

SPEC CPU2006	Idle	PADS	Overhead	PARSEC	Native	PADS	Overhead
400.perlbench	306	302	-1.32%	blackscholes	36.98	36.93	-0.12%
401.bzip2	396	389	-1.80%	bodytrack	29.88	30.44	1.88%
403.gcc	242	238	-1.68%	canneal	57.06	58.26	2.10%
429.mcf	234	211	-10.90%	dedup	13.73	14.02	2.11%
445.gobmk	374	371	-0.81%	facesim	94.45	96.28	1.94%
456.hmmer	327	325	-0.62%	ferret	63.64	64.77	1.77%
458.sjeng	405	403	-0.50%	fluidanimate	72.21	72.40	0.26%
462.libquantum	287	289	0.69%	freqmine	81.83	80.88	-1.17%
464.h264ref	419	417	-0.48%	netdedup	13.04	13.81	5.92%
471.omnetpp	292	292	0.00%	netferret	407.20	410.16	0.73%
473.astar	304	298	-2.01%	netstr.clust.	132.60	133.32	0.54%
483.xalancbmk	198	197	-0.51%	raytace	64.25	65.07	1.27%
433.milc	349	334	-4.49%	streamcluster	121.35	121.93	0.48%
444.namd	289	288	-0.35%	swaptions	45.33	44.95	-0.83%
447.dealII	214	213	-0.47%	vips	21.29	21.60	1.47%
450.soplex	195	194	-0.52%	x264	17.84	19.48	9.17%
453.povray	126	126	0.00%				
470.lbm	200	198	-1.01%				
482.sphinx3	400	396	-1.01%				
Average	292.47	288.47	-1.39%	Average	1272.69	1284.30	0.91%
Geo.Mean	279.59	275.64	-1.43%	Geo.mean	52.05	52.93	1.69%

(8 threads), 16 GB of memory. We compiled all SPEC CPU2006 C/C++ benchmarks with clang 3.4 and O3. To reduce noise we averaged over 3 runs using the ref dataset (the default configuration for a reportable run in SPEC CPU2006). We compiled PARSEC 3.0 in its default configuration and evaluate it using the native dataset and 16 threads.

Table 1 shows the performance results. In general, the overhead for HexPADS is negligible and in our experiment we observed a slight performance improvement for SPEC CPU2006 (likely due to cache variations and fluctuations of the scheduler placing benchmarks on different cores) and a slight performance degradation for PARSEC. The average and geometric mean is less than 2% and therefore likely noise for both benchmarks. The only infrequent false positives we measured were for CAIN on dedup/netdedup (see Sect. 6.4).

We conducted our experiments on an idle system with multiple cores. The SPEC CPU2006 benchmarks are single threaded but the PARSEC benchmarks are highly parallel. Most of the information is collected by low overhead performance counters and the HexPADS process sleeps most of the time. When observing HexPADS with the `htop` command it uses less than 1% of the CPU to continuously scan, measure, and analyze performance data. In comparison, ninja [7] detects illegal suid processes by scanning the process list at 1.5–2% overhead.

6.2   Case Study: Rowhammer

Rowhammer [25] is a DRAM vulnerability that causes bit flips in DRAM cells, triggered by frequent accesses to neighbouring cells. The DRAM accesses to the adjacent cells cause an interaction with the cell in between, resulting in random bit flips. The rowhammer attack executes cache flush instructions and accesses memory locations in a tight loop. In the attack scenario described by Google's P0 security group, they managed to cause bit flips in a Page Table Entry (PTE) that causes the PTE to point to a physical page under the control of the attacker. This hardware bug allows the attacker to escalate her privileges from user-space to the highest software level, side-stepping all hardware security layers, execution layers, and defense mechanisms.

While incredible powerful, the rowhammer attack is extremely noisy (on the memory bus) and long running. The attack only succeeds if a very large amount of adjacent DRAM accesses are executed in short order, i.e., between refresh intervals that negate all intermediate effects. The attack relies on a high bandwidth to the DRAM cells and therefore has limited interaction with the operating system, e.g., through the page fault handler that adds overhead, reducing the bandwidth for the attack. The overall amount of page faults (or page fault ratio) is therefore low.

Our rowhammer detector (see Fig. 2) measures cache accesses and cache misses of all running processes and checks if the cache miss rate is higher than 70 % (i.e., more than 70 % of all cache accesses are cache misses), the total amount of cache misses is significant, and the number of page faults is low. As rowhammer is a long running attack, our detector averages the cache misses over the sliding window of collected samples. In addition, the average page table miss rate must be low, otherwise the memory accesses would not happen fast enough. If the cache miss rate is too low then no bits are flipped. Using the

```
i ranges from 0 to NR_SAMPLES , not inclusive
cur = current iteration
prev = previous iteration
cache_access = sum(cache_access[i])/NR_SAMPLES
cache_miss = sum(cache_miss[i])/NR_SAMPLES
miss_rate = cache_miss / cache_access
fault_rate = page_faults[cur] / page_faults[prev]
if (
  miss_rate > 0.70 and
  cache_miss > 500,000 and
  fault_rate < 0.01
) cache_attack_detected();
```

Fig. 2. Pseudo code for rowhammer detector based on cache misses and page faults.

rowhammer prototype implementation[4] we always measured a cache miss rate of $> 90\%$ (more than 4,000,000 cache misses per iteration, the highest number of cache misses of a benign process was 101,000 cache misses per iteration) and the attack is detected immediately after the process starts up. Any successful rowhammer attack will always be noisy and the cache miss rate per instruction must be high for the attack to be successful. The default counter measure slows down an offending process for a configurable amount of time.

6.3 Case Study: Cache-Based CSCs

Cache CSCs are very similar in their cache access patterns to the rowhammer attack. Generally, a cache CSC uses one of three ways to communicate [8]: (i) evict and time (the attacker measures execution of the victim's code, evicts the cache, and measures the same code again), (ii) prime and probe (the attacker fills its own memory and measures through access times what data was evicted by the victim), or (iii) flush and reload (the attacker flushes shared memory and measures what memory was reloaded by the victim). All these attacks have in common that they result in a huge amount of cache misses in a short amount of time as large memory areas have to be flushed and read/written.

We have tested two cache covert channels: (i) cache template attacks [11] which is based on flush and reload and (ii) an enhanced version of C5 [18] which is based on prime and probe. The observed memory access pattern is very similar to rowhammer attacks with the difference that a cache CSC is only concerned about the cache itself and not if the memory is written back to DRAM. In our experiment, cache template attacks results in about 1,500,000 cache misses per iteration and C5 attacks in about 2,300,000 cache misses per iteration.

We therefore use the same detector as for rowhammer to detect cache CSCs. The covert channels described above rely on a combination of repeated flushing or filling of the cache and measuring timing. The cache flushing and filling is measurable through cache misses, indicating that a cache CSC is being used.

Our current detector does not distinguish between rowhammer and cache CSCs and successfully detects both attacks. Cache CSCs will always incur a high amount of cache misses, just like rowhammer attacks. If the attacker lowers the speed of the cache attack, the bandwidth will decrease alongside which results in additional noise on the channel. After a certain noise level is reached the attack becomes unrealistic.

6.4 Case Study: CAIN

CAIN (Cross VM ASL INtrospection) [2] leverages memory deduplication as a side channel to recover ASLR base addresses of loaded libraries in co-located virtual machines. For a successful attack, an attacker needs to execute user-space code on a virtual machine that is co-located with the target machine (i.e.,

---

[4] Google's prototype implementation is available at https://github.com/google/rowhammer-test.

runs on the same physical hardware). Memory deduplication searches for shared memory pages across virtual machines and coalesces any common pages. A write to a merged page results in a page fault caught by the VMM and triggers a copy-on-write operation, resulting in a timing side channel that allows the detection of specific memory pages in concurrently running virtual machines. Memory deduplication saves physical memory but causes performance degradation when pages are unmerged (e.g., when one virtual machine writes to the page). CAIN generates a large amount of page candidates for specific libraries, picking a page that is static except for a set of pointers relative to the library's base address. Each generated page candidate then has the probability of $\frac{1}{ASL\ entropy}$ of being present in the target virtual machine. CAIN uses all available memory to generate target pages and then waits for the memory deduplication mechanism to merge a candidate page and the target page. The correct target page is then detected by measuring timing when writing to the page (due to the copy-on-write it takes a longer time to write compared to an unmerged page).

CAIN behavior is naturally bursty and generates a large amount of page faults and cache misses in a short time whenever new candidate pages are generated. This behavior is easily detected by measuring the gradient of page faults and the amount of cache misses (for writing).

```
cur = current iteration
prev = previous iteration
page_faults = array of page fault measurements
cache_miss = array of cache miss measurements
page_miss_rate = page_faults[cur]/executed_instr

if (
 page_faults[prev] > 2.0 * page_faults[cur] and
 page_faults[cur] > 100000 and
 cache_miss[cur] > 10000 and
 page_miss_rate > 0.001
) CAIN_attack_detected();
if (
 page_faults[prev] + page_faults[cur] > 256000
) CAIN_attack2_detected();
```

Fig. 3. Pseudo code for CAIN detector based on cache misses and page faults.

Our detector (see Fig. 3) checks if (i) the amount of page faults in the current iteration is more than double the amount of page faults in the previous iteration (i.e., the amount of page faults doubled), there were more than 100,000 page faults, more than 10,000 cache misses in this iteration, and the page miss rate per executed instruction in the last interval was higher than 0.001 or (ii) the amount of page faults in the last two iteration is higher than 256,000 (which corresponds to 1024 MB of memory being initialized in a short interval).

The first part of the detector checks the increasing flank while the second part checks for a high amount of new memory that is allocated in a short burst. Our detector currently does not check for the ratio between read and write cache misses, for CAIN the amount of write cache misses would be much higher than the amount of read cache misses. Only the PARSEC dedup/netdeup benchmarks experienced false positives as this benchmark allocates a huge amount of memory during startup. For the complete evaluation, the first check results in 1 false positive and the second check in 24 false positives. CAIN attacks are not time critical, so for a future detector we will ensure that benign cases that continuously use the allocated memory do not trigger a detection.

The current detector measures the memory allocation pattern of a CAIN attack through page faults, cache misses, and the amount of allocated memory. CAIN attacks could mitigate the detection by allocating less memory, which would reduce the effectiveness of the attack. An extension of the detector could measure the absence of accesses after detection to detect the phase where CAIN is waiting for the VMM to merge individual pages.

6.5   Discussion, Limitations, and Future Work

The efficiency and success of HexPADS depends on the ability of the detectors to distinguish benign behavior from malicious behavior. The attacks evaluated in the case studies are fundamentally different from benign applications due to the underlying constraints of the attacks. With knowledge of the signatures (which will likely be widely distributed and analyzed), an attacker could launch some form of targeted Mimicry [30] attacks. Mimicry attacks hide the malicious behavior in benign behavior, thereby circumventing detection. HexPADS is not immune to Mimicry attacks and an attacker could, e.g., slow down the number of cache accesses to evade the rowhammer detection. But by slowing down the attack it becomes less efficient and more likely to fail, e.g., for rowhammer, if the attack does not achieve a sufficiently high number of memory accesses between memory refresh operations then the attack will fail. The design of effective detectors depends on a threshold where the attack is no longer successful, yet the amount of false positives remains low. We acknowledge the difficulty of finding such efficient thresholds, especially for programs with different program characteristics where the threshold must be conservative.

In the current version, the baseline behavior and the signatures are hard-coded. The current signatures are based on manual analysis of program executions. As future work we will look into ways of coming up with tighter and more precise signatures automatically, e.g., by collecting benign traces of a wide variety of applications and workloads and using machine learning to automatically extract a baseline pattern and classify the different samples into general signatures. In addition, we will look into aggregating performance measurements of child processes to mitigate an attacker that constantly spawns children to prevent detection. The current motivating examples and case-studies focus on memory attacks. In future work, HexPADS can either be extended to include other attack vectors (e.g., by sampling other performance events), or its concept can be integrated into other attack detection frameworks.

The current prototype implementation is limited to single host detection and does not coordinate information across different virtual machines (i.e., the detection mechanism must run on the same virtual machine as the attacker). This is merely an engineering limitation and the prototype can be extended through additional programming effort. A CAIN attack can only be observed on the same system, so the detector must either run on the attacker machine (e.g., in the case where the attacker controls only a user-space application) or at the level of the hypervisor. An advantage of the current implementation is that the daemon has negligible overhead and runs without any elevated privileges. Disadvantages of such an implementation are that (i) only effects on the system can be observed, attacks from non-monitored systems (virtual machines) are missed and (ii) the operating system is a part of the trusted computing base, any attacker with elevated privileges (administrator privileges) can disable the monitoring and detection mechanism.

7   Related Work

Related work for HexPADS exists in different areas. On one hand, prior work on CSC attacks is used as a motivation to develop our attack detection mechanism and we use different CSC mechanisms to evaluate our work. On the other hand, we compare our work against different existing CSC attack detection and mitigation mechanisms, showing key differences between our performance counter based approach and other approaches that focus on mitigation instead of detection. Last but not least, we compare against other existing intrusion detection mechanisms and explain why they detect attacks on a different abstraction level.

7.1   Covert and Side Channel Attacks

Last level caches are a prime target to extract information using CSC information leaks across processes or even across virtual machines. Sensitive information (e.g., cryptographic keys) can be extracted from unwilling sensitive processes [11,24,37,39] or two malicious processes can use the covert channel to communicate stealthily [18]. A challenge for these CSC attacks is the underlying hardware configuration as each CPU family can be different. Unfortunately, an automated exploration of the cache configuration is possible [11,19].

Other CSCs include, e.g., the last branch target buffer [1], the memory bus [36], memory deduplication mechanisms [2,14,26], and attacks against the underlying memory architecture [25].

7.2   Covert and Side Channel Attack Detection and Mitigation

A CSC attack detection mechanism may be implemented at the level of the hardware, the virtual machine monitor, the operating-system, or the application.

Hardware-based detection and mitigation mechanisms can be separated into approaches that partition resources [6,31,32] with the downside of potentially under-utilizing resources, randomizing accesses [32,33], or limiting the granularity of the timer [17].

On the hypervisor level, HomeAlone [38] detects cross-VM side channel attacks by monitoring cache misses and cache behavior. Other defense mechanisms in the hypervisor either partition resources to be used exclusively for a given virtual machine [15] (with the drawback that same-machine attacks are possible) or limit the timer granularity for virtual machines [28]. HexPADS in comparison measures fine-grained performance events on the process level and allows the identification of individual processes that cause the outlier.

Düppel [40] employs periodic cache flushing to introduce noise and to reduce the attacker's bandwidth. This is a pure mitigation mechanism that does not distinguish between benign behavior and attack behavior. HexPADS may use a mechanism to mitigate an ongoing attack as soon as it is detected with the advantage that cache flushing (and the associated overhead) only occurs during active attacks and not whenever a sensitive operation is executed.

7.3 Intrusion Detection and Mitigation

Network-based IDS like Bro [22] detect an intrusion by inspecting network packets. Host-based IDS observe system characteristics like system call patterns and parameters [13,20,34], log analysis [3], or file integrity checking (e.g., AFICK, Tripwire, or AIDE [3,10]) to detect malicious activity. Intrusion detection systems are either misuse-based or anomaly-based. A misuse-based IDS matches a set of patterns against the observed pattern [22,23,29]. An anomaly-based IDS detects deviations from a well known, good baseline [5,7,9,16,20,35].

HexPADS targets microarchitectural features and uses performance counters to collect fine-grained system information to detect attacks that are not directly observable by regular introspection methods but need support from hardware performance monitors (e.g., by measuring the amount of cache misses).

8   Conclusion

Intrusion detection and attack detection systems enable the detection of otherwise uncaught attacks (i.e., if all other defense mechanisms fail). We have presented the design and open-source implementation of HexPADS, a novel attack detection mechanism that leverages both core systems parameters and performance counter-based statistics on program execution to detect ongoing attacks. The core system measures a set of system parameters and performance characteristics like, e.g., cache misses, executed instructions, or page faults. Through a flexible plugin mechanism we can add dynamic detectors for individual attacks. In three case studies we have evaluated HexPADS and shown its effectiveness against rowhammer, cache-based covert and side channels, and CAIN attacks by implementing simple detectors that use cache accesses, cache misses, page faults, and number of executed instructions to detect attacks. The performance overhead of HexPADS is negligible (non-measurable) and the flexible design and plugin structure simplifies adding new detectors for other and future attacks.

Acknowledgments. We would like to thank Clémentine Maurice, Daniel Grauss, Antonio Barresi, Scott A. Carr, and Terry Ching-Hsiang Hsu for generous feedback on the paper. We also thank Clémentine and Daniel for providing access to the CSC implementation and Antonio for providing access to the CAIN implementation. This work was sponsored, in part, by NSF CNS-1513783.

References

1. Acıiçmez, O., Koç, Ç.K., Seifert, J.-P.: Predicting secret keys via branch prediction. In: Abe, M. (ed.) CT-RSA 2007. LNCS, vol. 4377, pp. 225–242. Springer, Heidelberg (2006)
2. Barresi, A., Razavi, K., Payer, M., Gross, T.R.: CAIN: silently breaking ASLR in the cloud. In: WOOT 2015: 9th Usenix Workshop on Offensive Technologies (2015)
3. Cid, D.B.: Ossec: open source host-based intrusion detection system (2015). http://ossec-docs.readthedocs.org/en/latest/
4. Corp, I.: Intel 64 and IA-32 Intel Architecture Software Developer's Manual Combined vols. 3A and 3B: System Programming Guide, Parts 1 and 2 (2015)
5. Denning, D.: An intrusion-detection model. IEEE Trans. Softw. Eng. **13**(2), 222–232 (1987)
6. Domnitser, L., Jaleel, A., Loew, J., Abu-Ghazaleh, N., Ponomarev, D.: Non-monopolizable caches: low-complexity mitigation of cache side channel attacks. ACM Trans. Archit. Code Optim. (2012)
7. Flo, T.R.: ninja process monitor (2010). http://forkbomb.org/ninja/
8. Fogh, A.: Cache side channel attacks (2015). http://dreamsofastone.blogspot.com/2015/09/cache-side-channel-attacks.html
9. Ghosh, A., Wanken, J., Charron, F.: Detecting anomalous and unknown intrusions against programs. In: Annual Computer Security Applications Conference (1998)
10. Grim, L., Vandenbrink, R.: Ids: File integrity checking. Technical report, SANS Institute (2014)
11. Gruss, D., Spreitzer, R., Mangard, S.: Cache template attacks: automating attacks on inclusive last-level caches. In: USENIX Security Symposium (2015)
12. Hiroaki, E., Kunikazu, Y.: ProPolice: improved stack-smashing attack detection. IPSJ SIG Notes **75**, 181–188 (2001)
13. Hofmeyr, S.A., Forrest, S., Somayaji, A.: Intrusion detection using sequences of system calls. J. Comput. Secur. **6**(3), 151–180 (1998)
14. Irazoqui, G., Inci, M.S., Eisenbarth, T., Sunar, B.: Wait a minute! a fast, cross-VM attack on AES. In: Stavrou, A., Bos, H., Portokalidis, G. (eds.) RAID 2014. LNCS, vol. 8688, pp. 299–319. Springer, Heidelberg (2014)
15. Kim, T., Peinado, M., Mainar-Ruiz, G.: Stealthmem: system-level protection against cache-based side channel attacks in the cloud. In: USENIX Security Symposium (2012)
16. Ko, C., Ruschitzka, M., Levitt, K.: Execution monitoring of security-critical programs in distributed systems: a specification-based approach. In: IEEE Symposium on Security and Privacy (1997)

17. Martin, R., Demme, J., Sethumadhavan, S.: Timewarp: rethinking timekeeping and performance monitoring mechanisms to mitigate side-channel attacks. In: International Symposium on Computer, Architecture (2012)

18. Maurice, C., Neumann, C., Heen, O., Francillon, A.: C5: cross-cores cache covert channel. In: Almgren, M., Gulisano, V., Maggi, F. (eds.) DIMVA 2015. LNCS, vol. 9148, pp. 46–64. Springer, Heidelberg (2015)

19. Maurice, C., Le Scouarnec, N., Neumann, C., Heen, O., Francillon, A.: Reverse engineering intel last-level cache complex addressing using performance counters. In: Bos, H., et al. (eds.) Raid 2015. LNCS, vol. 9404, pp. 48–65. Springer, Heidelberg (2015). doi:10.1007/978-3-319-26362-5_3

20. Mutz, D., Valeur, F., Vigna, G., Kruegel, C.: Anomalous system call detection. ACM Trans. Inf. Syst. Secur. 9(1) (2006)

21. PaX-Team. PaX ASLR (Address Space Layout Randomization) (2003). http://pax.grsecurity.net/docs/aslr.txt

22. Paxson, V.: Bro: a system for detecting network intruders in real-time. Comput. Netw. 31(23–24), 2435–2463 (1999)

23. Porras, P.A., Neumann, P.G.: Emerald: event monitoring enabling responses to anomalous live disturbances. In: Proceedings of the 20th National Information Systems Security Conference(1997)

24. Ristenpart, T., Tromer, E., Shacham, H., Savage, S.: Hey, you, get off of my cloud: exploring information leakage in third-party compute clouds. In: ACM Conference on Computer and Communication Security (2009)

25. Seaborn, M., Dullien, T.: Exploiting the dram rowhammer bug to gain kernel privileges (2015). http://googleprojectzero.blogspot.com/2015/03/exploiting-dram-rowhammer-bug-to-gain.html

26. Suzaki, K., Iijima, K., Yagi, T., Artho, C.: Memory deduplication as a threat to the guest OS. In: European Workshop on System Security (2011)

27. van de Ven, A., Molnar, I.: Exec shield (2004). https://www.redhat.com/f/pdf/rhel/WHP0006US_Execshield.pdf

28. Vattikonda, B.C., Das, S., Shacham, H.: Eliminating fine-grained timers in xen. In: ACM Cloud Computing Security Workshop (2011)

29. Vigna, G., Valeur, F., Kemmerer, R.A.: Designing and implementing a family of intrusion detection systems. In: European Software Engineering Conference (2003)

30. Wagner, D., Soto, P.: Mimicry attacks on host-based intrusion detection systems. In: ACM Conference on Computer and Communication Security (2002)

31. Wang, Z., Lee, R.B.: Covert and side channels due to processor architecture. In: Annual Computer Security Applications Conference (2006)

32. Wang, Z., Lee, R.B.: New cache designs for thwarting software cache-based side channel attacks. In: International Symposium on Computer, Architecture (2007)

33. Wang, Z., Lee, R.B.: A novel cache architecture with enhanced performance and security. In: International Symposium on Microarchitecture (2008)

34. Warrender, C., Forrest, S., Pearlmutter, B.: Detecting intrusion using system calls: alternative data models. In: IEEE Symposium on Security and Privacy (1999)

35. Wu, J., Ding, L., Wu, Y., Min-Allah, N., Khan, S.U., Wang, Y.: c^2 detector: a covert channel detection framework in cloud computing. Secur. Commun. Netw. 7(3), 544–557 (2014)

36. Wu, Z., Xu, Z., Wang, H.: Whispers in the hyper-space: high-speed covert channel attacks in the cloud. In: USENIX Security Symposium (2012)

37. Yarom, Y., Falkner, K.: Flush+reload: a high resolution, low noise, l3 cache side-channel attack. In: USENIX Security Symposium (2014)

38. Zhang, Y., Juels, A., Oprea, A., Reiter, M.K.: Homealone: co-residency detection in the cloud via side-channel analysis. In: IEEE Symposium on Security and Privacy (2012)
39. Zhang, Y., Juels, A., Reiter, M.K., Ristenpart, T.: Cross-VM side channels and their use to extract private keys. In: ACM Conference on Computer and Communication Security (2012)
40. Zhang, Y., Reiter, M.K.: Düppel: retrofitting commodity operating systems to mitigate cache side-channels in the cloud. In: ACM Conference on Computer and Communication Security (2013)

Analyzing the Gadgets
Towards a Metric to Measure Gadget Quality

Andreas Follner[1]([✉]), Alexandre Bartel[1], and Eric Bodden[2,3]

[1] Technische Universität Darmstadt, Darmstadt, Germany
{andreas.follner,alexandre.bartel}@cased.de
[2] Paderborn University, Paderborn, Germany
[3] Fraunhofer IEM, Paderborn, Germany
bodden@acm.org

Abstract. Current low-level exploits often rely on code-reuse, whereby short sections of code (*gadgets*) are chained together into a coherent exploit that can be executed without the need to inject any code. Several protection mechanisms attempt to eliminate this attack vector by applying code transformations to reduce the number of available gadgets. Nevertheless, it has emerged that the residual gadgets can still be sufficient to conduct a successful attack. Crucially, the lack of a common metric for "gadget quality" hinders the effective comparison of current mitigations.

This work proposes four metrics that assign scores to a set of gadgets, measuring quality, usefulness, and practicality. We apply these metrics to binaries produced when compiling programs for architectures implementing Intel's recent MPX CPU extensions. Our results demonstrate a 17 % increase in useful gadgets in MPX binaries, and a decrease in side-effects and preconditions, making them better suited for ROP attacks.

Keywords: ROP · Gadgets · Exploit · CFI · MPX · Metrics

1 Introduction

Several mitigation techniques guarding against control-flow attacks have been developed over the past 15 years. In contrast to modern-day attacks [6,7,10,11, 14,17,25,26,30,31], the attacks of the 90s [19] were simple. The latter typically exploited a stack-based buffer overflow vulnerability to overwrite a stack frame's return address with another that points to a location at which the attacker had previously injected malicious code. On returning from the compromised function, execution would consequently be redirected to the injected code block.

Since the early 2000s, the prevalent processor architectures have adopted the *No-eXecute* (NX bit) extensions. These allow an operating system to mark

---

The stamp on the top of this paper refers to an approval process conducted by the ESSoS artifact evaluation committee chaired by Alessandra Gorla and Jacques Klein. At the time this research was conducted Eric Bodden was at Fraunhofer SIT and TU Darmstadt.

J. Caballero et al. (Eds.): ESSoS 2016, LNCS 9639, pp. 155–172, 2016.
DOI: 10.1007/978-3-319-30806-7_10

memory pages that only contain data (namely the heap and stack) as being non-executable [18], thus stopping code-injection attacks. However, programs may need to be able to allocate executable memory, for example for just-in-time compilation [3]. For such cases, the operating system provides several API calls that can change the memory protection level of a memory area (e.g., VirtualProtect[1] on Windows). These API calls were quickly abused by attackers, who would leverage them to change the access privileges of a region of memory where they had previously injected their payload. To circumvent the NX bit protection and to execute the API calls which change the memory protection of the payload code to executable, current exploits reuse executable code snippets, or *gadgets*, comprising code from the running program and loaded libraries. Such attacks are known as code-reuse attacks, the most popular and widespread technique being Return-Oriented Programming (ROP) [24, 29].

The difficulty of staging a ROP attack in practice is subject to an attacker's concrete aims, the underlying environment, and the available gadgets. The latter, in particular, varies enormously between binaries. However, there is currently no established metric for quantifying the utility of gadgets within a given binary. Having such a metric would enable the comparison of gadgets in various kinds of transformed binaries, e.g., different optimization levels of compilers, or binaries that have been rewritten to add instructions for exploit mitigation. Currently, many tools that produce such binaries, even those meant to enhance a binary's security, do not take into account how their transformation affects ROP gadgets. Especially for exploit-mitigation techniques this is counterproductive: if a mitigation technique transforms code, how does one know that it does not in the end increase a binary's attack surface by adding useful gadgets?

This work presents four metrics based on practical exploit development, that are designed to aid researchers in the evaluation of mitigations. More generally, these metrics allow one to determine whether a binary transformation introduces gadgets that are better suited for ROP attacks than the original binary.

Since it is somewhat difficult to make statements about the usefulness of a set of gadgets without knowing the goal of the attacker and the underlying environment, the metrics cover two targeted, real-world exploitation scenarios, and two more general computations which reflect gadget variety and gadget usability. This work further applies the metrics to binaries protected by MPX (Memory Protection eXtensions) [23], a new buffer-overflow mitigation technique from Intel that adds instrumentation code to binaries through the compiler. As our evaluation shows, MPX-enabled binaries actually do contain more useful gadgets, and thereby increase the attack surface. This is particularly worrysome when running MPX-enabled binaries on legacy hardware that cannot benefit from the increased security that MPX is designed to offer. To summarize, our key contributions are:

- a definition of four metrics to measure gadget quality,
- GaLity, an open-source implementation to compute metrics on sets of gadgets, and

---

[1] https://msdn.microsoft.com/en-us/library/windows/desktop/aa366898%28v=vs.85%29.aspx.

- a case study using the metrics to determine how MPX affects gadgets on eight representative Windows x64 binaries.

The remainder of this paper is organized as follows. Section 2 motivates the necessity to evaluate gadget quality. Section 3 describes the proposed metrics. Section 4 explains the conducted case study on MPX. Section 5 covers related work. Finally, Sect. 6 concludes the paper.

2   Motivation

To the best of our knowledge, there exists no metric to assess the *quality* of a gadget or a set of gadgets. Such a metric, however, has a large variety of use cases. For example, it could be used to compare different control-flow integrity (CFI) [2] approaches. Today's CFI implementations [9,20,21,32–34] often use a metric which measures the reduction of gadgets (such as AIR, the average indirect target reduction [34], or DAIR, the dynamic average indirect target reduction [21]) to compare their results. For many approaches, this metric shows a reduction of over 99 %, yet this does not take into consideration the total number of gadgets, nor the quality of the remaining gadgets, limiting the metric's practical use. A DAIR of 90 % that leaves 50 gadgets with many side effects and preconditions intact is likely more secure than a DAIR of 99.5 %, that leaves intact exactly those 7 gadgets that an attacker requires to craft an exploit. Researchers using those metrics frequently acknowledge their limitations and the difficulty of developing a metric that measures gadget quality [6,21,32].

In general, attackers favour simple gadgets which have a minimum of side effects and preconditions. For example, consider a gadget that loads the value that rsp points to into rax. A clean and effective gadget for achieving this would be: pop rax; ret. In contrast, the gadget: pop rax; push rsp; pop rbp; mov [rdi+0x34fa], rsp; ret 0x2dbf1 will also achieve this aim, but will also have the side-effect of overwriting rbp. In addition, this gadget has the precondition that rdi+0x34fa has to point to writeable memory. Finally, ret 0x2dbf1 not only adds a large offset to rsp (which can be an issue if attacker-controlled memory is scarce, because it might set rsp to point outside of the allocated memory), it also disaligns the stack pointer, which is something normal programs do not do, hinting at a possible exploit execution. The next Section presents the four metrics we propose to compute gadget quality.

3   Metrics for Measuring Gadget Quality

In general, evaluating the quality of a set of gadgets is non-trivial. This stems primarily from the fact that an attacker's goal is potentially unknown, and that given sufficient gadgets, one can construct practically any program. In addition, the gadgets required for an attacker to achieve a goal vary by operating system and architecture. For example, on Windows x86, parameters to functions are usually passed on the stack, while on Windows x64, the first four parameters are passed through registers and all remaining ones are passed on the stack[2], leading to differences in gadget requirements. As a running example, we consider exploits targeting `VirtualProtect`, which is an API call that commonly serves as an avenue to bypassing NX protection on Windows 7 x64 [12,16,22]. We stress that our four metrics are not bound to evaluating this specific API call, as they consider the more general attack setup and execution procedures associated with ROP exploits. In addition, we perform an in-depth analysis of the various properties of gadgets with respect to their side effects, preconditions, usability, and usefulness.

3.1   Metric 1: Gadget Distribution

The *gadget distribution* metric is calculated by partitioning a given set of gadgets into twelve broad categories, with each category representing a class of operations, such as *arithmetic* and *data move*, as shown in Table 1. Gadgets are assigned to a category based on the first instruction of a gadget. For example, the gadget `add rax, 0x40; pop rcx; ret` would be assigned to the *arithmetic* category. We categorize on the basis of the first instruction as every suffix of a gadget is itself a gadget, and will be categorized separately. Note that gadgets containing privileged or sensitive instructions [1] are discarded and not considered in further steps because they trap in user mode, thereby making a gadget unusable.

Analyzing the distribution of frequencies of gadgets amongst categories is helpful as it allows comparing whether the distribution of gadgets in a transformed binary is similar to the one in the original binary, or if the number of gadgets in a category useful for an attacker has grown. Gadget quality and usefulness, however, are not measured and addressed by the remaining metrics.

While Table 1 does not contain all instructions of the x86–64 instruction set, it covers 99 % of the instructions found in gadgets of the binaries we used in the evaluation, i.e., a total of 20 MiB containing over one million instructions. Due to the large size of the x86–64 instruction set (over 700 instructions [1]), it would be a time-consuming, manual process to cover all existing instructions. However, the fact that we do not achieve 100 % coverage does not pose a threat to the metric, because all important and common instructions are categorized. The few we did not include do not have a big impact on the overall distribution. A manual inspection of uncategorized instructions in other binaries (we used several Windows 7 system libraries) revealed that there were many different instructions but in small numbers in any of the inspected binaries, which is what we expected.

Metric 1 allows to assess whether a transformed binary contains more gadgets in categories useful to an attacker.

---

[2] https://msdn.microsoft.com/en-us/library/windows/hardware/ff561499%28v=vs.85%29.aspx.

Table 1. Gadget categories

Category	Included instructions
Data move	pop, push, mov, xchg, lea, cmov, movabs
Arithmetic	add, sub, inc, dec, sbb, adc, mul, div, imul, idiv, xor, neg, not[a]
Logic	cmp, and, or, test
Control flow	call, sysenter, enter, int, jmp, je, jne, jo, jp, js, lcall, ljmp, jg, jge, ja, jae, jb, jbe, jl, jle, jno, jnp, jns, loop, jrcxz
Shift&Rotate	shl, shr, sar, sal, ror, rol, rcr, rcl
Setting flags	xlatb, std, stc, lahf, cwde, cmc, cld, clc, cdq
String	stosd, stosb, scas, salc, sahf, lods, movs
Floating point	divps, mulps, movups, movaps, addps, rcpss, sqrtss, maxps, minps, andps, orps, xorps, cmpps, vsubpd, vpsubsb, vmulss, vminsd, ucomiss, subss, subps, subsd, divss, addss, addsd, cvtpi2ps, cvtps2pd, cvtsd2ss, cvtsi2sd, cvtsi2ss, cvtss2sd, mulsd, mulss, fmul, fdiv, fcomp, fadd
Misc	wait, set, leave
MMX	pxor, movd, movq
NOP	nop
RET	ret

[a] It might appear peculiar that xor, neg, not are in the arithmetic category - however, this is how exploit developers often use these instructions. Since using nullbytes is sometimes prohibited by the environment, writing the negated or xor-ed value in memory, loading it to a register and then using the same operation on it again is used to bypass this restriction.

3.2 Metric 2: Gadget Environment Setup Capabilities

When constructing a ROP exploit, an attacker must be able to prepare the environment and operands for subsequent gadgets in a chain. For example, when attempting to perform a Windows API call via ROP, an attacker will generally require the ability to specify the call's arguments. The degree of ease with which an attacker may manipulate memory will affect the choice of gadgets that she uses. In this metric, we consider the most general case, whereby an attacker is able to inject arbitrary arguments into a target program's memory space at a known location. This could be possible due to, e.g., a browser with Javascript turned on, allowing heap sprays and Heap Feng Shui [31], and other vulnerabilities like information leaks [28]. We further assume the vulnerable program is running on a Windows 7x64 machine, which is a very common platform.

Consider the case whereby an attacker wants to invoke VirtualProtect, which takes four arguments. On the aforementioned target platform, the first four parameters are passed through registers (`rcx`, `rdx`, `r8`, `r9`). In such a scenario, an attacker needs to make sure that those registers contain the correct values before `VirtualProtect` can be invoked. To achieve that, three different kinds of gadgets are required, namely: (i) a *stack pivot* gadget which points `rsp` to the injected data, i.e., function arguments and addresses of gadgets, (ii) gadgets to load the arguments from memory to the appropriate registers, and (iii) a gadget that calls `VirtualProtect`.

This metric looks for gadgets that achieve these goals and distinguishes between gadgets that achieve only the required task or include other instructions. Of course, our tool reports gadgets only if the register that receives the argument is preserved, i.e., not overwritten by another instruction in the same gadget. In case the attacker wants to invoke an API that requires fewer arguments, like `VirtualAlloc`[3], fewer gadgets that load arguments are required.

A gadget is only useful in preparing a destination register r_d for use within a ROP chain if it does not destroy its value prior to returning. More concretely, consider a gadget consisting of a sequence of n instructions $i_0; i_1; \ldots i_{n-1}; \texttt{ret}$. If i_0 assigns the value to r_d, any subsequent instruction i_k with $k > 0$ that has r_d as a target operand and falls within the *data move, arithmetic,* or *shift and rotate* categories is tagged as being potentially destructive. A second refinement step is subsequently carried out, whereby the quirks of the target architecture are taken into account. For instance, instructions that output to a 32 bit subregister are handled differently than those that output to 16 or 8 bit subregisters. This is due to the behaviour that writing to a 32 bit subregister automatically zero-extends the value to fill the entire 64 bit register [1].

In the case of exploits making use of `VirtualProtect`, one finds that three of the four arguments that this API call takes (namely `lpAddress`, the start address of the memory region whose protection level is to be changed, `dsSize`, the size address of the memory region whose protection level is to be changed, and `lpflOldProtect`, an address where the old protection level will be stored) do not need to be precise. If `lpAddress` is a few bytes off an attacker can take this into account, just like a slightly smaller or larger size argument. `lpflOldProtect` is not used by an attacker and can therefore be written to any location. Therefore, the metric only deems two instructions destructive, namely `pop` and `mov` in 64 bit or 32 bit subregisters, as they overwrite the whole register. *Metric 2 allows one to assess whether a transformed binary contains gadgets typically required for an attack where the environment gives the attacker a lot of leeway.*

3.3 Metric 3: Gadget Environment Setup Capabilities - Restricted

In contrast to the previous metric, this metric considers the case where an attacker is restricted in the ways in which she can inject values into memory. In particular, we consider the scenario where an attacker may only inject data and

---

[3] https://msdn.microsoft.com/en-us/library/windows/desktop/aa366887%28v=vs.85%29.aspx.

hijack the control-flow via `strcpy`. This complicates the direct injection of values into memory because many parameters to API calls often contain null-bytes, which terminate strings, thus requiring that the arguments to be used for correctly invoking a function such as `VirtualProtect` be calculated dynamically at runtime. Imagine an attacker wants to indeed invoke `VirtualProtect`. By taking a look at the required parameters it becomes clear that many will contain null-bytes: `lpAddress` should point to the payload. Depending on the memory layout, this address may contain null-bytes (e.g., in a classic stack buffer overflow vulnerability on Windows, stacks are located at very low addresses making it very likely for the address to have its leftmost bytes set to null). `dwSize` must not be too large, i.e., `lpAddress + dwSize` must include only mapped pages. The value must also not be too small, as it has to cover the memory area where the payload is injected. Typically, the value is a couple of thousand bytes or smaller, which is a value that cannot be injected directly. `flNewProtect` is usually set to `0x40`, which cannot be injected directly because the leftmost bytes are null,and requires to be computed at runtime. `lpflOldProtect` will receive the old protection value, hence must point to writable memory, which may contain null-bytes. This example shows that in a scenario where the attacker is restricted, she will require various arithmetic and data-move gadgets in order to dynamically calculate parameters for API calls using gadgets.

The metric gauges the presence of gadgets that may be used to assist in evaluating values dynamically at runtime, specifically gadgets that move data between memory and registers and compute values: `pop`, `push`, `add`, `sub`, `adc`, `dec`, `inc`, `neg`, `not`, `mov`, `sbb`, `xchg`, `xor`. As in the case of *Metric 2*, a gadget is only considered if r_d is preserved. *Metric 3 allows one to assess whether a transformed binary contains gadgets typically required for an attack where the attacker has to make many calculations at runtime and cannot inject arbitrary data into a program.*

3.4   Metric 4: Gadget Quality

The aforementioned metrics do not measure the quality of a gadget per se, rather they provide an indication whether a specific attack can succeed given a set of gadgets. This metric focuses on assessing the quality of an individual gadget, whereby a high-quality gadget is defined as one having no preconditions or side-effects on other registers or memory. An example of a precondition is that a specific register has to point to writeable memory, e.g., in the gadget `pop rax; mov [rdi+0x34fa], rsp; ret`. To be usable, `rdi+0x34fa` must point to writeable memory. A side-effect is, for example, that data in another register is overwritten or the stack pointer is manipulated in a way that is difficult to undo, e.g., in the gadget `pop rax; mov rcx, 0xb0adffff; leave; ret`. This gadget overwrites the values in `rcx`, `rsp`, and `rbp`. To express gadget quality, a score is calculated for every gadget considered useful (see *Metric 3*). The score starts at 0 and is increased for side-effects and preconditions. Therefore, a higher score equals worse gadget quality. In the following we give a high-level overview of the two criteria we use to calculate the score for gadget quality.

Grading Instructions. To measure side-effects and preconditions, the metric inspects every instruction in a gadget. It reuses the categories introduced in Sect. 3.1 and assigns a score to each category, which reflects how destructive the instructions in the respective category are. Table 2 summarizes the scoring system. Depending on the destination of the instruction, we apply a modifier to the originally assigned score. The metric recognizes three possible kinds of destinations: rsp, which should ideally not be modified, because it is responsible for the control flow and always needs to point to the next gadget. Therefore, modifications of rsp usually have the largest influence on the overall score of a gadget. The second possible destination is r_d, the destination register in the first instruction of a gadget, for which we assume that this is also the register an exploit developer is interested in not being modified later on in the same gadget (in case a memory address is the target there is no active register; in case of an xchg instruction, both registers are active registers). Modifications of r_d are generally not desirable, but, depending on the modification, can be reversible, e.g., simple arithmetic. The third possible destination is any other general purpose register, except rsp and r_d, the metric considers all undesirable side effects and preconditions. Even if they do not affect rsp or r_d directly, they still negatively impact the final score.

Table 2. Rules for grading instructions. Category describes the category of the instruction (see Table 1). "RSP", "rd" and "Other" are possible targets for instructions, the stack pointer, the destination register of the first instruction of a gadget, or any of the other general purpose registers respectively. Categories not in the table generally do not affect the score, with some exceptions discussed in Sect. 3.4

Cat.	RSP	rd	Other	Notes
Data move	2	1	0.5	As opposed to all other instructions in this category, push does not affect the score of a gadget, since the only side effect it has is on rsp, and changes to rsp are covered by our rsp monitoring.
Arithmetic	2	1	0.5	Arithmetic instructions that modify a register other than rsp can be taken into account by the exploit developer. E.g., if r8 should contain 0x40, and a gadget like pop r9 ; add r8, 0x10 ; ret has to be executed as the last gadget, the developer can simply make sure r8 contains the value 0x30 before invoking the last gadget. Arithmetic instructions modifying rsp are covered by our rsp monitoring.
Shift & Rotate	3	2	0.5	These instructions are handled similarly to arithmetic instructions, however, they are more difficult to take into account, which is why they increase the score more than arithmetic instructions.

In a few cases grading all instructions in one category the same does not make sense and would result in false scoring, which is the reason for the following exceptions. *Exception #1:* Certain instructions that modify rsp need to be treated differently. This covers all instructions where we can statically determine the offset applied to `rsp`. Depending on how much `rsp` is changed, we adjust the overall score of the gadget. The details on this are covered in the next subsection. In case it is not possible to statically determine the offset (e.g., `leave` or `pop rsp`), the overall score of the gadget is increased depending on the category of the instruction, as presented in Table 2. *Exception #2:* `leave` is the only instruction in the miscellaneous category that needs to be taken into account, as it affects `rsp`. This is taken care of through our `rsp` monitoring. *Exception #3:* Remember from Sect. 3.1 that we do not cover all of the x86-64 instructions. This means that in very rare cases (less than 0.1 %) we cannot grade a gadget because it contains an instruction which we did not categorize. We discard these gadgets from the analysis. *Exception #4:* If an instruction uses a dereferenced register as destination its score is increased according to the rules in Table 2, because this poses a precondition - e.g., the gadget `pop r8; mov [rdx], 0xfffa; ret` has the precondition that `rdx` has to point to writable memory before the gadget can be used.

Monitoring `rsp` Offset. Modifications to `rsp` need to be tracked for each gadget, as explained in the previous paragraph. A short example will make clear why this is necessary. Assume the following gadget: `pop rax; add rbx, 0x10ff; push rcx; ret`. In this case, `rsp` will point to the value contained in `rcx` and jump to this address, which is not the injected address of the next gadget. For keeping track of the `rsp` offset the metric uses an SP-Score, *SPS*, which starts at 0, is increased for `pop` and decreased for `push` and `ret n` instructions. Of course, also arithmetic instructions on `rsp` are monitored and the respective value is added to or subtracted from *SPS*. When all instructions in a gadget have been analyzed and *SPS* is not 0 this means that `rsp` does not point to the next gadget, which might be problematic. Therefore, if *SPS* is negative, the overall score of the gadget will be increased by 2. Also, if *SPS* is large (more than 4 KiB) or not aligned, the score of the gadget will be increased by 1, because the former requires an attacker to be able to control more memory and the latter can be detected easily by exploit mitigation tools. If the instruction that operates on `rsp` takes a register and not an immediate (e.g., a `add rsp, rcx`), *SPS* is not changed but the gadget score will be increased by rules in Table 2. *Metric 4 allows one to assess the overall "quality" of a set of gadgets in respect to side-effects, preconditions, and usability.*

3.5   Discussion of the Metrics

We believe that metrics that measure the quality of a set of gadgets should focus on practical relevance rather than a theoretical concept such as Turing completeness [29]. Furthermore, they should also reflect whether real-world exploits can be constructed. Since at least Microsoft has seen a shift from classic, stack-based vulnerabilities to heap-related vulnerabilities [4], we believe that metrics should still consider both of these classes of attacks. Last but not least, the metrics should not be limited to well-defined and realistic attack scenarios, but also express overall gadget quality, i.e., side-effects and preconditions. To summarize, metrics as described above should:

- Be practical
- Measure if popular current attacks are possible with a given set of gadgets
- Measure if popular past attacks are possible with a given set of gadgets
- Measure gadget "quality"

The proposed metrics achieve all these goals. We would like to stress that our aim is assessing whether a binary contains gadgets suitable for today's ROP attacks. Recently, attacks that use longer and more complex gadgets have been proposed by researchers [7,11,13,14,26]. Such attacks are designed to bypass specific mitigation techniques, which are not used in the real world. Thus, in current environments, these complex attacks are cumbersome as they offer no advantage over using regular and simpler ROP gadgets, and we are not aware of any of these complex attacks being used in the wild.

Because of the lack of practical relevance, we decided not to treat gadgets potentially useful in such complex attacks differently than the other gadgets. Nevertheless, if new mitigations limiting the gadgets an attacker may use become widespread and attackers are forced to use more complex and longer gadgets and start using tools that assist in finding gadgets semantically rather than through simple pattern matching, our metrics will have to be updated to reflect this new environment. This is why we also plan to use a more abstract interpretation of gadgets and look into leveraging synergies created by combining gadgets in the future. Furthermore, we also leave an extension to jump-oriented programming (JOP) [5,8] for future work.

4   Evaluation

We have implemented the described metrics in a tool named GaLity, which takes a textfile containing gadgets as input and outputs the metrics we described in Sect. 3. We demonstrate that it is both practical and useful by applying it to binaries that are compiled to use MPX [23], Intel's latest mitigation technique against runtime exploits. MPX introduces new registers that contain the lower and upper bound of a pointer, and instructions that operate on those registers. This enables compilers to emit additional instructions (MPX and non-MPX) that tracks the sizes of buffers and accesses to those buffers at runtime, which can prevent buffer overflows. On processors which do not support MPX, MPX instructions execute as nop, making MPX compatible with older CPUs, but leaving those binaries unprotected by MPX. Given this observation one thus must wonder if the increased code size and thus increased availability of gadgets might actually decrease a binary's security on such systems. We then compare the results obtained by applying GaLity to binaries compiled with MPX support with the results obtained by applying GaLity to the same binaries compiled without MPX support, and determine which binaries contain more helpful gadgets for an attacker according to our metrics.

4.1   Implementation

We wrote GaLity in C#. GaLity takes a simple text file that contains gadgets as input and parses it in four passes, which correlate to the four metrics described in the previous section. While doing everything in one pass is certainly possible, we decided to use several passes, as this increases code readability, and performance was no issue (even large sets of gadgets containing hundreds of thousands of gadgets can be analyzed in less than 10 s on an Intel Core 2 Duo with 4 GiB RAM). Since current ROP attacks use rather simple gadgets we only reconstruct the semantic we require for our metrics. For example, GaLity recognizes the differences between instructions outputting to 64 bit, 32 bit, 16 bit, or 8 bit (sub)registers and treats them accordingly, but does not recognize that many instructions manipulate CPU flags. Knowledge about this would be required when utilizing more complex gadgets that use conditional branches. However, for current real-world attacks, the simpler but less error-prone approach is sufficient.

We looked into using an intermediate representation (IR) which makes side effects explicit, as this would allow more precise grading. However, we discovered that, as today's attacks use simple gadgets, there are few side effects that are relevant in our scenarios. Therefore, we leave designing an IR tailored to the very specific requirements of measuring gadget quality, that (1) can be reused and (2) recognizes more side effects, for future work.

4.2   Setup

To discover gadgets and write them to a file we used ROPgadget 5.4[4], with a maximum gadget length of 15. This might sound like a very high number, however, we did not want to risk potentially missing some useful gadgets. Also, our metrics ensure that gadgets that do not preserve r_d are discarded, i.e., not considered in the results, and gadgets that have many side effects have a bad score. Also, for this specific case study we decided to consider duplicate gadgets and not just unique gadgets, because if an important gadget exists several times in a binary, this binary is more attractive to an attacker than a binary which contains only one copy of that gadget. This matters, for example, in a scenario where a patch (security-related or not) or any other program modification removes said gadget. Furthermore, taking duplicate gadgets into account helps us measure, if the additional gadgets introduced by MPX are copies of useless or useful gadgets.

We compiled programs taken from SPEC2006, using Intel's latest GCC release with MPX support at the time of writing (5.0.0).[5] We decided to use the SPEC suite because it covers a wide range of application types, and present parts of real programs. MPX is still new and not integrated too well in build chains, which made compiling any program a challenge. However, we got the following eight programs to work properly: 401.bzip2, 403.gcc, 435.gromacs, 456.hmmer, 458.sjeng, 464.h264ref, 473.astar, 482.sphinx3. We compiled all binaries four times, with and without MPX and with and without optimizations

---

[4] https://github.com/JonathanSalwan/ROPgadget.

[5] https://software.intel.com/en-us/articles/intel-software-development-emulator.

(−O2). However, for our evaluation we only considered optimized binaries as this reflects real-world binaries.

4.3   Results

First of all, we noticed that MPX has a big influence on file size. With no optimizations, an MPX binary is, on average, almost 3 times as large as a non-MPX binary. With optimization level 2, which we used throughout our experiments, an MPX binary is still, on average, 86 % larger compared to a non-MPX binary. We noticed that, while the file size increases by a factor of almost two, the number of gadgets does not increase in the same way, MPX binaries contain, on average, only 23 % more gadgets than non-MPX binaries. This is because the number of gadgets is directly related to the number of ret instructions in a binary. MPX does not add many new functions but rather makes existing functions longer, therefore only few intended new ret instructions appear. Unintended ret instructions [24] might appear in some cases, however, since the new opcodes introduced by MPX do not contain a ret opcode, the possibility for this is rather low.

Table 3. Results for *Metrics 2, 3,* and *4.* Columns rcx, rdx, r8 and r9 denote the number of gadgets which load a value in the respective register, column pivot denotes the number of stack pivot gadgets. The first number denotes the number of gadgets without side-effects, the second number the number of gadgets with side-effects. Column call denotes the number of gadgets usable for indirect calls. These numbers are required for computing *Metric 2.* Column useful denotes the number of useful gadgets, calculated by *Metric 3.* Column Q denotes the number of gadgets with a score of 1 or lower, calculated by *Metric 4.*

	Metric 2						Metric 3	Metric 4
Program	rcx	rdx	r8	r9	pivot	call	useful	Q
h264ref no MPX	4 / 29	1 / 8	1 / 9	0 / 0	0 / 453	62	6,056	3,749
h264ref MPX	7 / 29	0 / 23	1 / 3	0 / 1	0 / 666	91	7,546	4,906
gromacs no MPX	228 / 320	39 / 135	0 / 2	0 / 0	0 / 1071	84	10,823	6,563
gromacs MPX	228 / 418	36 / 141	0 / 7	0 / 1	0 / 1214	155	13,002	8,170
hmmer no MPX	6 / 24	3 / 27	0 / 3	0 / 0	0 / 509	33	5,539	3,303
hmmer MPX	8 / 21	4 / 19	0 / 2	0 / 0	0 / 469	39	6,188	3,952
gcc no MPX	4 / 71	2 / 219	0 / 14	0 / 8	6 / 5295	588	50,766	32,949
gcc MPX	2 / 52	4 / 71	0 / 9	0 / 4	0 / 4337	763	59,522	39,342
sphinx3 no MPX	2 / 14	0 / 11	0 / 0	0 / 0	0 / 230	29	3,189	1,964
sphinx3 MPX	1 / 11	0 / 7	0 / 0	0 / 0	1 / 251	52	3,484	2,323
sjeng no MPX	1 / 3	0 / 3	0 / 0	0 / 1	0 / 122	72	1,444	983
sjeng MPX	1 / 4	0 / 5	0 / 0	0 / 0	0 / 137	76	1,982	1,414
astar no MPX	1 / 4	0 / 4	0 / 0	0 / 0	0 / 122	11	1,009	584
astar MPX	0 / 5	0 / 2	0 / 0	0 / 0	0 / 140	12	1,203	698
bzip2 no MPX	0 / 1	0 / 1	0 / 0	0 / 0	0 / 99	13	790	466
bzip2 MPX	0 / 1	0 / 1	0 / 0	0 / 0	0 / 112	16	987	605

Analyzing the increase or decrease of gadgets for each category due to MPX, illustrated in Fig. 1, shows that most categories gain gadgets. Arithmetic gadgets, which are helpful to an attacker, increase in both number and diversity. Data-move gadgets grow in numbers, but do not change a lot in respect to diversity. An interesting observation is that NOP-gadgets increase drastically, which is presumably due to the fact that the new MPX instructions are interpreted as multi-byte NOPs on hardware that does not support MPX. The categories flag, string and floating-point have a high standard deviation, indicating that changes in these categories are very application-specific. Gadgets in the miscellaneous category decrease both in diversity and number. Despite the large increase of nop gadgets, the overall distribution of gadgets remains roughly the same, as Fig. 2 shows. Overall we conclude that MPX binaries contain more gadgets in categories helpful to an attacker.

Next, we are interested in the two attack scenarios, i.e., *Metrics 2* and *3*. Regarding *Metric 2*, there is no big difference in the availability of gadgets. Gadgets that load arguments in r8 or r9 are rare in both MPX and non-MPX binaries, and sometimes the MPX binary and sometimes the non-MPX binary contains some. Regarding *Metric 3*, the number of useful gadgets increases in every binary and on average by 17 %, making MPX binaries a much more attractive target to attackers. We summarize the results in Table 3. Lastly, we determine overall gadget quality using *Metric 4*. In all eight binaries, the MPX versions contain more gadgets of high quality, i.e., with fewer side-effects and preconditions, as the last column of Table 3 shows.

By taking all four results into consideration we come to the conclusion, that binaries compiled with MPX support are favourable for an attacker. *Metric 1* shows an overall increase of gadgets in useful categories, further confirmed by *Metric 3*, which also shows that the additional gadgets in those categories are useful in practice. *Metric 2* gives no indication that MPX or non-MPX binaries contain more of the required gadgets. *Metric 4* gives the indication that MPX binaries tend to have more gadgets of higher quality, making them easier to use for an attacker.

5   Related Work

To the best of our knowledge, no previous work has been done on the topic of designing a metric to measure the quality of a set of gadgets, even though the metrics currently used to measure CFI strength are insufficient, exactly because gadget quality is not expressed by those metrics. Due to this lack of related work, we introduce the metrics that are currently used to evaluate CFI implementations, and discuss gadgets required for carrying out attacks against CFI.

Zhang et al. [34] propose using AIR which denotes how many gadgets are removed, because they are not acceptable targets of indirect branches. However, AIR does not take into account the quality of the remaining gadgets.

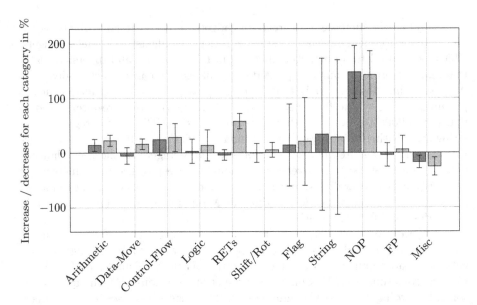

Fig. 1. This figure shows the average growth of gadgets for each category due to MPX across all eight applications. The blue bar represents the increase considering only unique gadgets, while the red bar represents the total increase of gadgets, i.e., also duplicate gadgets. We use the information about how the number of unique gadgets changes to infer if and how gadget variety is affected by a program transformation (Color figure online).

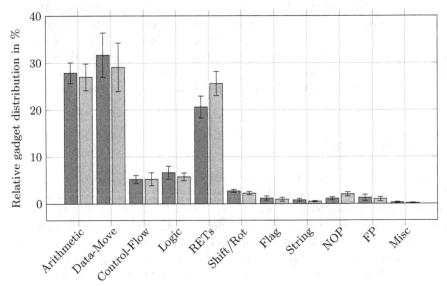

Fig. 2. This figure shows the average distribution of gadgets across all eight applications. The blue bar represents the non-MPX binaries, while the red bar represents the MPX binaries (Color figure online).

Payer et al. [21] propose DAIR, which works similarly to AIR but is dynamic, hence varies during program execution. Tice et al. [32] propose forward-edge AIR (fAIR), which is computed like AIR, but takes into account only forward-edge indirect control transfers, i.e., calls and jumps. All of the above metrics are limited to CFI though, and do not consider the quality of the remaining gadgets.

Carlini et al. [6] discuss the effectiveness of CFI implementations against ROP attacks and propose what they call a basic exploitation test (BET). BET consists of three generalized attack scenarios, namely arbitrary code execution, confined code execution and information leakage. They use a minimal program that allows exploitation, apply several CFI implementations, and evaluate, which of the described attack scenarios could be achieved. However, this process was done by a human, hence dependant on skill and knowledge of the exploit developer. Therefore, it can also not be used for mass-analyzing binaries.

In 2014 and 2015 many attacks targeting various CFI implementations, e.g., kBouncer [20], ROPecker [9], or CFI for COTS [34] have been published. As CFI places tight restrictions on indirect control-flow transfers, hence also gadgets, those attacks often incorporate gadgets that would rarely be used in real attacks. E.g., Carlini and Wagner [7], Davi et al. [11], and Göktaş et al. [13] discovered that long gadgets with few side effects are suitable for breaking heuristics-based mitigations. Such gadgets should consist of at least 20 instructions, preserve as many registers as possible, have few side-effects, and easily fulfillable preconditions. Gadgets of this length are generally not useful in today's attacks, and therefore GaLity does not treat them any different than other gadgets. Another kind of gadget commonly used in these attacks is an LBR-flushing gadget [7,26]. Recent CPUs have special registers which can be configured to store the addresses of up to the 16 most recent taken indirect branches [1], which is a feature kBouncer [20] and ROPecker [9] use. When certain, critical APIs are invoked, the LBR is inspected and, depending on whether the control-flow appears legitimate or not, an exception is raised. LBR-flushing gadgets are gadgets that naturally contain many indirect branches, present in the regular control flow, e.g., functions that call lots of sub-functions. By using such a gadget, the LBR is filled with legitimate addresses and there is no trace of irregular control flow, i.e., ROP, in the LBR.

Q [27] allows exploit developers to write a target program in the high-level language QooL and automatically builds a ROP chain that uses only gadgets from a provided binary. However, Q handles gadgets with side effects, which we call preconditions in this paper, very conservatively and discards such gadgets, potentially removing a large number of useful gadgets. Homescu et al. [15] present a Turing-complete set of gadgets, using only gadgets that are 3 byte or shorter. They find that all required gadgets appear very frequently in regular Linux binaries.

Lastly, there are many tools that assist exploit developers by finding and sorting gadgets, but none of them take into account the quality of gadgets. Some of these tools also attempt to automatically build a ROP exploit for one predefined scenario e.g., ROPgadget[6], Mona.py[7], or ropper[8], however, from our experience they are not very sophisticated and often fail, even if the necessary gadgets are available.

6 Conclusion

Return-Oriented Programming forms the cornerstone of many contemporary exploitation techniques, yet its viability hinges on the availability of useful gadgets. Program transformations, including exploit mitigation techniques, often do not take into consideration their impact on the quality and number of gadgets that they introduce into the binary to which they are applied. Evaluations usually concentrate on the security gained, but not the security that might be lost due to a set of gadgets that is now favourable for an attacker than in the original, unmodified binary.

This work addresses this issue and allows researchers to consider this important aspect, by developing a set of metrics that, by combining concrete attack scenarios and measuring overall gadget quality, cover a wide range of possible exploit scenarios. We implemented the described metrics in a tool called GaLity, and applied it to binaries compiled with MPX, a new buffer overflow prevention technique introduced by Intel. Our results show that MPX provides gadgets of higher quality, and also a favourable set of gadgets in a concrete attack scenario.

Acknowledgements. We want to express our thanks to the anonymous reviewers for their valuable comments. In particular, we want to thank our shepherd, Mathias Payer, who helped us give this paper its final form. This work was supported by the BMBF within EC SPRIDE, by the Hessian LOEWE excellence initiative within CASED, by the DFG Collaborative Research Center CROSSING, by the DFG Priority Program 1496 Reliably Secure Software Systems, and the project INTERFLOW.

References

1. Intel 64, ia-32 architectures software developer's manual combined volumes,: 1, 2a, 2b, 2c, 3a, 3b, and 3c, June 2015
2. Abadi, M., Budiu, M., Erlingsson, Ú., Ligatti, J.: Control-flow integrity. In: ACM Conference on Computer and Communication Security (CCS), Alexandria, VA, pp. 340–353, November 2005
3. Aycock, J.: A brief history of just-in-time. ACM Comput. Surv. (CSUR) **35**(2), 97–113 (2003)
4. Batchelder, D., Blackbird, J., Felstead, D., Henry, P., Jones, J., Kulkarni, A., Lambert, J., Lauricella, M., Malcolmson, K., Miller, M., Ng, N., Pecelj, D., Rains, T., Sekhar, V., Stewart, H., Thompson, T., Weston, D., Zink, T.: Microsoft Security Intelligence Report, vol. 16 (2013)

---

[6] http://shell-storm.org/project/ROPgadget/.

[7] https://www.corelan.be/index.php/2011/07/14/mona-py-the-manual/.

[8] https://scoding.de/ropper/.

5. Bletsch, T., Jiang, X., Freeh, V.W., Liang, Z.: Jump-oriented programming: a new class of code-reuse attack. In: Proceedings of the 6th ACM Symposium on Information, Computer and Communications Security, ASIACCS 2011, pp. 30–40. ACM, New York (2011)

6. Carlini, N., Barresi, A., Payer, M., Wagner, D., Gross, T.R.: Control-flow bending: On the effectiveness of control-flow integrity. In: 24th USENIX Security Symposium (USENIX Security 15), pp. 161–176. USENIX Association, Washington, D.C., August 2015

7. Carlini, N., Wagner, D.: ROP is still dangerous: Breaking modern defenses. In: 23rd USENIX Security Symposium (USENIX Security 14), pp. 385–399. USENIX Association, San Diego, August 2014

8. Checkoway, S., Davi, L., Dmitrienko, A., Sadeghi, A.-R., Shacham, H., Winandy, M.: Return-oriented programming without returns. In: CCS 2010, pp. 559–572. ACM (2010)

9. Cheng, Y., Zhou, Z., Yu, M., Ding, X., Deng, R.H., ROPecker: A generic and practical approach for defending against ROP attacks (2014)

10. Conti, M., Crane, S., Davi, L., Franz, M., Larsen, P., Negro, M., Liebchen, C., Qunaibit, M., Sadeghi, A.-R.: Losing control: On the effectiveness of control-flow integrity under stack attacks. In: Proceedings of the 22nd ACM SIGSAC Conference on Computer and Communications Security, CCS 2015, pp. 952–963. ACM, New York (2015)

11. Davi, L., Sadeghi, A.-R., Lehmann, D., Monrose, F.: Stitching the gadgets: On the ineffectiveness of coarse-grained control-flow integrity protection. In: Proceedings of the 23rd USENIX Conference on Security, SEC 2014, pp. 401–416. USENIX Association, Berkeley (2014)

12. Ducklin, P.: Anatomy of an exploit - inside the CVE-2013-3893 internet explorer zero-day-part 2, October 2013

13. Göktaş, E., Athanasopoulos, E., Polychronakis, M., Bos, H., Portokalidis, G.: Size does matter: Why using gadget-chain length to prevent code-reuse attacks is hard. In: Proceedings of the 23rd USENIX Conference on Security Symposium, SEC 2014, pp. 417–432. USENIX Association, Berkeley (2014)

14. Göktas, E., Athanasopoulos, E., Bos, H., Portokalidis, G.: Out of control: Overcoming control-flow integrity. In: Proceedings of the IEEE Symposium on Security and Privacy, SP 2014, pp. 575–589. IEEE Computer Society, Washington, DC (2014)

15. Homescu, A., Stewart, M., Larsen, P., Brunthaler, S., Franz, M.: Microgadgets: Size does matter in turing-complete return-oriented programming. In: Presented as part of the 6th USENIX Workshop on Offensive Technologies. USENIX, Berkeley (2012)

16. Jurczyk, M.: One font vulnerability to rule them all #2: Adobe reader RCE exploitation, August 2015

17. Li, X., Szor, P.: Emerging stack pivoting exploits bypass common security, May 2013

18. Microsoft. Data execution prevention

19. One, A.: Smashing the stack for fun and profit. Phrack 7(49), 14–16 (1996)

20. Pappas, V., Polychronakis, M., Keromytis, A.D.: Transparent rop exploit mitigation using indirect branch tracing. In: Proceedings of the 22Nd USENIX Conference on Security, SEC 2013, pp. 447–462. USENIX, Berkeley (2013)

21. Payer, M., Barresi, A., Gross, T.R.: Fine-grained control-flow integrity through binary hardening. In: Almgren, M., Gulisano, V., Maggi, F. (eds.) DIMVA 2015. LNCS, vol. 9148, pp. 144–164. Springer, Heidelberg (2015)

22. Pi, P.: Unpatched flash player flaw, more POCs found in hacking team leak, July 2015
23. Ramakesavan, R., Zimmerman, D., Singaravelu, P.: Intel memory protection extensions (intel mpx) enabling guide, April 2015
24. Roemer, R., Buchanan, E., Shacham, H., Savage, S.: Return-oriented programming: Systems, languages, and applications. ACM Trans. Inf. Syst. Secur. **15**(1), 2:1–2:34 (2012)
25. Schuster, F., Tendyck, T., Liebchen, C., Davi, L., Sadeghi, A.-R., Holz, T.: Counterfeit object-oriented programming: On the difficulty of preventing code reuse attacks in C++ applications. In: 36th IEEE Symposium on Security and Privacy (Oakland), May 2015
26. Schuster, F., Tendyck, T., Pewny, J., Maaß, A., Steegmanns, M., Contag, M., Holz, T.: Evaluating the effectiveness of current anti-ROP defenses. In: Stavrou, A., Bos, H., Portokalidis, G. (eds.) RAID 2014. LNCS, vol. 8688, pp. 88–108. Springer, Heidelberg (2014)
27. Schwartz, E.J., Avgerinos, T., Brumley, D.: Q: Exploit hardening made easy. In: Proceedings of the 20th USENIX Conference on Security, SEC 2011, pp. 25–25. USENIX Association, Berkeley (2011)
28. Serna, F.J.: The info leak era of software exploitation (2012)
29. Shacham, H.: The geometry of innocent flesh on the bone: return-into-libc without function calls (on the x86). In: Proceedings of the 14th ACM Conference on Computer and Communications Security, CCS 2007, pp. 552–561. ACM, New York (2007)
30. Snow, K.Z., Monrose, F., Davi, L., Dmitrienko, A., Liebchen, C., Sadeghi, A.-R.: Just-in-time code reuse: On the effectiveness of fine-grained address space layout randomization. In: Proceedings of the IEEE Symposium on Security and Privacy, SP 2013, pp. 574–588 (2013)
31. Sotirov, A.: Heap feng shui in javascript (2007)
32. Tice, C., Roeder, T., Collingbourne, P., Checkoway, S., Erlingsson, Ú., Lozano, L., Pike, G.: Enforcing forward-edge control-flow integrity in GCC & LLVM. In: 23rd USENIX Security Symposium (USENIX Security 14), pp. 941–955. USENIX Association, San Diego, August 2014
33. van der Veen, V., Andriesse, D., Göktaş, E., Gras, B., Sambuc, L., Slowinska, A., Bos, H., Giuffrida, C.: Practical context-sensitive CFI. In: Proceedings of the 22nd ACM SIGSAC Conference on Computer and Communications Security, CCS 2015, pp. 927–940. ACM, New York (2015)
34. Zhang, M., Sekar, R.: Control flow integrity for COTS binaries. In: Proceedings of the 22Nd USENIX Conference on Security, SEC 2013, pp. 337–352. USENIX Association, Berkeley (2013)

Empirical Analysis and Modeling of Black-Box Mutational Fuzzing

Mingyi Zhao[(⊠)] and Peng Liu

College of Information Sciences and Technology,
Pennsylvania State University, State College, USA
{muz127,pliu}@ist.psu.edu

Abstract. Black-box mutational fuzzing is a simple yet effective method for finding software vulnerabilities. In this work, we collect and analyze fuzzing campaign data of 60,000 fuzzing runs, 4,000 crashes and 363 unique bugs, from multiple Linux programs using CERT Basic Fuzzing Framework. Motivated by the results of empirical analysis, we propose a stochastic model that captures the long-tail distribution of bug discovery probability and exploitability. This model sheds light on practical questions such as what is the expected number of bugs discovered in a fuzzing campaign within a given time, why improving software security is hard, and why different parties (e.g., software vendors, white hats, and black hats) are likely to find different vulnerabilities. We also discuss potential generalization of this model to other vulnerability discovery approaches, such as recently emerged bug bounty programs.

Keywords: Mutational fuzzing · Software vulnerability · Empirical analysis · Stochastic modeling

1 Introduction

Software vulnerability is the root cause of many security breaches. However, it has also been observed that discovering software vulnerability is hard. While software companies invest heavily to eliminate vulnerabilities, other parties including white hats [32] and black hats [27] are frequently able to find new vulnerabilities, even when endowed with less resources (e.g. computing power, manpower, information). In addition, investment in software security exhibits diminishing returns [15], which has also been discussed in the field of software reliability growth [6].

Understanding these phenomena has important theoretical and practical implications. Existing work on the economy of security usually involves models of software vulnerability discovery [16,18,29]. Such models can be improved by empirical analysis of real vulnerability discovery data. The effort of studying vulnerability discovery also help practitioners. For example, software companies can

---

The stamp on the top of this paper refers to an approval process conducted by the ESSoS artifact evaluation committee chaired by Alessandra Gorla and Jacques Klein.

© Springer International Publishing Switzerland 2016
J. Caballero et al. (Eds.): ESSoS 2016, LNCS 9639, pp. 173–189, 2016.
DOI: 10.1007/978-3-319-30806-7_11

make better decisions on the level of security investment and the extent of collaboration with outside security researchers (e.g., white hats) [32]. Cyber-insurance organizations might also be able to assess the security of customers more accurately [5].

In this work, we conduct an empirical analysis and propose models for black box mutational fuzzing. Introduced in early 1990s [21], black box mutational fuzzing remains an effective method for discovering real world vulnerabilities [9,17]. Its basic idea is very simple. Given a program and a set of diverse seed files, the fuzzing tool randomly mutates the files and use the program to process them. Once the program crashes, a triaging tool identifies the underlying bug and determines its properties such as exploitability. This simplicity makes black-box mutational fuzzing not only easy to use, but also easy to analyze and model.

We first apply black-box mutational fuzzing to multiple Linux programs and collect data from each fuzzing campaign, based on the CERT Basic Fuzzing Framework (BFF) [14] (Sect. 3). Our dataset contains 60,000 fuzzing runs, 4,000 crashes and 363 unique bugs. Then, we empirically analyze the data and discuss the long-tail distribution of discovery probability (Sect. 4), as well as the distribution of exploitability of bugs (Sect. 6.3). Motivated by the empirical analysis, we propose a stochastic model of black-box mutational fuzzing (Sect. 5.1). The model is derived from software reliability growth models [4,6,10,23]. However, one unique contribution of our model is that we assume the arrival rates of individual bugs follow a power law distribution, which is consistent with our data. Together with a simulation model (Sect. 5.2), we attempt to explain phenomena discussed at the beginning of this section. First, we provide a method to estimate the expected discovery outcome, which sheds light on the diminishing return of security investment (Sect. 6.1). Next, we explain why it is hard for software companies to eliminate the vulnerability stockpile of black hats (Sect. 6.2). Finally, we discuss several potential directions for future work, including the generalization of this model to other vulnerability discovery mechanisms (Sect. 7). All scripts and data are published online[1] for reproducible research.

2   Related Work

Black-box mutational fuzzing has been widely used in software vulnerability discovery since early 1990s, when Miller et al. surprisingly found out that random inputs crash 25 % – 33 % of Unix utilities [21]. Since then, black-box mutational fuzzing has been used to find numerous real world bugs and security vulnerabilities in various programs [11,14,22]. Compared with other forms of more sophisticated fuzzing approaches, such as generational fuzzing [20], whitebox fuzzing [12], taint-based fuzzing [30], etc., black-box mutational fuzzing is simpler and easier to use, but is usually inferior in terms of code coverage.

More recently, various methods were proposed to improve the effectiveness of black-box mutational fuzzing. Householder and Foote studied the problem of selecting seeds and fuzzing ratio using BFF [14]. The basic idea is to have more selection weight on parameter values that yield higher crash density in

---

[1] http://github.com/movingname/fuzzingModel.

the past. Woo et al. considered a similar scheduling problem in which the target program of each fuzzing run is also selected on the fly [31]. They designed several online scheduling algorithms and showed an average of 1.5× improvement over the one used in BFF. Rebert et al. designed and evaluated 6 seed selection algorithms [28]. The motivation of our work is different from but complementary to these work. Instead of optimizing the fuzzing process, we want to understand the fuzzing process better, based on empirical analysis and theoretical modeling.

The stochastic model built in this paper is derived from software reliability models [4,6,10,23], since fuzzing or vulnerability discovery in general is one particular approach to improve the software reliability. However, different from existing work, we assume that the arrival rates of the individual bugs follow power law distribution. This enables us to obtain similar observations on the difficulty of software reliability growth [4,6], or in other words, diminishing returns of software vulnerability discovery. In addition, we also uniquely use the power law-based stochastic model to explain why other parties (e.g. black hats) seem always being able to discovery unique vulnerabilities in Sect. 6.2. We further analyze the exploitability of bugs in Sect. 6.3, which is missing from software reliability growth models.

This paper assumes that the discovery probability of bugs follows power law distribution. Such long-tail distributions have been observed and discussed in various cyber security domains recently. Allodi showed that vulnerability exploitation in several common programs may follow power law distribution [2], which can be used for vulnerability prioritization. Maillart and Sornette showed that the sizes of personal identify theft follow power law distribution [19]. Finally, Edwards et al. found that data breach size is log-normally distributed while the daily frequency of breaches can be described by a negative binomial distribution [8]. These results can be used to predict data breaches and their associated cost.

3   BFF and Data Collection

Figure 1 shows the workflow of black-box mutational fuzzing. We have created several Python scripts for seed collection, code coverage, seed selection and data analysis. The fuzzing tool and triaging tool is from the CERT Basic Fuzzing Framework (BFF) [14]. BFF is shown to be effective in finding real vulnerabilities in various programs, and has been used in previous work on improving black-box mutational fuzzing [28,31] as well. Next, we outline the details of our experiment.

Step 1. Target Selection. By combining the lists of target programs used in the literature [7,14,28,31], we have collected 18 programs that handle various types of video, audio, graphical, and document inputs. Table 1 list all 9 programs in which BFF has successfully found bugs. We have also tried to apply fuzzing to the following programs: a2mp3, eog, gifsicle, mplayer, mp3blaster, mpg123, moc, Outside In Viewer 8.5.2, and pdf2svg. However, for any of these programs, BFF triggers less than 3 or even 0 crashes. We therefore exclude them from the following analysis.

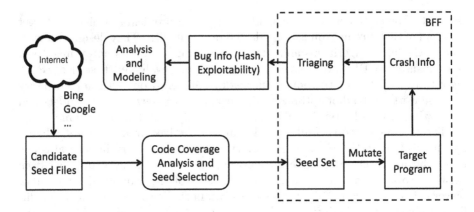

Fig. 1. The fuzzing experiment workflow.

Step 2. Seed Collection and Selection. We have collected thousands of candidate seeds files, including pdf documents, mp3 files, videos and images, from search engines like Bing and Google. The # cand columns of Table 1 shows the number of candidate seed files for each program. We then write a script to collect the basic blocks (bbls) covered by each seed using the Intel Pin framework. In general, the higher coverage of seeds, the more vulnerabilities will be found in fuzzing [28]. Next, we select 50 seed files to form the final seed set for each program, using a simple greedy algorithm that maximizes the coverage in each iteration. Table 1 shows that the final seed sets still achieve similar levels of coverage (% bbls column).

Step 3. Fuzzing. We use BFF as the fuzzing tool and use its default *fuzzing configuration*. The main configuration parameter is the seed used in each fuzzing run, and the *fuzzing ratio*, which indicates how many bits in the seed will be flipped. We use the default probability-based parameter selection method implemented in BFF [14]. The outcome of a fuzzing run is either a crash or nothing, while the result of a fuzzing campaign is a sequence of crashes caused by software bugs in the program. Since multiple crashes could correspond to the same bug, we need a triaging step to map a crash to the corresponding bug.

Step 4. Triaging. Once a crash is encountered, BFF will run the triaging step, which calculates the hash for the underlying bug based on the stack trace[2], minimizes the input that triggers the crash, and determines whether the bug is exploitable or not. Similar to other triaging tools such as the !exploitable for Windows OS and CrashWrangler for Mac OS X, the CERT Triage Tools in BFF assigns one of the following exploitability levels to each crash: unknown, not_exploitable, probably_not_exploitable, probably_exploitable and exploitable.

---

[2] The method used to generate the hash is an extension of the fuzzy stack hash method proposed in the literature [24].

Step 5. Data Analysis. At the end, we know the, seed file, configuration, and outcome of each fuzzing run, as well as the hash and exploitability of each bug discovered. We then analyze the data and show statistics of the results in Table 1. We present our main analysis results in the next section.

Table 1. Seed selection and fuzzing statistics of selected programs. # cands is the number of candidate seed files we collected from the Internet. # bbls is the number of unique basic blocks recored when parsing the candidate seed files. % bbls is the percentage of basic blocks covered by the final seed set.

Program	Seed Selection			Fuzzing			
	# cand	# bbls	% bbls	# runs	# crashes	# bugs	max_freq
xpdf 3.02-2	2,161	188,023	93.1%	4,303	185	37	73
mupdf[a]	-	-	-	9,900	201	25	61
convert 5.2.0[b]	-	-	-	79,636	32,161	134	3,197
ffmpeg 0.8	787	121,875	86.7%	16,055	3,872	96	863
autotrace 0.31.1	149[c]	-	100%	29,729	2,548	23	593
jpegtran 1.2.0	320	6,837	99.4%	303,898	116	33	31
gif2png 2.5.4-2	1,084	12,772	99.8%	136,768	2,305	7	34
feh 2.2	1,332	56,266	94.8%	5,209	159	5	51
mp3gain 1.5.2	214	7,224	99.9%	1,369	1,451[d]	7	861

[a] mupdf and xpdf share same seeds.
[b] We use the seeds provided in the default BFF vm image for convert.
[c] Since the size is small, we use all of the seeds in fuzzing.
[d] Here, # crashes is actually larger than # runs. This is caused by a stack corruption bug that confuses triaging process to correlate the same crash into different bugs [28].

4   The Long-Tail Distribution of Bugs

The major goal of fuzzing and any bug discovery effort is to find as many bugs as possible. Moreover, it has been observed that the easiness of discovering different bugs is different. In black-box mutational fuzzing, we can quantify easiness of discovering bug i as its *discovery probability* (λ_i):

$$\lambda_i = \frac{c_i}{t} \tag{1}$$

where c_i is the number of crashes caused by bug i, and t is the number of fuzzing runs in the fuzzing campaign. Then the question is, what is the distribution for bug discovery probability?

In Fig. 2, we plot the empirical probability distribution of bugs for all 6 programs with more than 20 bugs discovered. We see that these distributions all have *the long-tail shape*; that is, a few bugs trigger a large number of crashes,

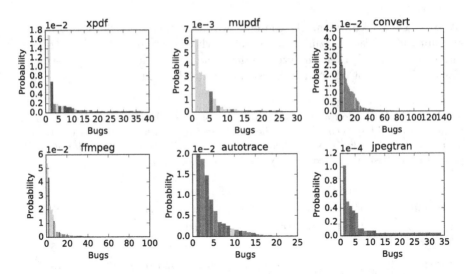

Fig. 2. Probability distributions of bugs triggered in fuzzing campaigns. The color of a bar represents the exploitability of the bug. The meaning of colors are: exploitable (red), probably_exploitable (blue), probably_not_exploitable (yellow), not_exploitable (green), unknown (grey) (Color figure online).

while most bugs have only triggered crashes a few times. Many such distributions [26], including vulnerability exploitation [2], have been proposed to follow the power law distribution. We thus propose the following hypothesis:

Hypothesis 1. *The discovery probability of bugs in a program follows a power law distribution.*

More specifically, we assume the following discrete power law distribution [26]:

$$P(\text{discover bug } i \text{ in a fuzzing run}) = \lambda_i = \frac{i^{-\alpha}}{\zeta(\alpha)} \qquad (2)$$

where α is the scaling factor of the power law distribution, $\zeta(\alpha)$ is the Riemann ζ-function as the normalizer. As we will show in Sect. 6, a smaller α leads to more bugs discovered in the same number of fuzzing runs. i is the *rank id* of the bug among all bugs sorted by their discovery probability inside the program. A bug with a larger rank id (lower rank) has lower probability to be discovered, as Eq. 2 tells. To complete the probability distribution, we also use λ_1 to represent the probability of no crash. We can think about no crash as a special bug, and it has the highest probability in these 6 fuzzing campaigns.

We next need to estimate the scaling factor α of a power law distribution from the empirical distribution. The most common approach is to use Maximum Likelihood Estimators (MLEs) [3,26]. However, we could not apply these estimators because *we do not know the true rank id of a bug discovered in fuzzing.*

We only know a bug's rank among all discovered bugs. For example, the 20th bug in the empirical data could have the true rank id of 100.

We propose a simulation method to estimate α. We could think a fuzzing campaign with t runs as generating t values form the corresponding power law distribution. We then choose the α that minimizes the difference between the number of unique bugs discovered in the experiment and the number of unique values generated from the distribution. Table 2 shows the estimates of α.

Table 2. Estimates of α.

Program	α
xpdf	2.39
mupdf	2.88
convert	2.38
ffmpeg	2.21
autotrace	3.25
jpegtran	3.53

Because we do not know the true rank id of bugs discovered, it is also difficult to apply goodness-of-fit tests, either through bootstrapping or by comparing with alternative distributions [2,3]. In this work, we will test the estimates of α by comparing the predicted number of bugs discovered with the actual number of bugs discovered in Sect. 6.1. More rigorous methods of estimating α and testing the goodness-of-fit are left as future work. In the following sections, we will show that this power law hypothesis enables us to answer some interesting questions related to vulnerability discovery and software security.

5   Modeling a Fuzzing Campaign

We then build models for a fuzzing campaign. First, we propose a stochastic model based on existing software reliability literature [4,6,10,23], in Sect. 5.1. Although expressive, this stochastic model has two assumptions that might not be realistic. We remove one assumption by proposing a simulation model in Sect. 5.2 (Table 3).

5.1   A Stochastic Model

Since each fuzzing run is independent from other runs, and the outcome of a fuzzing run is either 1 (crashed) or 0 (not crashed), it is natural to consider the fuzzing process as a Poisson Process $\{N(t), t \geq 0\}$, where $N(t)$ is the number of

Table 3. Notations.

Variable	Explanation
$N(t)$	The Poisson process for number of crashes in a fuzzing campaign
$N_i(t)$	The Poisson process corresponding to bug i
λ_i	The rate for $N_i(t)$ and the discovery probability of bug i
n	Total number of bugs in the program
$N'(t)$	The non-homogeneous Poisson process for number of unique bugs
$\mathcal{D}(t)$	The set of discovered bugs by time t.
$D(t)$	Number of discovered bugs by time t. $D(t) = |\mathcal{D}(t)|$.
$\mathcal{L}(t)$	The set of remaining bugs by time t

crashes seen till time t. Furthermore, since crashes are caused by different bugs, we can expressed $N(t)$ as:

$$N(t) = \sum_{i=2}^{\infty} N_i(t) \tag{3}$$

Here, i is the rank id of a bug and t is the number of fuzzing runs. $\{N_i(t), t \geq 0\}$ is the corresponding Poisson process for the i-th bug, and $N_i(t)$ is the number of crashes for the i-th bug we have seen till time t. We can see that the discovery probability of the i-th bug we have discussed in the previous subsection is actually the rate λ_i of the Poisson process $N_i(t)$. A larger λ_i means that bug i causes crashes more frequently.

In a fuzzing campaign, we are mostly interested in the first crash of a bug. This is equivalent to the assumption in the software reliability models that a bug is found and instantly fixed, while the fix does not influence the discovery of other remaining bugs [4]. We define $\mathcal{D}(t)$ as the set of bugs that have already been found by time t, and $\mathcal{L}(t)$ as the set of remaining bugs. So we have:

$$\lambda'(t) = \sum_{i \in \mathcal{L}(t)}^{\infty} \lambda_i = \sum_{i=2}^{\infty} \lambda_i - \sum_{i \in \mathcal{D}(t)}^{\infty} \lambda_i \tag{4}$$

Therefore, we obtain a new non-homogeneous Poission process, $N'(t)$, for the discovery of unique bugs. $\lambda'(t)$ is the arrival rate of new bugs, and the expected time to discover the next bug is $1/\lambda'(t)$.

We currently do not know how to solve Eq. 4 analytically. Thus when doing calculation, we replace ∞ with n, in order to obtain an approximate result. Intuitively, we assume there are n bugs in total inside the program. By choosing a larger n, we can further approximate the true result. In our following analysis, we set $n = 1000$. The probability that $i > 1000$ is only 1.3e-4 for ffmpeg ($\alpha = 2.21$), and 9.03e-9 for jpegtran ($\alpha = 3.53$).

In addition, this stochastic model relies on the following two assumptions:

Assumption 1. *In one fuzzing run, multiple bugs can be triggered.*

However, in BFF, each fuzzing run stops at the first crash, which is then triaged to one bug. Thus, with Assumption 1, the model will slightly overestimate the number of bugs discovered, as we will see in Sect. 6.1. But we expected that this effect is small because most bugs have low discovery probability (Fig. 2), and the chance that multiple bugs are triggered in the same fuzzing run is even lower.

Assumption 2. *The discovery probability distribution is the same for all fuzzing runs in a fuzzing campaign.*

This assumption also oversimplifies the reality. Since the fuzzing seeds and fuzzing ratio are different among fuzzing runs, each fuzzing run will explore a unique input space and be able to trigger a different subset of all latent bugs. We will discuss this more in Sect. 7.

Improving this stochastic model by relaxing these two assumptions is challenging, which is left as a future work. In the next sub section, we propose a simulation model that removes Assumption 1.

5.2 A Simulation Model

Similar to the discussion in Sect. 4, we could think a fuzzing campaign with t runs as generating t values form the corresponding power law distribution. Algorithm 1 returns a simulated bug discovery sequence as well as unique bugs discovered, given α and t as the inputs. Step 1 and 2 can be implemented using existing software package [3]. In step 5, we add the condition $id > n$ because we will compare the simulation model with the stochastic model.

Algorithm 1. Simulate a fuzzing campaign.

---

   input : α of the bug distribution, and t, the number of fuzzing runs
   output: Simulated bug discovery sequence and unique bugs discovered
1 dist = powerlaw(α, xmin=1, discrete=True) ;
2 seq = dist.gen_random(t);
3 bugs = {};
4 **foreach** $id \in bugs$ **do**
5   **if** $id == 1$ *or* $id > n$ **then**
6 | continue;
7   **if** $id \notin bugs$ **then**
8 | bugs.add(id);
9 **return** seq, bugs;

In this simulation model, we remove Assumption 1 since each fuzzing run only yields at most one bug discovery. In Sect. 6.1, we will compare the predicted numbers of bugs discovered by these two models, and the actual number of bugs discovered.

6   Analysis Results

We present 3 analysis results in this section. We first use the models presented in the last section to calculate the expected number of bugs discovered, and discuss the diminishing returns in software security. We then examine the order of bug discovery to explain why different parties are likely to find different bugs. Finally, we empirically study the exploitability of bugs and discuss its implications.

6.1   Expected Number of Bugs Discovered

The first question is, what is the expected number of unique bugs find by time t? Under the stochastic model proposed in Sect. 5.1, we know that the time of the first occurrence of each bug follows the exponential distribution with parameter λ_i. Therefore, the probability of bug i undiscovered by time t is $e^{-\lambda_i t}$, and the expected number of undiscovered bug at time t is $\sum_{i=2}^{n} e^{-\lambda_i t}$. We then know that the number of expected bugs discovered by time t is:

$$E[D(t)] = n - \sum_{i=2}^{n} e^{-\lambda_i t} = n - \sum_{i=2}^{n} e^{-\frac{i-\alpha}{\zeta(\alpha)} t} \tag{5}$$

We also use the simulation model proposed in Sect. 5.2 to obtain $E[D(t)]$. We repeat the simulation 10 times and take the average number of bugs discovered by time t as $E[D(t)]$. As we have discussed in Sect. 5.1, we set $n = 1000$ for both models.

In Fig. 3, we show the plots of expected bugs discovered based on the Poisson process and the simulation, and the real trajectory, of 6 fuzzing campaigns. We see that the predicted curves from both models are close to the real curve, except for autotrace. We suspect that the large prediction error for autotrace is due to a poor fit of power law to its empirical distribution. We plan to further investigate this in the future. In addition, the curve of the stochastic model is generally above the other two. This can be partly explained by Assumption 1, as we have discussed in Sect. 5.1. In general, the simulation model gives more accurate prediction for the 6 fuzzing campaigns than the stochastic model.

The concave shape of all curves show the *diminishing returns*: as the fuzzing campaign enters the long tail, the rate of discovery ($\lambda'(t)$) decreases, and the number of bugs discovered in the same amount of time reduces. This diminishing of return is consistent with our experience of fuzzing and software reliability growth [10,23]. A software company can use the two models to decide how long the fuzzing campaign shall run. First, the company need to run a fuzzing campaign for a limited amount of time, in order to estimate α. Then, the company needs to define the reliability and security utility gain of finding a bug, and the fuzzing cost, which might include computing resource consumption, delayed product release, etc. Next, the company can generate the accumulated utility curve and the accumulated cost curve based on the curve of expected bug discovery ($E[D(t)]$) proposed in this section. At the point when the utility of fuzzing is below the cost, the fuzzing campaign should be terminated.

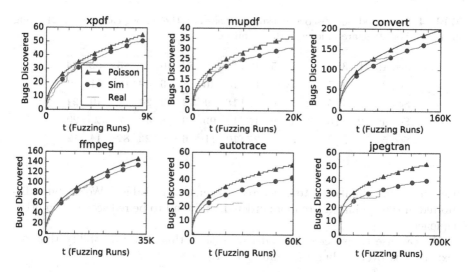

Fig. 3. Plots of expected number of bugs discovered ($E[D(t)]$) and actual number of bugs discovered (D) overtime. We have doubled the number of fuzzing runs in order to observe how two models predict.

6.2 The Order of Bug Discovery

The diminishing of return discussed in Sect. 6.1 might appear to be a good thing for security. If there is a strong order of bug discovery, then bugs with larger discovery probability will almost always be eliminated first. Thus, as long as the software company invests more resources than other parties, including black hats [27] and white hats [32], in vulnerability discovery, these other parties will not likely to find new vulnerabilities.

However, in reality, we see that many vulnerabilities of famous software have been discovered by outside parties, many of whom are just individuals [1,13,25,32]. There are multiple reasons to explain this. In this work, we propose one explanation based on the power law hypothesis. The basic idea is that the order of bug discovery is weak in the long-tail part of the distribution.

To further explain this, we first define *the order of bug discovery*. At the end of a fuzzing campaign, the expected sequence of rank id (S) of discovered k bugs is $2, 3 \ldots k$, because a higher ranked bug has higher discovery probability, and thus is expected to be discovered earlier. However, due to the randomness, the actual id sequence ($\hat{S}$) would be different from the expected sequence S. We can calculate the edit distance $D(S, \hat{S})$ between these two sequences, and define the order of bug discovery as:

$$order(\hat{S}) = k - D(S, \hat{S}) \tag{6}$$

Intuitively, the bug discovery is strongly/weakly ordered if the distance between S and $\hat{S}$ is small/large. However, since we do not know the true rank id of bugs discovered, we cannot calculate the order of empirical sequences directly.

Table 4. Simulated bug discovery sequence based on the **ffmpeg** case ($\alpha = 2.21$). Bug ids in the bold font are unique to that sequence.

Seq 1:	**19**	3	2	9	4	5	12	14	6	**84**	10	7	**85**	**95**	**24**
Seq 2:	2	3	7	4	5	**17**	10	13	**40**	8	6	**49**	12	11	9
Seq 3:	2	4	5	**28**	3	6	7	**18**	9	12	13	20	11	10	**21**
Seq 4:	2	5	6	3	4	9	**15**	12	**99**	10	8	**46**	7	**225**	20
Seq 5:	3	2	4	7	8	5	**27**	10	11	6	9	**23**	**82**	14	12

Instead, we run simulation to generated 5 sequences in Table 4. We see bugs discovered in the beginning are more ordered, and tend to be rediscovered in other sequences.

We can use the stochastic model to explain this. The probability that the next new discovery is bug i (assuming $i \in \mathcal{L}(t)$) is:

$$P(\text{bug } i \text{ is the next one after time } t) = \frac{\lambda_i}{\left(\sum_{j \in \mathcal{L}(t)} \lambda_j\right)} \propto \lambda_i \propto i^{-\alpha} \qquad (7)$$

For bug i and bug $i + 1$ (assuming $i + 1 \in \mathcal{L}(t)$), we have:

$$P_i - P_{i+1} \propto i^{-\alpha} - (i+1)^{-\alpha} \qquad (8)$$

which decreases to 0 as $i \to \infty$. This means that when i is small (the fuzzing process is in the "head part" of the distribution), a bug with higher discovery probability is much more likely to be discovered first, and the fuzzing process has a stronger order. However, as i increases and the fuzzing process enters the long-tail, which vulnerability will come next is harder to predict. In addition, a smaller α will make the fuzzing outcome less ordered, while a larger α makes the process more ordered.

To understand its implication, we consider a "fuzzing competition" between a software company and a black hat. Both sides run fuzzing and try to find as many bugs as possible. We assume that the software company has a resource advantage A over the black hat. That is, while the black hat can conduct a fuzzing campaign with t runs, the company can do At runs, by having a larger fuzzing server farm. We want to know how many unique bugs can the black hat find.

We simulate 10,000 fuzzing runs for the black hat, and simulate $10,000 \times A$ runs for the software company. The two curves in Fig. 4 show the number of unique bugs found by the black hat for two programs. We observe that although in the beginning, the software company can quickly reduce the bug pool of the black hat by investing more resources, the return of investment quickly diminishes as A further grows. When the software company has 30 times more fuzzing resources, the black hat is still able to find 2 unique bugs for ffmpeg and 1 unique bug for xpdf on average. Intuitively, it means that when the fuzzing enters the long-tail, the outcome is more random, so the company is less capable of interfering the black hat's outcome. This partly explains why in the reality, outside

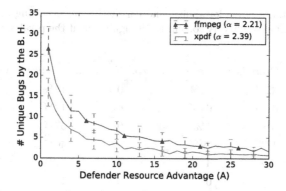

Fig. 4. Simulated number of unique bugs discovered in two programs by the black hat under different resource advantage A of the software company.

parties such as black hats and white hats are able to find security holes, despite software companies have already spent significant effort in software security. From Fig. 4, we can also see that when α is smaller, more unique bugs can be found by the black hat.

In summary, the power law hypothesis favors attackers, since they are able to find vulnerabilities even if the defender has much more resources. In addition, there is an asymmetry between attackers and defenders: the attackers only need to find a few exploitable bugs to succeed, while the defenders have to patch all holes. On the other hand, this result also encourages software companies to collaborate with outside benign white hats, through vulnerability disclosure and bug bounty programs [32]. We will discuss this more in Sect. 7.2. But before that, we need to ask one more question: are these unique bugs discovered by the black hat exploitable?

6.3   Exploitability

Table 5 shows the distribution of bug exploitability in the data. We see that a significant portion of the bugs are either exploitable or probably exploitable.

Then, we further ask the question: is there any correlation between discovery probability and exportability? If there is a positive correlation, then it means that harder to be discovered bugs are harder to be exploited, which favors the software company side. To answer this question, we calculate the Pearson correlation between the logarithm of discovery probability, and the exploitability which is mapped to a 1–4 scale, with 1 meaning not exploitable and 4 means exploitable. We exclude bugs of unknown exploitability. The result is shown in Table 5. We see that although 5 out of 6 programs have a weak negative correlation (i.e., harder to be discovered bugs are easier to be exploited), there is only one that is statistically significant (xpdf). We thus propose the following hypothesis:

Hypothesis 2. *Bug discovery probability and exploitability do not have a strong correlation.*

Table 5. Percentages of exploitability, and correlation between bug discovery probability (log) and exploitability. A correlation is significant if the p-value is less than 0.1.

Program	Exp.	Prob. Exp.	Prob. Not Exp.	Not Exp.	Unknown	Corr.	p-value
xpdf	27%	32%	22%	0	19%	−0.35	**0.06**
mupdf	24%	0	64%	0	12%	−0.21	0.35
convert	33%	3%	7%	0	57%	−0.05	0.68
ffmpeg	8%	17%	29%	0	46%	−0.16	0.25
autotrace	39%	4%	4%	0	52%	0.02	0.95
jpegtran	79%	6%	0	0	15%	−0.03	0.87

Hypothesis 2 has several implications. First, it indicates that the next bug to be found could be exploitable, no matter how many runs have been conducted before. This gives an additional advantage for black hats, who not only are likely to find unique bugs, but are also able to find exploitable ones. Second, by assuming the independence between discovery probability and exploitability, one can predict the exploitability of the next bug based on the empirical exploitability distribution in Table 5. For example, in the case of xpdf and $A = 30$ in Fig. 4, we can predict that the 1 unique bug discovered by the black hat has roughly 25% probability of being exploitable. By combining the vulnerability discovery models and the exploitability distribution, the software company can thus better forecast potential attacks and allocate defense resources accordingly.

7   Discussion and Future Work

7.1   Apply Our Analysis to Larger Datasets

Although our dataset includes most of the programs studied in previous work [7,14,28,31], it is still not enough to fully test the hypotheses we proposed. Therefore, an important future work is to increase the scope of analysis to other programs, other platforms (e.g., Microsoft Windows and Mac OS), and other fuzzing frameworks [20]. It would also be helpful to run the fuzzing campaign for much longer time.

Another important direction is to apply our analysis to different fuzzing configurations, which include the selection of fuzzing ratio, seeds, etc. It is possible that the same bug's discovery probability might be significantly different in different configurations. This diversity gives an additional explanation to why other parties are likely to find unique bugs, in addition to our discussion in Sect. 6.2. That is, different parties tend to have different configurations, and thus the discovery probability distribution is distinct to each of them. However, although the discovery probability of a bug might be different under different fuzzing configurations, we hypothesize that the discovery probability distribution will still be a power law distribution:

Hypothesis 3. *The bug discovery probability under different fuzzing configurations follow power law distributions.*

7.2 Generalization to Other Vulnerability Discovery Approaches

We choose to study black-box mutational fuzzing first because it is probably the simplest vulnerability discovery method. However, black-box mutational fuzzing is just one method in the vulnerability discovery toolbox. Other methods include code review, static analysis, symbolic execution, dynamic analysis, etc. We hypothesize that these approaches might resemble fuzzing, and thus the bug discovery "easiness", a generalization of the discovery probability, could also follow the power law distribution. Collecting empirical data from these approaches and applying similar analysis would be an interesting future work.

Some other vulnerability discovery paradigms also share similarities with black box mutational fuzzing. For example, many companies today collaborate with a large number of outside security researchers (or white hats) through vulnerability disclosure and bug bounty programs [32]. Actually, our discussion in Sect. 6.2 provides one explanation of why such collaboration is necessary. In addition, these white hats, with diverse background and skill levels, will often test different parts of the system, or using various testing payload. This is similar to the seed mutation in a black-box mutational fuzzing, although the distribution of inputs might be more complex than random bit flipping. Therefore, we could possibly generalize the proposed models to understand and analyze data from these bug bounty programs.

8 Conclusion

Understanding the process of vulnerability discovery and why software security is hard has important practical implications. In this work, we have collected empirical data of black-box mutational fuzzing. We show that the fuzzing process can be modeled as a non-homogeneous Poisson process with the rates of individual bugs following a power law distribution. We then show how to calculate the expected outcome of a fuzzing campaign. We further show that once the vulnerability discovery enters the long-tail, there will be significant diminishing returns, and less order in the bug arrival. These effects pose challenge for the software companies that try to eliminate vulnerabilities before the black hats, and call for collaboration with white hats. Finally, we show that the model can potentially be extended to other vulnerability discovery mechanisms, such as bug bounty programs, that have diversity and randomness.

Acknowledgment. We sincerely thank our shepherd and the anonymous reviewers for their valuable comments and suggestions on early versions of this paper. This work was supported by ARO W911NF-13-1-0421 (MURI), NSF CCF-1320605, NSF CNS-1422594, NSF CNS-1505664, ARO W911NF-15-1-0576, and NIETP CAE Cybersecurity Grant.

References

1. Algarni, A., Malaiya, Y.: Software vulnerability markets: discoverers and buyers. Int. J. Comput. Appl. Technol. Inf. Sci. Eng. **8**(3), 71–81 (2014)
2. Allodi, L.: The heavy tails of vulnerability exploitation. In: Piessens, F., Caballero, J., Bielova, N. (eds.) ESSoS 2015. LNCS, vol. 8978, pp. 133–148. Springer, Heidelberg (2015)
3. Alstott, J., Bullmore, E., Plenz, D.: Powerlaw: a python package for analysis of heavy-tailed distributions. PLoS ONE **9**, e85777 (2014)
4. Bishop, P., Bloomfield, R.: A conservative theory for long-term reliability-growth prediction [of software]. IEEE Trans. Reliab. **45**(4), 550–560 (1996)
5. Böhme, R., Schwartz, G.: Modeling cyber-insurance: towards a unifying framework. In: The Workshop on the Economics of Information Security (WEIS) (2010)
6. Brady, R.M., Anderson, R., Ball, R.C.: Murphy's law, the fitness of evolving species, and the limits of software reliability. Number 471. University of Cambridge, Computer Laboratory (1999)
7. Cha, S.K., Woo, M., Brumley, D.: Program-adaptive mutational fuzzing. In: 36th IEEE Symposium on Security and Privacy (2015)
8. Edwards, B., Hofmeyr, S., Forrest, S.: Hype, heavy tails: a closer look at data breaches. In: The Workshop on the Economics of Information Security (WEIS) (2015)
9. Evans, C., Moore, M., Ormandy, T.: Fuzzing at scale. Google Online Security Blog
10. Fenton, N., Bieman, J.: Software metrics: a rigorous and practical approach. CRC Press, Boca Raton (2014)
11. Forrester, J.E., Miller, B.P.: An empirical study of the robustness of windows nt applications using random testing. In: Proceedings of the 4th USENIX Windows System Symposium, Seattle, pp. 59–68 (2000)
12. Godefroid, P., Levin, M.Y., Molnar, D.A., et al.: Automated whitebox fuzz testing. In: The Network and Distributed System Security Symposium, vol. 8, pp. 151–166 (2008)
13. Hafiz, M., Fang, M.: Game of detections: how are security vulnerabilities discovered in the wild? Empirical Software Engineering, pp. 1–40 (2015)
14. Householder, A.D., Foote, J.M.: Probability-based parameter selection for black-box fuzz testing. In: CERT (2012)
15. W. Jackson. Has secure software development reached its limits? GCN
16. Johnson, B., Laszka, A., Grossklags, J.: Games of timing for security in dynamic environments. In: Khouzani, M.H.R., et al. (eds.) GameSec 2015. LNCS, vol. 9406, pp. 57–73. Springer, Heidelberg (2015). doi:10.1007/978-3-319-25594-1_4
17. Jurczyk, M., Coldwind, G.: Ffmpeg and a thousand fixes. Google Online Security Blog
18. Laszka, A., Grossklags, J.: Should cyber-insurance providers invest in software security? In: Pernul, G., Ryan, P.Y.A., Weippl, E. (eds.) ESORICS 2015, Part I. LNCS, vol. 9326, pp. 483–502. Springer, Heidelberg (2015)
19. Maillart, T., Sornette, D.: Heavy-tailed distribution of cyber-risks. Eur. Phys. J. B **75**(3), 357–364 (2010)
20. McNally, R., Yiu, K., Grove, D., Gerhardy, D.: Fuzzing: the state of the art. Technical report, DTIC Document (2012)
21. Miller, B.P., Fredriksen, L., So, B.: An empirical study of the reliability of unix utilities. Commun. ACM **33**(12), 32–44 (1990)
22. Miller, C.: Babysitting an army of monkeys. In: CanSecWest (2010)

23. Miller, D.R.: Exponential order statistic models of software reliability growth. IEEE Trans. Softw. Eng. **1**, 12–24 (1986)
24. Molnar, D., Li, X.C., Wagner, D.: Dynamic test generation to find integer bugs in x86 binary linux programs. In: USENIX Security Symposium, vol. 9 (2009)
25. Naraine, R.: Teenager hacks google chrome with three 0day vulnerabilities. ZDNet
26. Newman, M.: Power laws, Pareto distributions and Zipf's law. Contemp. Phys. **46**(5), 323–351 (2005)
27. Radianti, J.: Eliciting information on the vulnerability black market from interviews. In: Proceedings of the SECURWARE, pp. 154–159 (2010)
28. Rebert, A., Cha, S.K., Avgerinos, T., Foote, J., Warren, D., Grieco, G., Brumley, D.: Optimizing seed selection for fuzzing. In: Proceedings of the USENIX Security Symposium, pp. 861–875 (2014)
29. Rue, R., Pfleeger, S.L.: Making the best use of cybersecurity economic models. IEEE Secur. Priv. **4**, 52–60 (2009)
30. Wang, T., Wei, T., Gu, G., Zou, W.: Taintscope: a checksum-aware directed fuzzing tool for automatic software vulnerability detection. In: IEEE Symposium on Security and Privacy (2010)
31. Woo, M., Cha, S.K., Gottlieb, S., Brumley, D.: Scheduling black-box mutational fuzzing. In: ACM Conference on Computer and Communications Security (2013)
32. Zhao, M., Grossklags, J., Liu, P.: An empirical study of web vulnerability discovery ecosystems. In: ACM Conference on Computer and Communications Security (2015)

On the Security Cost of Using a Free and Open Source Component in a Proprietary Product

Stanislav Dashevskyi[1,3], Achim D. Brucker[2,3(✉)], and Fabio Massacci[1]

[1] University of Trento, Trento, Italy
{stanislav.dashevskyi,fabio.massacci}@unitn.it
[2] Department of Computer Science, The University of Sheffield, Sheffield, UK
[3] SAP SE, Walldorf, Germany
a.brucker@sheffield.ac.uk

Abstract. The work presented in this paper is motivated by the need to estimate the security effort of consuming Free and Open Source Software (FOSS) components within a proprietary software supply chain of a large European software vendor. To this extent we have identified three different cost models: centralized (the company checks each component and propagates changes to the different product groups), distributed (each product group is in charge of evaluating and fixing its consumed FOSS components), and hybrid (only the least used components are checked individually by each development team). We investigated publicly available factors (e. g., development activity such as commits, code size, or fraction of code size in different programming languages) to identify which one has the major impact on the security effort of using a FOSS component in a larger software product.

Keywords: Free and open source software usage · Free and open source software vulnerabilities · Security maintenance costs

1 Introduction

Whether Free and Open Source Software (FOSS) is more or less secure than proprietary software is a heavily debated question [8,9,22].

We argue that, at least from the view of a software vendor who is consuming FOSS, this question is not the right question to ask. First, there may be just no alternative to use FOSS components in a software supply chain, because FOSS components are the de-facto standard (e. g., Hadoop for big data). Second, FOSS may offer functionalities that are very expensive to re-implement and, thus, using FOSS is the most economical choice.

A more interesting question to ask is which factors are likely to impact the "security effort" of a selected FOSS component.

A.D. Brucker—Parts of this research were done while the author was a Security Testing Strategist and Research Expert at SAP SE in Germany.

J. Caballero et al. (Eds.): ESSoS 2016, LNCS 9639, pp. 190–206, 2016.
DOI: 10.1007/978-3-319-30806-7_12

As the security of a software offering depends on all components, FOSS should, security-wise, be treated as one's own code. Therefore, software companies that wish to integrate FOSS into their products must tackle two challenges: the selection of a FOSS product and its maintenance. In order to meet the FOSS selection challenge, large business software vendors perform a thorough security assessment of FOSS components that are about to be integrated in their products by running static code analysis tools to verify the combined code base of a proprietary application and a FOSS component in question, and by performing a thorough audit of the results. The security maintenance problem is not easier: when a new security issue in a FOSS component becomes publicly known, a business software vendor has to verify whether that issue affects customers who consume software solutions where that particular FOSS component was shipped by the vendor. In ERP systems and industrial control systems this event may occur years after deployment of the selected FOSS product.

Addressing either problem requires expertise about both the FOSS component and software security. This combination is usually hard to find and resources must be allocated to fix the problem in a potentially unsupported product. It is therefore important to understand which characteristics of a FOSS component (number of contributors, popularity, lines of code or choice of programming language, etc.) are likely to be a source of "troubles". The number of vulnerabilities of a FOSS product is only a part of a trouble: a component may be used by hundreds of products.

Motivated by the need to estimate the efforts and risks of consuming FOSS components for proprietary software products of a large European software vendor – SAP SE, we investigate the factors impacting three different cost models:

1. The *centralized model*, where vulnerabilities of a FOSS component are fixed centrally and then pushed to all consuming products (and therefore costs scale sub-linearly in the number of products)
2. The *distributed model*, where each development team fixes its own component and effort scales linearly with usage
3. The *hybrid model*, where only the least used FOSS components are selected and maintained by individual development team

In the rest of the paper we describe the FOSS consumption in SAP (Sect. 2), introduce our research question and the three security effort models (Sect. 3), and discuss related works (Sect. 4). Then we present the data sources used for analyzing the impact factors (Sect. 5), describe each variable in detail and discuss the expected relationships between them (Sect. 6). Next we (Sect. 7) discuss the statistical analysis of the data. Finally we conclude and outline future work (Sect. 8).

2   FOSS Consumption at SAP

SAP's product portfolio ranges from small (mobile) applications to large scale ERP solutions that are offered to customers on-premise as well as cloud solutions.

Table 1. Our sample of FOSS projects and their historical vulnerability data

(a) Languages

(b) Distribution of vulnerability types

Language	Portion	Vulnerability	Portion	Vulnerability	Portion
Java	40%	DoS	30.8%	Gain Privileges	3.1%
C++	30%	Code execution	20.3%	Directory Traversal	2.4%
PHP	13%	Overflow	16.6%	Memory Corruption	2.2%
C	10%	Bypass Something	10.3%	CSRF	0.9%
JavaScript	5%	Gain Information	7.1%	HTTP response splitting	0.3%
Other	1%	XSS	5.9%	SQL injection	0.1%

This wide range of options requires both flexibility and empowerment of the (worldwide distributed) development teams to choose the software development model that fits their needs best while still providing secure software.

While, overall, SAP is using a large number of FOSS components, the actual number of such components depends heavily on the actual product. For example, the traditional SAP systems in ABAP usually do not contain a lot of FOSS components; the situation is quite the opposite for recent cloud offerings based on OpenStack (http://www.openstack.com) or Cloud Foundry (https://www.cloudfoundry.org/).

For each vulnerability that is published for a consumed FOSS component, an assessment is done to understand whether the vulnerability makes the consuming SAP product insecure. In this case, a fix needs to be developed and shipped to SAP customers. For example, in 2015 a significant number of SAP Security Notes (i. e., patches) fixed vulnerabilities in consumed FOSS components.

Overall, this results in additional effort both for the development teams as well as the teams that work on the incident handling (reports from customers). Thus, there is a need for approaches that support SAP's development teams in estimating the effort related to maintain the consumed FOSS components.

To minimize the effort associated with integrating FOSS components as well as to maximize the usability of the developed product, product teams consider different factors. Not all of them are related to security: e. g., the compatibility of the license as well as requests from customers play an important role as well. From a effort and security perspective, developer teams currently consider:

– How widely a component is used within SAP? Already used components require lower effort as licensing checks are already done and internal expertise can be tapped. Thus, effort for fixing issues or integrating new versions can be shared across multiple development teams.
– Which programming languages and build systems are used? If a development team has already expertise in them, a lower integration and support effort can be expected.
– What maintenance lifecycle is used by the FOSS components? If the security maintenance provided by the FOSS community "outlives" the planned maintenance lifecycle of the consuming SAP product, only the integration of minor releases in SAP releases would be necessary.

– How active is the FOSS community? Consuming FOSS from active and well-known FOSS communities (e. g., Apache) should allow a developer team to leverage external expertise as well as externally provided security fixes.

Table 1 illustrates the characteristics of a selection of FOSS components used within SAP. We have chosen the most popular 166 components used by at least 5 products.

3 Research Question and Cost Models

Considering the above discussion we can summarize our research question:

RQ. *Which factors have significant impact on the security effort to manage a FOSS component in centralized, distributed, and hybrid cost models?*

A key question is to understand how to capture effort in broad terms. In this respect, there are three critical activities that are generated by using FOSS components in a commercial software company [26,29] and specifically at SAP: the analysis of the licenses, security analysis, and maintenance. Licensing is out of scope for this work, and we focus on the other two stages.

In the previous section we have already sketched some of the activities that the security team must perform in both stages. A development team can be assigned to a maintenance which includes several tasks, security maintenance being only one of them. Therefore, it is close to impossible to get analytical accounting for security maintenance to the level of individual vulnerabilities. Further, when a FOSS component is shared across different consuming applications, each development team can differ significantly in the choice of the solution and hence in the effort to implement it.

Therefore, we need to find a proxy for the analysis of our three organizational models. Preliminary discussion with developers and company's researchers suggested the combination of vulnerabilities of the FOSS component itself and the number of company's products using it. A large number of vulnerabilities may be the sign of either a sloppy process or a significant attention by hackers and may warrant a deeper analysis during the selection phase or a significant response during the maintenance phase. This effort is amplified when several development teams are asking to use the FOSS component as a vulnerability which eschewed detection may impact several hundred products and may lead to several security patches for different products.

We assume that the effort structure has the following form

$$e = e_{\text{fixed}} + \sum_{i=1}^{m} e_i \tag{1}$$

where e_i is a variable effort that depends on the i-th FOSS component, and e_{fixed} is a fixed effort that depends on the security maintenance model (e. g., the initial set up costs). For example, with a distributed security maintenance approach an organization will have less communication overhead and more free-dom for developers in distinct product teams, but only if a small number of teams are using a component.

Let $|vulns_i|$ be the number of vulnerabilities that have been cumulatively fixed for the i-th FOSS component and let $|products_i|$ be the number of propri-etary products that use the component:

1. In the `centralized model` a security fix for all instances of a FOSS compo-nent is issued once by the security team of the company and then distrib-uted between all products that are using it. This may happen when, as a part of FOSS selection process, development teams must choose only com-ponents that have been already used by other teams and are supported by the company. To reflect this case the effort for security maintenance scales logarithmically with the number of products using a FOSS component.

$$e_i \propto \log(|vulns_i| * |products_i|) \tag{2}$$

2. The `distributed model` covers the case when security fixes are not central-ized within a company, so each development team has to take care of security issues in FOSS components that they use. In this scenario the effort for secu-rity maintenance increases linearly with the number of products using a FOSS component.

$$e_i \propto |vulns_i| * |products_i| \tag{3}$$

3. The `hybrid model` combines the two previous models: security issues in the least consumed FOSS components (e. g., used only by lowest quartile of prod-ucts consuming FOSS) are not fixed centrally. After this threshold is reached and some effort linearly proportional to the threshold of products to be con-sidered has been invested, the company fixes them centrally, pushing the changes to the remaining products.

$$e_i \propto \begin{cases} |vulns_i| * |products_i| & \text{if } |products_i| \leq p_0 \\ p_0 * |vulns_i| + \log(|vulns_i| * (|products_i| - p_0)) & \text{otherwise} \end{cases} \tag{4}$$

As shown in Fig. 1, the hybrid model is a combination of the distributed model and centralized model, when centralization has a steeper initial cost. The point V_0 is the switching point where the company is indifferent between the cen-tralized and distributed cost models. The hybrid model captures the possibility of a company to switch models after (or before) the indifference point. The fixed effort of the centralized model is obviously higher than the one of a distributed model (e. g., setting up a centralized vulnerability fixing team, establishing and communicating a fixing process, etc.).

Hence, we extend the initial function after the threshold number of products p_0 is reached so that only a logarithmic effort is paid on the *remaining* products. This has the advantage of making the effort e_i continuous in $|products_i|$. An

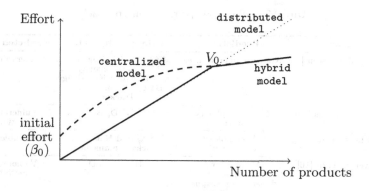

Fig. 1. Illustration of the three cost models

alternative would be to make the cost logarithmic in the overall number of products after $|products_i| > p_0$. This would create a sharp drop in the effort for the security analysis of FOSS components used by several products after p_0 is reached. This phenomenon is neither justified on the field, nor by economic theory. In the sequel, we have used for p_0 the lowest quartile of the distribution of the selected products.

We are not aiming to select a particular model – we consider them as equally possible scenarios. Our goal is to see which of the FOSS characteristics can have impact on the effort when such models are in place, keeping in mind that this impact could differ from one model to another.

We now define the impact that the characteristics of the i-th FOSS component have on the expected effort e_i as a (not necessarily linear) function f_i of several variables and a stochastic error term ϵ_i:

$$e_i = f(x_{i1}, \ldots, x_{il}, y_{il+1}, \ldots, y_{im}, d_{im+1}, \ldots, d_n) + \epsilon_i \tag{5}$$

The variables $x_{ij}, j \in [1, l]$ impact the effort as scaling factors, so that a percentage change in them also implies a percentage change in the expected effort. The variables $y_{ij}, l \in [l+1, m]$ directly impact the value of the effort. Finally, the dummy variables $d_{ij}, j \in [m+1, n]$ denote qualitative properties of the code captured by a binary classification in $\{0, 1\}$.

For example, in our list the 36-th component is "Apache CXF" and the first scaling factor for effort is the number of lines of code written in popular programming languages so that $x_{i,1} \doteq locsPopular_i$, and $x_{36,1} = 627,639$.

Given the above classification we can further specify the impact equation for the i-th component as follows

$$\log(e_i) = \beta_0 + \log(\prod_{j=1}^{l}(x_{ij}+1)^{\beta_j}) + \sum_{j=l+1}^{m} \beta_i * e_{ij}^y + \sum_{j=m+1}^{n} \beta_i * d_{ij} + \epsilon_i \tag{6}$$

where β_0 is the initial fixed effort for a specific security maintenance model.

Table 2. Vulnerability prediction approaches

Paper	Predictors	Vulnerability data	Predicted vars
Massacci & Nguyen [14]	Known vulnerabilities	MFSA, NVD, Bugzilla, Microsoft Security Bulletin, Apple Knowledge Base, Chrome Issue Tracker	Vulnerabilities
Shin & Williams [25]	Code complexity metrics	MFSA, NVD, Bugzilla	Vulnerable functions
Shin et al. [24]	Code complexity, Developer activity	MFSA, Red Hat Linux package manager	Vulnerable files
Nguyen & Tran [16]	Member and Component dependency graphs, Code metrics	MFSA, NVD	Vulnerable functions
Walden & Doyle [27]	SAVD, SAVI	NVD	Post-release vulnerabilities
Scandriato et al. [21]	Frequencies of terms	Fortify SCA warnings	Vulnerable files
Walden et al. [28]	Code metrics, Frequencies of terms	NVD	Vulnerable files

These models focus on technical aspects of security maintenance of consumed FOSS components putting aside all organizational aspects (e. g., communication overhead). For organizational aspects please see ben Othmane et al. [4].

4   Related Work

An extensive body of research explores the applicability of various metrics for estimating the number of vulnerabilities of a FOSS component.

The simplest metric is time (since release), and the corresponding model is a Vulnerability Discovery Model. Massacci and Nguyen [14] provide a comprehensive survey and independent empirical validation of several vulnerability discovery models. Several other metrics have been used: code complexity metrics [16, 24, 25], developer activity metrics [24], static analysis defect densities [27], frequencies of occurrence of programming constructs [21, 28], etc. We illustrate some representative cases in Table 2.

Shin and Williams [25] evaluated software complexity metrics for identifying vulnerable functions. The authors collected information about vulnerabilities in Mozilla JavaScript Engine (JSE) from MFSA[1], and showed that nesting complexity could be an important factor to consider. The authors stress that their approach has few false positives, but several false negatives. In a follow-up work, Shin et al. [24] also analyzed several developer activity metrics showing that poor developer collaboration can potentially lead to vulnerabilities, and that code complexity metrics alone are not a good vulnerability predictor.

Nguyen and Tran [16] built a vulnerability prediction model that represents software with dependency graphs and uses machine learning techniques to train

---

[1] https://www.mozilla.org/en-US/security/advisories/.

the predictor. They used static code analysis tools to compute several source code metrics and tools for extracting dependency information from the source code, adding this information to the graphs that represent an application. To validate the approach, the authors analyzed Mozilla JSE. In comparison to [25], the model had a slightly bigger number of false positives, but less false negatives.

Walden and Doyle [27] used static analysis for predicting web application security risks. They measured the static-analysis vulnerability density (SAVD) metric across version histories of five PHP web applications, which is calculated as the number of warnings issued by the Fortify SCA[2] tool per one thousand lines of code. The authors performed multiple regression analyses using SAVD values for different severity levels as explanatory variables, and the post-release vulnerability density as the response variable showing that SAVD metric could be a potential predictor for the number of new vulnerabilities.

Scandriato et al. [21] proposed to use a machine learning approach that mines source code of Android components and tracks the occurrences of specific patterns. The authors used the Fortify SCA tool: if the tool issues a warning about a file, this file is considered to be vulnerable. However, it may not be the case as Fortify can have many false positives, and authors verified manually only the alerts for 2 applications out of 20. The results show that the approach had good precision and recall when used for prediction within a single project. Walden et al. [28] confirmed that the vulnerability prediction technique based on text mining (described in [21]) could be more accurate than models based on software metrics. They have collected a dataset of PHP vulnerabilities for three open source web applications by mining the NVD and security announcements of those applications. They have built two prediction models: (1) a model that predicts potentially vulnerable files based on source code metrics; and (2) a model that uses the occurrence of terms in a PHP file and machine learning. The analysis shows that the machine learning model had better precision and recall than the code metrics model, however, this model is applicable only for scripting languages (and must be additionally adjusted for languages other than PHP).

Choosing the right source of vulnerability information is crucial, as any vulnerability prediction approach highly depends on the accuracy and completeness of the information in these sources. Massacci and Nguyen [13] addressed the question of selecting the right source of ground truth for vulnerability analysis. The authors show that different vulnerability features are often scattered across vulnerability databases and discuss problems that are present in these sources. Additionally, the authors provide a study on Mozilla Firefox vulnerabilities. Their example shows that if a vulnerability prediction approach is using only one source of vulnerability data - MFSA, it would actually miss an important number of vulnerabilities that are present in other sources such as MFSA and NVD. Of course, the same should be true also for the cases when only the NVD is used as the ground truth for predicting vulnerabilities.

To the best of our knowledge, there is no work that predicts the effort required to resolve security issues in consumed third-party products.

---

[2] http://www8.hp.com/us/en/software-solutions/static-code-analysis-sast/.

5   Data Sources

We considered the following public data sources to obtain the metrics of FOSS projects that could impact the security effort in maintaining them:

1. **National Vulnerability Database (NVD)** – the US government public vulnerability database, we use it as the main source of public vulnerabilities (https://nvd.nist.gov/).
2. **Open Sourced Vulnerability Database (OSVDB)** – an independent public vulnerability database. We use it as the secondary source of public vulnerabilities to complement the data we obtain from the NVD (http://osvdb.org).
3. **Black Duck Code Center** – a commercial platform for the open source governance can be used within an organization for the approval of the usage of FOSS components by identifying legal, operational and security risks that can be caused by these components. We use SAP installation to identify the most popular FOSS components within SAP.
4. **Open Hub (formerly Ohloh)** – a free offering from the Black Duck that is supported by the online community. The Open Hub retrieves data from source code repositories of FOSS projects and maintains statistics that represent various properties of the code base of a project (https://www.openhub.net/).
5. **Coverity Scan Service website** – in 2006 Coverity started the initiative of providing free static code scans for FOSS projects, and many of the projects have registered since that time. We use this website as one of the sources that can help to infer whether a FOSS project is using SAST tools (https://scan.coverity.com/projects)
6. **Exploit Database website** – the public exploit database that is maintained by the Offensive Security[3] company. We use this website as the source for the exploit numbers (https://www.exploit-db.com/).
7. **Core Infrastructure Initiative (CII) Census** – the experimental methodology for parsing through data of open source projects to help identify projects that need some external funding in order to improve their security. We use a part of their data to obtain information about Debian installations (https://www.coreinfrastructure.org/programs/census-project).

6   FOSS Project Metrics Selection

Initially we considered SAP installation of the Black Duck Code Center repository as the source of metrics that could impact the security maintenance effort when using FOSS components. We also performed a literature review and a survey of other repositories to identify potentially interesting variables not currently used in the industrial setting, clustering them by the following four categories:

---

[3] https://www.offensive-security.com/.

Security Development Lifecycle (SDL) – metrics that characterize how the SDL is implemented within a FOSS project. It includes indicators whether a project encourages to report security issues privately, is using one or more static analysis tools during development, etc.

Implementation – various implementation characteristics such as the main programming language and the type of a project.

Popularity – metrics that are relevant to the overall popularity of a FOSS project (e. g., user count and age in years).

Effort – we use these variables as the proxy for the desired response variable - the effort required by companies to update and maintain their applications that are using FOSS components.

Table 3 shows the initial set of metrics that we considered, describing the rationale for including them and references to the literature in which the same or similar metrics were used.

The age of a project, its size and the number of developers (`years`, `locsTotal`, and `contribs`) are traditionally used in various studies that investigate defects and vulnerabilities in software [7,31], the software evolution [3,5] and maintenance [32]. We consider security vulnerabilities to be a specific class of software defects, which are likely to be impacted by these factors.

Several studies considered the popularity of FOSS projects as being relevant to their quality and maintenance [19,20,32] - we used `userCount` from Open Hub and `debianInst` from CII Census as measures of popularity for a project. Many studies investigated whether frequent changes to the source code can introduce new defects [11,15,24,32,34] - we intended to capture this with `locsAdded`, `locsRemoved`, and `commits` metrics from Open Hub.

The presence of security coding standards as a taxonomy of common programming errors [10,23] that caused vulnerabilities in projects should reduce the amount of vulnerabilities and the effort as well. We could not find references to how the presence of security tests could impact the effort.

Wheeler [29] suggested that successful FOSS projects should use SAST tools, which should at least reduce the amount of "unforgivable" security issues discussed by Christey [6].

Numbers of vulnerabilities and exploits have a strong correlation (in our dataset: rho = 0.71, p < 0.01) because security researchers can create exploits to test published vulnerabilities and, alternatively, they can create exploits to prove that a vulnerability indeed exists (so that it will be published as a CVE entry *after* an exploit was disclosed). We tested both values without finding significant differences and for simplicity we report here the `vulns` variable as the proxy for effort.

After obtaining the values of these metrics for a sample of 50 projects we understood that only variables that could be extracted automatically and semi-automatically are interesting for the maintenance phase. Gathering the data manually introduces bias and limits the size of a dataset that we can analyze, and, therefore, the validity of the analysis at all. Thus, we removed the

Table 3. FOSS project metrics

Variable	Category	Source	Collection method	Rationale	Ref.
noSecTest	SDL	Project website, code repository	Manual	The test suite neither contains tests for past vulnerabilities (regression) nor for tests for security functionality.	[29] [1] [6]
privateVulns	SDL	Project website	Manual	There is a possibility to report security issues privately.	
secList	SDL	Project website	Manual	There is a list of past security vulnerabilities.	
noSast	SDL	Project website, code repository, Coverity website	Manual	A project is not using static code analysis tools.	[29] [1]
noManagedLang	Implement.	Open Hub	Automatic	Parts of the project are written in a language without automatic memory management.	[32] [30]
scriptingLang	Implement.	Open Hub	Automatic	Parts of a project are written in a scripting language.	[32]
noCodingStand	Implement.	Project website	Manual	A project has no coding standards for potential contributors.	[1]
locsTotal	Implement.	Open Hub	Automatic	Total size of code bases (LoC).	[5] [7] [31] [3] [33]
locsPopular	Implement.	Open Hub	Automatic	Total size of Java, C, C++, PHP, JavaScript, and SQL code (LoC).	
locsBucket	Implement.	Open Hub	Automatic	Size of the code bases for other programming languages (LoC).	
locsAdded, locsRemoved	Implement.	Open Hub	Automatic	The development activity of a project.	[15] [7] [34] [24] [11] [32]
userCount	Popularity	Open Hub	Automatic	The user count of a project (Open Hub).	[19] [18] [1] [32] [20] [30]
debianInst	Popularity	CII Census	Semi-automatic	The number of installations of a Debian package that corresponds to a project.	
commits	Popularity	Open Hub	Automatic	Number of total commits.	[15] [24] [32]
contribs	Popularity	Open Hub	Automatic	Total number of contributors.	[15] [24] [32]
years	Popularity	Open Hub	Automatic	Age of a project in years.	[17] [32] [11]
vulns	Effort	NVD, OSVDB, and other vuln. databases	Semi-Automatic	The total number of publicly disclosed vulnerabilities.	
exploits	Effort	Exploit DB website	Semi-automatic	The total number of publicly available exploits.	
blackduck	Effort	Black Duck Code Center	Automatic	The total number of requests for a FOSS component within SAP.	

manual variables and expanded the initial dataset up to 166 projects (at least 5 consuming products in SAP Black Duck repository).

We also tried to find commonalities between FOSS projects and to cluster them. However, this process would introduce significant human bias. For example, the "Apache Struts 2" FOSS component is used by SAP as a library in one project, and as a development framework in another one (indeed, it can be considered to be both a framework and a set of libraries). If we "split" the "Apache Struts 2" data point into another two instances marked as "library" and "framework", this would introduce dependency relations between these data points. Assigning arbitrarily only one category to such data points would also be inappropriate.

A comprehensive classification of FOSS projects would require to perform a large number of interviews with developers to understand the exact nature of the usage of a component and the security risk. However, it is unclear what would be the added value *to developers* of this classification.

Below we describe relations between explanatory and response variable:

1. **locsPopular** and **locsBucket** (x_{ij}) – the more there are lines of code, the more there are potential vulnerabilities (that are eventually disclosed publicly). We use these two variables instead of just having **locsTotal** because almost every project in our dataset is written in multiple programming languages, including widely-used languages (**locsPopular**), and rarely-used ones (**locsBucket**). Therefore, different ratios between these two variables could impact the effort differently.

2. **locsEvolution** (y_{ij}) shows how the code base of a project was changed for the whole period of its life. We compute this metric by obtaining the sum of the total number of added and removed lines of code divided by **locsTotal**. Figure 2 shows that we could not use added and deleted lines of code as the measure of global changes as they correlate with each other and with **locsTotal**, however **locsEvolution** has no correlations with **locsTotal** and can be used as an independent predictor.

3. **userCount** and **debianInst** (y_{ij}) – the more there are users, the more potential security vulnerabilities will be reported. **debianInst** provides an alternative measure for **userCount**, however, the two measures are not exactly correlated as some software is usually downloaded from the Web (e. g., Wordpress) so it is very unlikely that someone would install it from the Debian repository, even if a corresponding package exists. On the other hand, some software may be distributed only as a Debian package.

4. **years** (y_{ij}) – more vulnerabilities could be discovered over time.

5. **commits** (y_{ij}) – many commits introduce many atomic changes that can lead to more security issues.

6. **contribs** (y_{ij}) – many contributors might induce vulnerabilities as they might not have exhaustive knowledge on the project and can incidentally break some features they are unaware of.

7. **noManagedLang** (d_{ij}) – parts written in programming languages without built-in memory management could have more security vulnerabilities (e. g., DoS, Sensitive information disclosure, etc.).
8. **scriptingLang** (d_{ij}) – software including fragments in scripting languages could be prone to code injection vulnerabilities.

In spite of their intuitive appeal we excluded dummy variables related to the programming language from our final analysis because we realized that essentially all projects have components of both, therefore, all regression equations would violate the independence assumption.

Table 4 shows the descriptive statistics of response and explanatory variables selected for the analysis.

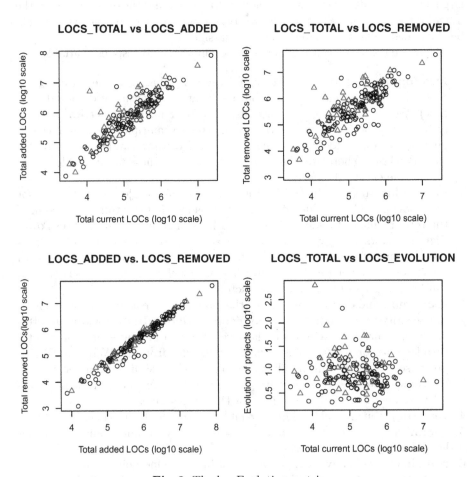

Fig. 2. The locsEvolution metric

Table 4. Descriptive statistics

Variable			Statistic			
	Min	1st Quartile	Median	Mean	3rd Quartile	Max
effort_centralized	0.69	3.64	4.43	4.81	5.75	10.13
effort_distributed	2.00	38.24	84.50	706.60	316.50	25020.00
effort_hybrid	2.00	2540.00	44.00	210.10	139.10	4554.00
years	1.00	7.00	10.00	10.27	13.75	28.00
userCount	0.00	9.00	52.00	258.00	178.00	9390.00
debianInst	0.00	42.75	1407.00	21970.00	12390.00	175900.00
contribs	1.00	15.00	32.00	115.20	101.20	1433.00
commits	18.00	1160.00	4365.00	9785.00	8806.00	174803.00
locsPopular	0.00	32350.00	110700.00	345700.00	310700.00	13830000.00
locsBucket	58.00	5216.00	32770.00	195600.00	128000.00	9372000.00
locsEvolution	1.70	4.85	7.10	15.18	12.60	638.10

7 Analysis

To analyze the statistical significance of the models and identify the variables that impact security effort, we employ a least-square regression (OLS). Our reported R^2 values (0.21, 0.34, 0.39) and F-statistic values (5.30, 10.13, 12.41) are acceptable considering that we have deliberately run the OLS regression with all variables of interest, as our purpose is to see which variables have no impact. The results of estimates for each security effort model are given in Table 5.

Zhang et al. [31] demonstrated a positive relationship between the size of a code base (LOC) and defect-proneness. Zhang [33] evaluated the LOC metrics for defect prediction and concluded that larger modules tend to have more defects. Security vulnerabilities are a subclass of software defects, and our results show that this effect only holds for particular programming languages: the **locsPopular** variable has a positive impact on the effort (it is statistically significant for the **distributed** and **hybrid** models), the **locsBucket** is essentially negligible as a contribution (10^{-5}).

Table 5. Ordinary least-square regression results

	Centralized model		Distributed model		Hybrid model	
Intercept	$9.40 \cdot 10^{-1}$	$(4.29 \cdot 10^0)^{\S}$	$1.89 \cdot 10^0$	$(2.33 \cdot 10^0)^{\dagger}$	$-1.83 \cdot 10^0$	$(-1.96)^*$
log(locsPop+1)	$2.68 \cdot 10^{-2}$	$(1.32 \cdot 10^0)$	$1.46 \cdot 10^{-1}$	$(1.94 \cdot 10^0)^*$	$0.25 \cdot 10^0$	$(2.89)^{\ddagger}$
log(locsBucket+1)	$-4.01 \cdot 10^{-5}$	$(-2.00 \cdot 10^{-3})$	$1.28 \cdot 10^{-2}$	$(1.86 \cdot 10^{-1})$	$9.42 \cdot 10^{-2}$	(1.20)
locsEvolution	$1.83 \cdot 10^{-4}$	$(0.31 \cdot 10^0)$	$8.62 \cdot 10^{-4}$	$(3.98 \cdot 10^{-1})$	$9.31 \cdot 10^{-4}$	(0.38)
years	$2.07 \cdot 10^{-2}$	$(2.66 \cdot 10^0)^{\ddagger}$	$8.47 \cdot 10^{-2}$	$(2.94 \cdot 10^0)^{\ddagger}$	$9.47 \cdot 10^{-2}$	$(2.87)^{\ddagger}$
commits	$-1.01 \cdot 10^{-6}$	$(-0.46 \cdot 10^0)$	$-5.96 \cdot 10^{-6}$	$(-7.26 \cdot 10^{-1})$	$-1.11 \cdot 10^{-5}$	(-1.19)
contribs	$-7.50 \cdot 10^{-5}$	$(-0.47 \cdot 10^0)$	$-3.31 \cdot 10^{-4}$	$(-0.57 \cdot 10^0)$	$5.66 \cdot 10^{-4}$	(0.84)
userCount	$8.77 \cdot 10^{-5}$	$(2.22 \cdot 10^0)^{\dagger}$	$5.49 \cdot 10^{-4}$	$(3.74 \cdot 10^0)^{\S}$	$5.81 \cdot 10^{-4}$	$(3.46)^{\S}$
debianInst	$1.70 \cdot 10^{-6}$	$(2.39 \cdot 10^0)^{\dagger}$	$8.99 \cdot 10^{-6}$	$(3.41 \cdot 10^0)^{\S}$	$1.10 \cdot 10^{-5}$	$(3.64)^{\S}$
N	166		166		166	
Multiple R^2	0.21		0.34		0.39	
Adjusted R^2	0.17		0.30		0.36	
F-statistic	5.30 (p < 0.01)		10.13 (p < 0.01)		12.41 (p < 0.01)	

Note, t-statistics are in parentheses.   Signif.codes: * 5%, † 1%, ‡ 0.1%, § 0.001%

The **locsEvolution, commits** and **contribs** variables do not seem to have an impact. We expected the opposite result, as many works (e. g., [7,15,24]) suggest a positive relation between number (or frequency) of changes and defects. However, these works assessed changes with respect to distinct releases or components or methods, while we are using the cumulative number of changes for all versions in a project; we may not capture the impact because of this.

The study by Li et al. [12] showed that the number of security bugs can grow significantly over time. Also, according to the vulnerability discovery process model described by Alhazmi et al. [2], the longer is the active phase of a software the more attention it will attract, and more hackers will get familiar with it to break it. Massacci and Nguyen [13] illustrated this model by showing that the vulnerability discovery rate was the highest during the active phase of Mozilla Firefox 2.0. We find that the age of a project – **years** has a significant impact in all our effort models, thus supporting those models.

It is a folk knowledge that "Given enough eyeballs, all bugs are shallow" [19], meaning that FOSS projects have the unique opportunity to be tested and scrutinized not only by their developers, but by their user community as well. We found that in our models the number of external users (**userCount** and **debian-Inst**) of a FOSS component has small but statistically significant impact. This could be explained by the intuition that only a major increase of the popularity of a FOSS project could result in finding and publishing new vulnerabilities: not every user would have enough knowledge in software security for finding vulnerabilities (or motivation for reporting them).

8   Conclusions

In this paper we have investigated the publicly available factors that can impact the effort required for performing security maintenance process within large software vendors that have extensive consumption of FOSS components. We have defined three security effort models – `centralized`, `distributed`, and `hybrid`, and selected variables that may impact these models. We automatically collected data on these variables from 166 FOSS components currently consumed by SAP products and analyzed the statistical significance of these models.

As a proxy for security maintenance effort of consumed FOSS components we used the combination of the number of products using a these components, and the number of known vulnerabilities in them. As the summary of our findings, the main factors that influence the security maintenance effort are the amount of lines of code of a FOSS component and the age of the component. We have also observed that the external popularity of a FOSS component has statistically significant but small impact on the effort, meaning that only large changes in popularity will have a visible effect.

As a future work we plan collecting a wider sample of FOSS projects, assessing other explanatory variables and investigating our models further. Using the data for prediction of the effort is also a promising direction for the future work.

Acknowledgments. This work has been partly supported by the European Union under agreement no. 285223 SECONOMICS, no. 317387 SECENTIS (FP7-PEOPLE-2012-ITN), the Italian Project MIUR-PRIN-TENACE, and PON - Distretto Cyber Security attività RI.4.

References

1. Aberdour, M.: Achieving quality in open-source software. IEEE Softw. **24**(1), 58–64 (2007)
2. Alhazmi, O., Malaiya, Y., Ray, I.: Security vulnerabilities in software systems: a quantitative perspective. In: Jajodia, S., Wijesekera, D. (eds.) Data and Applications Security 2005. LNCS, vol. 3654, pp. 281–294. Springer, Heidelberg (2005)
3. Beecher, K., Capiluppi, A., Boldyreff, C.: Identifying exogenous drivers and evolutionary stages in floss projects. J. Syst. Softw. **82**(5), 739–750 (2009)
4. ben Othmane, L., Chehrazi, G., Bodden, E., Tsalovski, P., Brucker, A.D., Miseldine, P.: Factors impacting the effort required to fix security vulnerabilities: An industrial case study. In: López, J., Mitchell, C.J. (eds.) ISC 2015. LNCS, vol. 9290, pp. 102–119. Springer, Heidelberg (2015)
5. Capiluppi, A.: Models for the evolution of os projects. In: Proceedings of International Conference on Software Maintenance (2003)
6. Christey, S.: Unforgivable vulnerabilities. Black Hat Briefings (2007)
7. Gegick, M., Williams, L., Osborne, J., Vouk, M.: Prioritizing software security fortification throughcode-level metrics. In: Proceedings of the 4th ACM Workshop on Quality of Protection (2008)
8. Hansen, M., Köhntopp, K., Pfitzmann, A.: The open source approach opportunities and limitations with respect to security and privacy. Comput. Secur. J. **21**(5), 461–471 (2002)
9. Hoepman, J.-H., Jacobs, B.: Increased security through open source. Commun. ACM **50**(1), 79–83 (2007)
10. Jones, R.L., Rastogi, A.: Secure coding: Building security into the software development life cycle. Inf. Syst. Secur. **13**(5), 29–39 (2004)
11. Kamei, Y., Shihab, E., Adams, B., Hassan, A.E., Mockus, A., Sinha, A., Ubayashi, N.: A large-scale empirical study of just-in-time quality assurance. IEEE Trans. Softw. Eng. **39**(6), 757–773 (2013)
12. Li, Z., Tan, L., Wang, X., Lu, S., Zhou, Y., Zhai, C.: Have things changed now?: An empirical study of bug characteristics in modern open source software. In: Proceedings of the 1st Workshop on Architectural and System Support for Improving Software Dependability (2006)
13. Massacci, F., Nguyen, V.H.: Which is the right source for vulnerability studies?: an empirical analysis on mozilla firefox. In: Proceedings of the 6th International Workshop on Security Measurements and Metrics (2010)
14. Massacci, F., Nguyen, V.H.: An empirical methodology to evaluate vulnerability discovery models. IEEE Trans. Softw. Eng. **40**(12), 1147–1162 (2014)
15. Nagappan, N., Ball, T.: Use of relative code churn measures to predict system defect density. In: Proceedings of 27th International Conference on Software Engineering (2005)
16. Nguyen, V.H., Tran, L.M.S.: Predicting vulnerable software components with dependency graphs. In: Proceedings of the 6th International Workshop on Security Measurements and Metrics (2010)

17. Ozment, A., Schechter, S.E.: Milk or wine: Does software security improve with age? In: Proceedings of Usenix Security Symposium (2006)
18. Polančič, G., Horvat, R.V., Rozman, T.: Comparative assessment of open source software using easy accessible data. In: Proceedings of 26th International Conference on Information Technology Interfaces (2004)
19. Raymond, E.: The cathedral and the bazaar. Knowl. Technol. Policy **12**(3), 23–49 (1999)
20. Sajnani, H., Saini, V., Ossher, J., Lopes, C.V.: Is popularity a measure of quality? an analysis of maven components. In: Proceedings of IEEE International Conference on Software Maintenance and Evolution (2014)
21. Scandariato, R., Walden, J., Hovsepyan, A., Joosen, W.: Predicting vulnerable software components via text mining. IEEE Trans. Softw. Eng. **40**(10), 993–1006 (2014)
22. Schryen, G.: Is open source security a myth? Commun. ACM **54**(5), 130–140 (2011)
23. Seacord, R.C.: Secure coding standards. In: Proceedings of the Static Analysis Summit, NIST Special Publication (2006)
24. Shin, Y., Meneely, A., Williams, L., Osborne, J., et al.: Evaluating complexity, code churn, and developer activity metrics as indicators of software vulnerabilities. IEEE Trans. Softw. Eng. **37**(6), 772–787 (2011)
25. Shin, Y., Williams, L.: An empirical model to predict security vulnerabilities using code complexity metrics. In: Proceedings of the Second ACM-IEEE International Symposium on Empirical Software Engineering and Measurement (2008)
26. Stol, K.-J., Ali Babar, M.: Challenges in using open source software in product development: A review of the literature. In: Proceedings of the 3rd International Workshop on Emerging Trends in Free/Libre/Open Source Software Research and Development (2010)
27. Walden, J., Doyle, M.: Savi: Static-analysis vulnerability indicator. IEEE Secur. Priv. J. **10**(3), 32–39 (2012)
28. Walden, J., Stuckman, J., Scandariato, R.: Predicting vulnerable components: Software metrics vs text mining. In: Proceedings of IEEE 25th International Symposium on Software Reliability Engineering (2014)
29. Wheeler, D.A.: How to evaluate open source software/free software (oss/fs) programs (2005). http://www.dwheeler.com/oss_fs_eval.html
30. Wheeler, D.A., Khakimov, S.: Open source software projects needing security investments (2015)
31. Zhang, D., El Emam, K., Liu, H., et al.: An investigation into the functional form of the size-defect relationship for software modules. IEEE Trans. Softw. Eng. **35**(2), 293–304 (2009)
32. Zhang, F., Mockus, A., Zou, Y., Khomh, F., Hassan, A.E.: How does context affect the distribution of software maintainability metrics? In: Proceedings of 29th IEEE International Conference on Software Maintenance (2013)
33. Zhang, H.: An investigation of the relationships between lines of code and defects. In: Proceedings of IEEE International Conference on Software Maintenance (2009)
34. Zimmermann, T., Nagappan, N., Williams, L.: Searching for a needle in a haystack: Predicting security vulnerabilities for windows vista. In: Proceedings of Third International Conference on Software Testing, Verification and Validation (2010)

Idea: Usable Platforms for Secure Programming – Mining Unix for Insight and Guidelines

Sven Türpe[✉]

Fraunhofer Institute for Secure Information Technology SIT,
Darmstadt, Germany
sven.tuerpe@sit.fraunhofer.de

Abstract. Just as security mechanisms for end users need to be usable, programming platforms and APIs need to be usable for programmers. To date the security community has assembled large catalogs of dos and don'ts for programmers, but rather little guidance for the design of APIs that make secure programming easy and natural. Unix with its setuid mechanism lets us study usable security issues of programming platforms. Setuid allows certain programs to run with higher privileges than the user or process controlling them. Operating across a privilege boundary entails security obligations for the program. Obligations are known and documented, yet developers often fail to fulfill them. Using concepts and vocabulary from usable security and usability of notations theory, we can explain how the Unix platform provokes vulnerabilities in such programs. This analysis is a first step towards developing platform design guidelines to address human factors issues in secure programming.

1 Introduction

When humans, while interacting with technology, run into the same kind of problem often enough for us to see a pattern, the technology is often at fault: its design does not sufficiently take into account human factors and human capabilities. Programming is no exception, "programmers are people, too" [2]. We know numerous vulnerability patterns and collect them in databases like CWE (http://cwe.mitre.org), but the security community is only starting to pay attention to the human factors involved in secure programming and the usability of programming platforms [4,6,12,15,16].

A classical example of vulnerability-inducing platform design is the set-user-id/set-group-id (setuid/setgid) mechanism of Unix. Setuid lets Unix processes under certain conditions change their identity (persona) and thus their privileges, allowing users to run particular programs with elevated privileges.

While useful and even necessary sometimes, setuid is also an inexhaustible source of vulnerabilities. Hundreds of vulnerability reports related to setuid can be found in the U.S. National Vulnerability Database (NVD, http://nvd.nist.gov); new instances continue to appear [8]. While usability issues in the immediate setuid API have been addressed in the literature [5,7,14], setuid causes a

© Springer International Publishing Switzerland 2016
J. Caballero et al. (Eds.): ESSoS 2016, LNCS 9639, pp. 207–215, 2016.
DOI: 10.1007/978-3-319-30806-7_13

cariety of challenges elsewhere, which are thus far only covered by secure programming guides [1, 3, 10].

This paper proposes to analyze the guidelines for writing secure setuid programs and the underlying API design from a usability perspective. While the
secure coding guides are technically sound, programmers apparently have difficulties following them consistently. Are there features in the design of Unix APIs
that make it hard to put secure coding advice into practice? If so, how could
these APIs be improved to make it easier for programmers to write secure code?

The ultimate aim is a collection of applicable design principles for APIs and
programming platforms that facilitate secure programming. Setuid and the Unix
API constitute an ideal starting point for such an investigation: they have been
widely deployed and used, so that a large body of accessible code exists. Setuid
also facilitates comparison, as a flip of a bit changes the security context of a
program without any change in its code.

After briefly describing the setuid mechanism, this paper applies parts of
existing usability and usable security theory to a toy program, demonstrating
how such an analysis can yield insight.

2 Setuid in a Nutshell

Unix processes access kernel functions and system resources through the kernel's syscall API. The kernel enforces two kinds of policies there. First, files
and file-like resources (device files, named pipes, sockets, etc.) are subject to
discretionary access control. Second, some functions in the syscall API require
superuser privileges to be called at all or with unrestricted parameter values.

The kernel makes its access control decisions based on the persona associated
with a process. A process persona comprises an effective user id (EUID) and
one or more group IDs. File access control uses these IDs together with a file's
ownership (user and group) to select the set of permission bits to evaluate before
granting or denying access. The superuser (root, user ID 0) can override these
permission checks and access any file. A process with effective user ID 0 is also
the only way to get unrestricted access to the syscall API.

2.1 Setuid Mechanism

A regular process inherits its persona from its parent. This behavior corresponds
with the intuition of a user session: upon login, a user obtains a shell process
with the appropriate persona, and whatever is being run from there has the same
persona and privileges [9, 13]. The setuid mechanism allows some programs to
run with a different persona; it has two parts:

1. The setuid/setgid permission bits, when set on program files, override identity
   inheritance. When a program with one of these bits is executed, the child
   process runs with the effective user or group ID determined by the file owner,
   rather than those inherited from the parent. Inherited IDs are also preserved,
   so that the process can switch privileges.

2. The `setuid()`/`setgid(` family of system calls allows processes to manipulate their own persona, subject to a number of constraints. A process with EUID = 0 can take on any persona; this is also used to configure the persona of a login shell after user authentication. Processes with other effective user IDs cannot normally change their persona. However, in conjunction with the setuid permission bits, a process can drop and regain privileges.

For detail on setuid, its pitfalls, and proposed design improvements see the literature on setuid [5,7,14] and Unix programming [10,13].

2.2 Uses

Setuid is a versatile and useful mechanism and allows programs to handle cases not covered by the semantics and granularity of file access control. Setuid is used, for example, in these cases:

- The login program, running with root privileges, uses the `setuid()`/`setgid()` API to personalize the shell process for an authenticated user.
- File access control cannot enforce finer-grained policies. Unix password files, for example, need line-by-line access control so that non-**root** users can change only their own passwords. A setuid program can enforce arbitrary policies on resources accessible for the program but not for its users.
- Some programs need to make privileged system calls but should nevertheless be started and controlled by a regular user. The standard ports for HTTP (80) and HTTPS (443), for example, are privileged. Otherwise, however, a web server is a regular program that needs no special privileges.

In principle, setuid may be used with any user identity. However, setuid **root** is the most common and also the most critical use.

3 Security Obligations and Programming Rules

3.1 Security Obligations

Setuid places a process at a privilege boundary. Program code is being executed with elevated privileges while input is controlled and output is received by a user holding at most a subset of these privileges. Input includes a program's standard input stream, files read by the program, environment variables, signals, and possibly interactive commands. Output includes the standard output stream, error or log messages, and files written or manipulated.

On the one hand, setuid allows designated programs to refine and extend the access control policy enforced by a system. The **passwd** program, for example, which allows users to change their own password but not those of other users, enforces a policy on lines of the password file, whereas file access control can only enforce permissions on entire files. On the other hand, a program at a privilege boundary becomes a guardian of the higher privilege. A program meant to attain privileges by the setuid mechanism needs to make sure that

1. It enforces the required policy completely and correctly. Any failure to do so defeats its purpose.
2. No matter what the caller does to inputs and outputs, the program does not support any operation not part of its intended purpose.

The latter is the harder problem. Consider just some of the things that should not happen across a privilege boundary:

- Write user-controlled data to a user-selected file
- Execute user-specified commands or code with elevated privileges
- Read files and forward information about their content to the user.

Data and control flows must be carefully constrained across the entire input and output space of the program. Due to the purpose of the program – extending and refining access control – this burden rests with the program alone.

3.2 Programming Rules

The abstract obligations of a setuid program translate into a larger set of rules for the programmer. Bishop [3] developed an early set of rules, including items like:

- "Close all but necessary file descriptors before calling exec." (The exec call loads and runs a new program within the process, replacing the one currently running. Open files remain open.)
- "Check the environment in which the process will run." (The process environment is inherited from the parent. It contains a number of user-controlled variables and parameters, which influence the behavior of library functions and programs.)
- "Make only safe assumptions about recovery of errors." (Attempts at error recovery that might be helpful in a regular program can become dangerous in conjunction with setuid.)

Such rules have their roots in design subtleties that can be exploited in a setuid setting. The process environment, for example, is passed on silently in the background and controls critical behaviors of programs and libraries – how program files are searched, how new files are created, and so on.

Garfinkel et al. [10] later offered advanced design guidelines, advising programmers, for example, to bracket code sections that actually need elevated privileges between code that restores privileges before and drops them after a the respective calls. Chen et al. [5] propose a revision of the setuid API that makes this idiom easier to use and more robust.

4   Example: A Good Program Turning Vulnerable

4.1   Hello, World!

Listing 1 outlines a "Hello, world!" program, which instead of just printing its message, sends an email to the address specified as the first command line argument. After some declarations, the program creates a command string of the form `mail <email_addr>` in the string buffer `cmd` (line 8), executes this command through a `popen()` call (line 9), and writes the message to the pipe thus opened (line 10). All error handling has been omitted for brevity, but should be straightforward: verify that `argv[1]` is present (let the `mail` program care about syntax) and check return values after each call.

Listing 1. This program has multiple vulnerabilities when executed setuid `root`.

```
1   #include <stdio.h>
2   #include <stdlib.h>
3
4   int main(int argc, char *argv[]) {
5       char cmd[256] ="";
6       FILE *mail = NULL;
7       /* ... */
8       snprintf(cmd, 256, "mail %s", argv[1]);
9       mail = popen(cmd, "w");
10      fputs("Hello, world!\n", mail);
11      pclose(mail);
12      return 0;
13  }
```

Apart from the intentional omission of error handling, the program in Listing 1 exhibits two reasonable design decisions. First, the program reuses the existing `mail` command, which in turn takes care of the complexities of email sending. Second, our "Hello, world!" program uses `popen()` to call this subprogram. This, too, hides complexity from the programmer and takes care of many details. Using lower-level calls, such as `fork()` and `exec()`, to implement the same functionality would require a lot more code; `popen()` together with the file API of the C standard library offers a convenient abstraction. As `popen()` uses the Unix shell to execute commands and child processes inherit environments, programs executed this way also follow platform conventions, such as searching for programs as specified by the `PATH` variable or honoring the `LANG` and `LC_*` environment variables controlling internationalization.

4.2 Some Vulnerabilities

As soon as the setuid mechanism is applied to run the program in Listing 1 with `root` privileges while keeping an unprivileged user in control of its inputs, the program becomes a ragbag of vulnerabilities:

- Line 8 embeds user input in a command string to be executed by a command interpreter. The caller can use a variety of separators and other mechanisms to sneak in arbitrary commands to be executed with elevated privileges.
- Line 8 does not specify an absolute path for the `mail` command. The shell invoked by the `popen()` call in line 9 will hence search for an executable file named `mail` in the directories specified in the `PATH` environment variable; the

caller can manipulate the search path so that an arbitrary program named `mail` is found first.
- Line 9 executes the prepared command as a child process, passing on all environment variables. These variables may influence the operation of the mail command or the hidden shell used to execute this command.

As an immediate mitigation, the programmer might (1) specify an absolute path to the mail command [10] and (2) use lower-level calls to spawn the `mail` subprocess without executing a shell command or searching files along an environment-specified path [1]. It is also recommended to (3) sanitize environment variables [3, 10] and to (4) assure open file descriptors do not leak across privilege boundaries [3].

To reduce the risk from buffer overflow and similar defects, which can occur anywhere in a program, and as a general matter of hygiene, it is further recommended to employ either of two patterns dependent on how often elevated privileges are required: (a) carry out all privileged work early, then drop privileges permanently, or (b) drop privileges temporarily whenever they are not needed. This reduces the amount of code actually running with elevated privileges. Getting privilege changes right can be a challenge of its own [5, 14].

5   The API Usability Perspective

For every possible functional specification we can think of two different programming tasks T_R and T_S, where T_R is the task of writing a program that approximates the specified behavior closely enough, and T_S is the same task with the additional requirement that the result also be secure. Translated into Unix with its setuid mechanism, T_R is the task of writing a regular program that runs with the privileges of the user controlling its inputs, and T_S is the task of writing a functionally equivalent program suitable for setuid `root` use.

Ideally, task T_S of writing a secure program should not be harder to accomplish than task T_R of writing an equivalent regular program. If we can identify factors in platform and API design that systematically make T_S harder to accomplish than the corresponding T_R, then the platform leaves room for usability improvement. The ideal "don't care" situation may not be attainable, but perhaps security mechanisms and APIs can be redesigned to make it easier for programmers to fulfill their remaining security duties.

To identify factors that complicate secure programming we can apply usable security principles [17] and general API usability guidelines [11]. The following two subsections will illustrate this for subsets of the respective criteria.

5.1   Usable Security Principles

Yee [17] proposes ten principles of user interaction design for secure systems. Although programming tasks differ in important respects from interactive use of a program or security mechanism, some of these principles can be applied to programming environments. Two examples:

Path of Least Resistance. "The most natural way to do any task should also be the most secure way" [17]. This is a different way of putting the ideal outlined above, where the programmer just does not have to care about security. Many secure programming rules imply that programmers should replace short and straightforward pieces of code with longer and more complicated ones. Apple's secure programming guide [1], for example describes a supposedly secure alternative to the popen() call. This recommended alternative would vastly increase the length of the example in Listing 1, introduce some potential for new defects, and require the programmer to deal with lower-level APIs. Requiring such programming games [11] clearly violates the path of least resistance principle.

Explicit Authorization. Originally referring to transfers of a user's authority to others, explicit authorization can be required for any critical aspect. The setuid mechanism violates this principle by placing programs in a security-critical context without asking for the programmer's consent. Rather than letting the programmer acquire privileges when needed, the platform forces programmers to drop privileges when they do not need them. From the programmer's point of view, running with elevated privileges is the default rather than an explicitly authorized exception.

5.2   Cognitive Dimensions

The cognitive dimensions framework [11] provides a vocabulary to discuss usability properties of programming languages, APIs, and other information artefacts. A discussion by the cognitive dimensions merely describes properties of a notation; how these properties affect usability depends on the kind of task to be accomplished. Programming as an interactive design task imposes high demands on the notation. The cognitive dimensions include aspects like the following:

Hard Mental Operations. Some operations, such as Boolean logic, are inherently hard to carry out for the human mind. A notation requiring such operations to be understood therefore becomes hard to use. Secure programming in a setuid scenario requires the programmer to keep track of data flows between the two privilege levels, regular and elevated, and make sure the program cannot be abused to read or write data with elevated privileges beyond its intended purpose and policy. However, this is nearly impossible even for small programs. The example in Listing 1 is only a toy program without subroutines, yet it contains already two indirections: to understand the fputs() call in line 10, one has to track the file handle mail to the popen() call before, which in turn depends on a command assembled in line 8 using user input (argv[1]). This becomes hopeless rather quickly as programs grow.

Visibility. To write secure setuid programs, the programmer has to follow numerous rules, but the API does not give any hints as to which rules to apply when and where. There are no defined markers for safe or unsafe functions or for data that could or should not be used in certain ways. Secure programming rules rely entirely on information in the programmer's head. The programmer needs to know the inner workings of functions like popen() to understand the risks, contrary to the idea that functions should hide implementation detail and rather adhere to an explicit contract.

Hidden Dependencies. Hidden dependencies occur if actions in one place have a non-obvious effect elsewhere. In the case of popen(), a hidden dependency exists between the process environment and the behavior of popen(). Environment variables are being passed on down the process tree. The programmer can intervene, but as a default, environment variables are hidden from the programmer rather than passed explicitly.

6   Outlook

Although incomplete, the preceding analysis already suggests some directions for API redesign. An improved version of the Unix API could for example:

- Let programmers acquire privileges and corresponding responsibilities through an explicit call,
- Offer safe alternatives to unsafe functions, so that secure alternatives do not require writing more code, and
- Detect inappropriate security contexts inside critical functions and return an error when a functions is being called where it shouldn't.

The second step of research after analysis of the existing API is therefore improvement. Tradeoffs will likely appear between the different design goals, so even if we know what to aim for, devising an improved API remains a challenge. Finally, any proposed improvement needs to be tested with real programmers. This may be the hardest part. Research prefers small, controlled lab experiments, whereas real programming takes place in large projects and code bases and is done by programmers that acquire skills and habits over time as they use and reuse platforms. As an alternative, once a set of usability principles has been established, other platforms and their program vulnerbility patterns can be analyzed to see whether the principles explain the patterns.

For the first two steps, a vast amount of data is freely available for research. Vulnerability databases are full of reports of defect instances. Many of those concern open source software and can be reviewed. Open source platform implementations – Linux and *BSD – facilitate experimentation, the more so as extensions like Linux capabilities, SELinux, and Capsicum exist, which address the same set of issues from a technical rather than from a human factors perspective.

References

1. Apple Inc.: Secure Coding Guide, 2014-02-11 edn. (2006–2014). https://developer.apple.com/library/mac/documentation/Security/Conceptual/SecureCodingGuide/
2. Arnold, K.: Programmers are people, too. ACM Queue **3**(5), 54–59 (2005)
3. Bishop, M.: How to write a setuid program. Login **12**(1), 5–11 (1987)
4. Cappos, J., Zhuang, Y., Oliveira, D., Rosenthal, M., Yeh, K.C.: Vulnerabilities as blind spots in developer's heuristic-based decision-making processes. In: Proceedings of New Security Paradigms Workshop, NSPW 2014, pp. 53–62. ACM, New York, NY, USA (2014)
5. Chen, H., Wagner, D., Dean, D.: Setuid demystified. In: USENIX Security Symposium, pp. 171–190 (2002)
6. Crandall, J.R., Oliveira, D.: Holographic vulnerability studies: vulnerabilities as fractures in interpretation as information flows across abstraction boundaries. In: Proceedings of New Security Paradigms Workshop, NSPW 2012, pp. 141–152. ACM, New York, NY, USA (2012)
7. Dittmer, M.S., Tripunitara, M.V.: The unix process identity crisis: a standards-driven approach to setuid. In: Proceedings of the 2014 ACM SIGSAC Conference on Computer and Communications Security, CCS 2014, pp. 1391–1402. ACM, New York, NY, USA (2014)
8. Esser, S.: OS X 10.10 DYLD_PRINT_TO_FILE local privilege escalation vulnerability. https://www.sektioneins.de/blog/15-07-07-dyld_print_to_file_lpe.html (2015)
9. Free Software Foundation Inc: The GNU C Library Reference Manual, glibc 2.22 edn, August 2015. https://www.gnu.org/software/libc/manual/
10. Garfinkel, S., Spafford, G., Schwartz, A.: Practical UNIX and Internet Security, 3rd edn. O'Reilly Media, Sebastopol (2003)
11. Green, T.R.G., Petre, M.: Usability analysis of visual programming environments: a 'cognitive dimensions' framework. J. Vis. Lang. Comput. **7**(2), 131–174 (1996)
12. Oliveira, D., Rosenthal, M., Morin, N., Yeh, K.C., Cappos, J., Zhuang, Y.: It's the psychology stupid: how heuristics explain software vulnerabilities and how priming can illuminate developer's blind spots. In: Proceedings of 30th Annual Computer Security Applications Conference, ACSAC 2014, pp. 296–305. ACM, New York, NY, USA (2014)
13. Stevens, W.R.: Advanced Programming in the UNIX Environment. Addison-Wesley Publishing Company, Reading (1992)
14. Tsafrir, D., Da Silva, D., Wagner, D.: The murky issue of changing process identity: revising "setuid demystified". Login **33**(3), 55–66 (2008)
15. Türpe, S.: Point-and-shoot security design: can we build better tools for developers? In: Proceedings of New Security Paradigms Workshop, NSPW 2012, pp. 27–42. ACM, New York, NY, USA (2012)
16. Wurster, G., van Oorschot, P.C.: The developer is the enemy. In: Proceedings of New Security Paradigms Workshop, NSPW 2008, pp. 89–97. ACM, New York, NY, USA (2008)
17. Yee, K.-P.: User interaction design for secure systems. In: Deng, R.H., Qing, S., Bao, F., Zhou, J. (eds.) ICICS 2002. LNCS, vol. 2513, pp. 278–290. Springer, Heidelberg (2002). doi:10.1007/3-540-36159-6_24

AppPAL for Android
Capturing and Checking Mobile App Policies

Joseph Hallett[(✉)] and David Aspinall

School of Informatics, University of Edinburgh, Edinburgh, UK
s1361467@sms.ed.ac.uk

Abstract. It can be difficult to find mobile apps that respect one's security and privacy. Businesses rely on employees enforcing company mobile device policies correctly. Users must judge apps by the information shown to them by the store. Studies have found that most users do not pay attention to an apps permissions during installation [19] and most users do not understand how permissions relate to the capabilities of an app [30]. To address these problems and more, we present AppPAL: a machine-readable policy language for Android that describes precisely when apps are acceptable. AppPAL goes beyond existing policy enforcement tools, like Kirin [16], adding delegation relationships to allow a variety of authorities to contribute to a decision. AppPAL also acts as a "glue", allowing connection to a variety of local constraint checkers (e.g., static analysis tools, packager manager checks) to combine their results. As well as introducing AppPAL and some examples, we apply it to explore whether users follow certain intended policies in practice, finding privacy preferences and actual behaviour are not always aligned in the absence of a rigorous enforcement mechanism.

1 Introduction

Finding the right apps can be tricky. Users need to discover which are not going to abuse their data. This can be difficult as it isn't obvious how apps use the data each has access to. Consider a user attempting to buy a flashlight app. By searching the Play store the user is presented with a long list of apps. Clicking through each one they can find the permissions each requests but not the reasons why each was needed. They can see review scores from users but not from tools to check apps for problems and issues like SSL misconfigurations [17]. If they want to use the app at work will it break their employers rules for mobile usage?

App stores give some information about their apps; descriptions, screenshots and review scores. Android apps show a list of permissions when they're first installed. In Android Marshmallow apps will display permissions requests when the app first tries to access sensitive data (such as contacts or location information). Users do not understand how permissions relate to their device [19,46]. Ultimately the decision of which apps to use and which permissions to grant must be made by the device user.

© Springer International Publishing Switzerland 2016
J. Caballero et al. (Eds.): ESSoS 2016, LNCS 9639, pp. 216–232, 2016.
DOI: 10.1007/978-3-319-30806-7_14

Some apps are highly undesirable. Many potentially unwanted programs (PUP) are being propagated for Android devices [45, 47]. Employees are increasingly using their own phones for work. An employer may restrict which apps their employees can use. The IT department may set a mobile device policy—a series of rules describing what kinds of apps may be used and how—to prevent information leaks. Some users worry that apps will misuse their personal data—sending their address book or location to an advertiser without their permission. Such a user avoids apps which can access their location, or address book; they may apply their own personal security policies when downloading and running apps.

These policies can only be enforced by the users continuously making the correct decision when prompted about apps. An alternative is to write the policy down and make the computer enforce it. To implement this we propose using a logic of authorization—a language designed to express rules about permissible actions.

We present AppPAL, an instantiation of Becker et al.'s SecPAL [6] with constraints (statements checkable using information external to the language such as the time of day or static analysis tools) and predicates that allow us to decide which apps to run or install. The language allows us to reason about apps using statements from third parties. AppPAL allows us to enforce the policies on a device. We can express trust relationships amongst these parties and use constraints to do additional checks, such as using security checks. This lets us enforce more complex policies than existing tools such as Kirin [16] which are limited to permissions checks. Policies can be enforced by the stores selling the apps, on the devices installing apps or by third-parties providing app vetting services.

Consider the following example: a user, Alice, may have rules she has to follow when using apps for work and her own policies when using apps at home in her private life. Using AppPAL we can write policies for work and home, and decide which policy to enforce using a user's location, or the time of day:

```
' alice '  says App isRunnable          ' alice '  says App isRunnable
  if'home-policy' isMetBy(App)            if'work-policy' isMetBy(App)
  where at('work') = false.               where beforeHourOfDay('17') = true.
```

We can delegate policy specification to third parties or roles, and assign principals to roles:

```
' alice '  says'it-department' can-say'work-policy' isMetBy(App).
' alice '  says'alice ' can-act-as'it-department'.
```

We can write policies specifying which permissions an app must or must not have by its app store categorization. For example, it would be okay allowing a photography app access to the camera, but not to allow access to location data if the user doesn't want their photos geotagged.

```
' alice '  says App isRunnable
  if' permissions-policy' isMetBy(App).
' alice '  says'permissions-policy' isMetBy(App)
  if App isAnApp
```

```
where
  category(App,'Photography'),
  hasPermission(App,'LOCATION') = false,
  hasPermission(App,'CAMERA') = true.
```

There has been much work developing app analysis tools for Android. Tools such as Stowaway [18] detect over-privileged apps. TaintDroid [15] and Flow-Droid [1,33] can do taint and control flow analysis; sometimes even between app components. Other tools like QUIRE [11] can find privilege escalation attacks between apps. ScanDAL [31] and SCanDroid [21] help detect privacy leaks. Appscopy [20] searches for specific kinds of malware. Tools like DroidRanger [49] scan app markets for malicious apps. Various tools such as AppGuard [3], Dr. Android and Mr. Hide [29] or AppFence [27] can control the permissions or data an app can get. MalloDroid [17] looks for apps configured to use SSL incorrectly (for instance by not verifying hostnames or certificates).

AppPAL can act as a *"glue"* between static analysis tools and the app installation policies device owners are trying to enforce. This avoids creating tools with hard-coded fixed policies. For example a store might not want to sell apps with SSL errors or apps flagged by an anti-virus tool. Using AppPAL we can combine tools for checking apps to implement the store's policies.

```
'play-store' says App isSellable
  if App isAnApp
  where mallodroidCheck(App) = true,
        mcafeeAVCheck(App) = true.
```

No additional attempt is made to ensure these static analysis tools are sound. The policy designer must be aware of the tool's limitations. Black or whitelisting may have to be used to avoid some false positives or negatives.

2   Enforcing a Policy at Work

An employee *Alice* works for *Emma*. Emma allows Alice to use her personal phone as a work phone but has some specific concerns.

- Alice shouldn't run any apps that can track her movements. Alice's workplace is at a secret location and it mustn't be leaked.
- Apps should come from a reputable source, such as the Google Play Store.
- Emma uses an anti-virus (AV) program by McAfee. It should check all apps before they're installed.

To ensure this policy is met Alice promises to follow it. She might even sign a document promising never to break the rules within the policy. This is error-prone—what if she makes a mistake or misses an app that breaks her policy? Alternatively Emma's policy could be partially enforced using existing tools. *Google's Device Policy for Android* [23] could configure Alice's device to disallow apps from outside the Google Play Store and let Emma set the permissions granted to each app [40].

We could implement Emma's policy using existing tools (such as an AV checker, and a taint analysis tool like Flowdroid [1,33]) but it is a clumsy solution—they are not flexible. Each has to be configured separately to implement only part of the policy. If Emma changes her policy or Alice changes jobs she must recheck her apps and then alter or remove the software on her phone to ensure compliance. It isn't clear what an app must do to be run, or what checks have been done if it is already running on the phone. The relationship between Alice (the user), Emma (the policy setter) and the tools Emma trusts to implement her policy isn't immediately apparent.

What happens when Alice goes home? Emma shouldn't be able to overly control what Alice does in her private life. Alice might not be allowed to use location tracking apps at work but at home she might want to (to meet friends, track jogging routes or find restaurants for example). Some mobile OSs, such as iOS and the latest version of Android, allow app permissions to be enabled and disabled at run time. Can we enforce different policies at different times or locations?

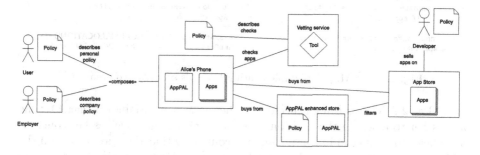

Fig. 1. Ecosystem of devices and stores with AppPAL.

We propose using the mobile ecosystem shown in Fig. 1. People have policies which are enforced by AppPAL on their devices. They can be composed with policies from employers or others to create enhanced devices that ensure apps meet the policies of their owners. The device can make use of vetting services which run tools to infer complex properties about apps. Users can buy from enhanced stores which ensure the only apps they sell are the apps which meet the store's explicit policies, or ones requested by users. Developers could decide which stores to sell their apps in on the basis of policies about stores.

3   Expressing Policies in AppPAL

In Sect. 2, Alice and Emma had policies they wanted to enforce but no means to do so. Instead of using several tools to enforce Emma's policy disjointedly, we could use an authorization logic. In Fig. 2 we give an AppPAL policy implementing Emma's app concerns on Alice's phone.

SecPAL is a logic of authorization for access control decisions in distributed systems. It has a clear and readable syntax, as well as rich mechanisms for delegation and constraints. SecPAL has already been used as a basis for other policy languages in areas such as privacy preferences [7] and data-sharing [2]. We present AppPAL as a modified form of SecPAL, aimed at mobile apps.

Other access control languages, such as XACML 3.0 [37], could also have been used as the basis for AppPAL. SecPAL however can capture distribution and delegation relationships between principles and serves as a simplified model of a more complex system like XACML, and has a well defined semantics, and decidability.

```
1   'alice' says 'emma' can-say inf        15   'emma' says 'anti-virus-policy' isMetBy(App)
2       App isRunnable.                     16   if App isAnApp
3                                           17   where
4   'emma' says App isRunnable             18      mcafeeAVCheck(App) = true.
5       if 'no-tracking-policy' isMetBy(App), 19
6          'reputable-policy' isMetBy(App),  20   'emma' says 'no-location-permissions'
7          'anti-virus-policy' isMetBy(App). 21   can-act-as 'no-tracking-policy'.
8                                           22
9   'emma' says                            23   'emma' says
10      'reputable-policy' isMetBy(App)    24      'no-location-permissions' isMetBy(App)
11          if App isBuyable.              25      if App isAnApp
12                                         26      where
13  'emma' says 'google-play' can-say      27         hasPermission(App, 'COARSE_LOCATION')=false,
14      App isBuyable.                     28         hasPermission(App, 'FINE_LOCATION')=false.
```

Fig. 2. AppPAL policy implementing Emma's security requirements.

In line 2 Alice lets Emma specify whether an `App` (a variable) `isRunnable`; she allows her to delegate the decision (`can-say inf`). Emma specifies her concerns as policies to be met in line 4: if Emma is convinced that these are met then she will say the `App` `isRunnable`. In line 10 and line 14 Emma specifies that an app meets the `reputable-policy` if the `App` `isBuyable`; with 'google-play' deciding of what is buyable or not. Google is not allowed to delegate the decision further, i.e. Google is not allowed to specify Amazon as a supplier of apps as well. Emma specifies the 'anti-virus-policy' in line 15 using a constraint. When checking the policy the `mcAfeeVirusCheck` should be run on the `App`. Only if this returns *false* will the policy be met. To specify the 'no-tracking-policy' Emma says that the 'no-location-permissions' rules implement the 'no-tracking-policy' (line 21). Emma specifies this in line 24 by checking the app is missing two permissions.

Alice wants to install a new app (`com.facebook.katana`) on her phone. She collects statements to show the app meets the `isRunnable` predicate.

- 'google-play' *says*'com.facebook.katana' isReputable. Required to convince Emma that the app came from a reputable source.
- 'emma'*says*'anti-virus-policy' isMetBy('com.facebook.katana'). She can obtain this by running the AV program on her app.
- 'emma'*says*'no-locations-permissions' isMetBy('com.facebook.katana'). Needed to show the App meets Emma's no-tracking-policy. Emma will say this if the app has no location permissions.

These last two statements require the checker to do some extra checks to satisfy the constraints. To get the second statement AppPAL must run the AV program on her app and check the result. The results from the AV program may change with time as its signatures are updated; so the checker must re-run this check every time it wants to obtain the statement connected to the constraint. For the third statement the AppPAL checker needs to examine the permissions of the app. It could do this by looking in the MANIFEST.xml inside the app itself, or through the Android package manager if it is running on a device.

We could also imagine Emma wanting a personalised app store where all apps sold meet her policy. With AppPAL this can be implemented by taking an existing store and selectively offering only the apps which will meet the user's policy. This gives us a *filtered store* which, from an existing set of apps, we get a personalised store that only sells apps that meet a policy.

4   AppPAL

AppPAL is implemented as a library for Android and Java. The parser is implemented using ANTLR4. AppPAL's syntax is inherited from SecPAL [6] (shown in Fig. 3).

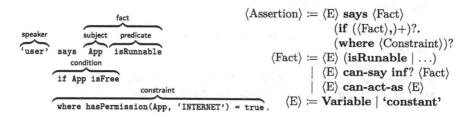

Fig. 3. Structure and simplified grammar of an AppPAL assertion.

In SecPAL the precise nature of predicates and constraints is left open. In instantiating SecPAL, AppPAL makes the predicates and constraints explicit. AppPAL policies can make use of the predicates and constraints in Table 1. Additional predicates can be created in the policy files, however constraints must be implemented individually. For example, on Android the hasPermission constraint uses the Android package manager to check what permissions an app requests, but the Java version uses the Android platform tools to check.

Splitting the decision about whether an app is runnable into a series of policies that must be met gives us flexibility in how the decision is made. It allows us to describe multiple means of making the same decision, and provide backup routes when one fails. Some static analysis tools are not quick to run. Even taking minutes to run a battery draining analysis can be undesirable: if a user wants to download an app quickly they may not be willing to wait to check that a policy is met. In that case, it may be preferable to delegate to an online database.

Table 1. AppPAL predicates and constraints.

Name	Description
App `isRunnable`	Says an app can be run
App `isInstallable`	Says an app can be installed
App `isAnApp`	Tells AppPAL that an app exists
Policy `isMetBy`(App)	Used to split policies into smaller components
`hasPermission`(App, Permission)	Constraint to check if an app has a permission
`beforeHourOfDay`(time)	Constraint used to check the time
`ToolCheck`(App, Property)	Constraint to run an analysis tool on an app

In Sects. 2 and 3 we described a *no-tracking-policy* to prevent a user's location being leaked. In Emma's policy we checked this using the app's permissions; if the app couldn't get access to the GPS sensors (using the permissions) then it meets this policy. Some apps may want to access this data, but may not leak it. We could use a taint analysis tool to detect this (e.g. FlowDroid [1,33]). Our policy becomes:

```
'emma' says'no−locations−permissions'
  can-act-as'no−tracking−policy'.

'emma' says'no−locations−permissions' isMetBy(App)
  if App isAnApp
  where
    hasPermission(App,'ACCESS_FINE_LOCATION') = false,
    hasPermission(App,'ACCESS_COARSE_LOCATION') = false.

'emma' says'location−taint−analysis'
  can-act-as'no−tracking−policy'.

'emma' says'location−taint−analysis' isMetBy(App)
  if App isAnApp
  where
    flowDroidCheck(App,'Location','Internet') = false.
```

Sometimes we might want to use location data. For instance Emma might want to check that Alice is at her office. Emma might track Alice using a location tracking app. Provided the app only talks to Emma, and it uses SSL correctly (using MalloDroid [17]) she is happy to relax the policy.

```
'emma' says'relaxed−no−tracking−policy' canActAs'no−tracking−policy'.
'emma' says'relaxed−no−tracking−policy' isMetBy(App)
  if App hasCategory('tracking')
  where
    mallodroidSSLCheck(App) = false,
    connectionsCheck(App,'[https://emma.com]') = true.
```

This gives us four different ways of satisfying the *no-tracking-policy*: with permissions, with taint analysis, with a relaxed version of the policy, or by Emma directly saying the app meets it. When we come to check the policy if any of these ways give us a positive result we can stop our search.

4.1 Policy Checking

AppPAL has the same policy checking rules as SecPAL [6]. AppPAL uses an assertion context of known facts and rules, as well as facts deduced while checking. While Becker et al. used a DatalogC based checking algorithm, we have implemented the rules directly in Java as no DatalogC library is currently available for Android. Pseudo-code is shown in Fig. 4.

On a mobile device memory is at a premium. We want to keep the assertion context as small as possible. For some assertions (like isAnApp) we derive them by checking the arguments at evaluation time. This gives us greater control of the evaluation and how the assertion context is created. For example, when checking the isAnApp predicate; we can fetch the assertion that the subject is an app based on the app in question. When delegating we will also be able to request facts from the delegated party dynamically (although this is not yet implemented).

```
def evaluate(ac, rt, q, d)
  return rt[q, d] if rt.contains q, d
  p = cond(ac, rt, q, d)
  if p.isValid then
    return (Proven, rt.update q, d, p)
  p = canSay_CanActAs(ac, rt, q, d)
  if p.isValid then
    return (Proven, rt.update q, d, p)
  else
    return (Failure, rt.update q, d, Failure)


def cond(ac, rt, q, d)
  ac.add q.fetch if q.isFetchable
  ac.assertions.each do |a|
    if (u = q.unify a.consequent) &&
      (a = u.sub a).variables == none
      return checkConditions ac, rt, a, d
  return Failure


def canSay_CanActAs(ac, rt, q, d)
  ac.constants.each do |c|
    if c.is_a :subject
      p = canActAs ac, rt, q, d
      return Proven if p.isValid
    elsif c.is_a :speaker
      p = canSay ac, rt, q d
      return Proven if p.isValid
  return Failure


def checkConditions(ac, rt, a, d)
  getVarSubs(a,ac.constants).each do |s|
    sa = s.sub a
    if sa.antecedents.all
      { |a| evaluate(ac, rt, a, d).isValid }
      p = evaluateC sa.constraint
      return Proven if p.isValid
  return Failure
```

Fig. 4. Partial-pseudocode for AppPAL evaluation.

4.2 Benchmarks

When AppPAL runs on a mobile phone, apps should be checked as they are installed. Since policy checks may involve inspecting many rules and constraints one may ask whether the checking will be acceptably fast. Downloading and installing an app takes about 30 seconds on a typical Android phone over wifi. If checking a policy delays this even further a user may become annoyed and disable AppPAL.

The policy checking procedure is at its slowest when having to delegate repeatedly; the depth of the delegation tree is the biggest factor for slowing the search. Synthetic benchmarks were created to check that the checking procedure performed acceptably. Each benchmark consisted of a chain of delegations. The *1 to 1* benchmark consists of a repeated delegation between all the principals. In the *1 to 2* benchmark each principal delegated to 2 others and in the *1 to 3* benchmark each principal delegated to 3 others. These benchmarks are reasonable as they model the slowest kinds of policies to evaluate—though worse ones could be designed by delegating even more or triggering an expensive constraint check.

For each benchmark we controlled the number of principals in the policy file: as the number of principals increased so did the size of the policy. The results are shown in Fig. 5. We have only used a few delegations per decision when describing hypothetical user policies. We believe the policy checking performance of AppPAL is acceptable as unless a policy consists of hundreds of delegating principals the overhead of checking an AppPAL policy is negligible.

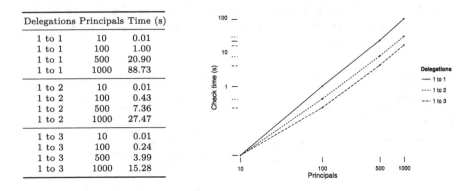

Delegations	Principals	Time (s)
1 to 1	10	0.01
1 to 1	100	1.00
1 to 1	500	20.90
1 to 1	1000	88.73
1 to 2	10	0.01
1 to 2	100	0.43
1 to 2	500	7.36
1 to 2	1000	27.47
1 to 3	10	0.01
1 to 3	100	0.24
1 to 3	500	3.99
1 to 3	1000	15.28

Fig. 5. Benchmarking results on a Nexus 4 Android phone.

5   Measuring Policy Compliance

Throughout we have asserted that users often have informal policies and that there is a need for policy enforcement tools. Corporate mobile security bring your own device (BYOD) policies have started appearing and NIST have issued recommendations for writing them [41,44]. In a study of 725 Android users, Lin et al. found four patterns that characterise user privacy preferences for apps [35] demonstrating a refinement of Westin's privacy segmentation index [32]. Using app installation data from Carat [12,38] we used AppPAL to find the apps satisfying each policy Lin et al. identify and measure the extent that each user was following a policy.

Lin et al. identified four types of user. The *Conservative* (C) users were uncomfortable allowing an app access to any personal data for any reason.

The *Unconcerned* (U) users felt okay allowing access to most data for almost any reason. The *Advanced* (A) users were comfortable allowing apps access to location data but not if it was for advertising. Opinions in the largest cluster, *Fencesitters* (F), varied but were broadly against collection of personal data for advertising. We wrote AppPAL policies to describe each of these behaviours as increasing sets of permissions. These simplify the privacy policies identified by Lin et al. as we do not take into account the reason each app might have been collecting each permission (we could write more precise rules if we could determine why each permission was requested). Lin et al. used Androguard [13] as well as manual analysis to determine the precise reasons for each permission [35].

Policy	C	A	F	U
GET_ACCOUNTS	✗	✗	✗	✗
ACCESS_FINE_LOCATION	✗	✗	✗	
READ_CONTACT	✗	✗	✗	
READ_PHONE_STATE	✗	✗		
SEND_SMS	✗	✗		
ACCESS_COARSE_LOCATION	✗			

It is also interesting to discover when people install apps classified as malware. McAfee classify malware into several categories, and provided us with a dataset of apps classified as malware and PUPs. The *malicious* and *trojan* categories describe traditional malware. Other categories classify PUP such as aggressive adware. Using AppPAL we can write policies to differentiate characterising users who allow dangerous apps and those who install poor quality ones.

```
'user' says'mcafee' can-say
 'malware' isKindOf(App).
'mcafee' says'trojan ' can-act-as'malware'.
'mcafee' says'pup' can-act-as'malware'.
```

If a user is enforcing a privacy policy we might also expect them to install less malware. We can check this by using AppPAL policies to measure the number of malwares each user had installed.

We now want to test how closely user behavior follows policies. Installation data was taken from a partially anonymized[1] database of installed apps captured by Carat [38]. By calculating the hashes of known package names we see who installed what. The initial database has over 90,000 apps and 55,000 users. On average each Carat user installed around 90 apps each; 4,300 apps have known names. Disregarding system apps (such as com.android.vending) and very common apps (Facebook, Dropbox, Whatsapp, and Twitter) we reduced the set to an average of 20 known apps per user. To see some variation in app type, we considered only the 44,000 users who had more than 20 known apps.

---

[1] Users are replaced with incrementing numbers, app names are replaced with hashes to protect sensitive names.

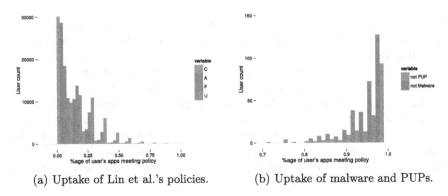

(a) Uptake of Lin et al.'s policies.   (b) Uptake of malware and PUPs.

Fig. 6. Policy compliance graphs. Each histogram shows the number of users who followed a policy to a certain extent. Users who installed no malware have been omitted from Fig. 6(b).

Using this data, and the apps themselves taken from the Google Play Store and Android Observatory [4], we checked which apps satisfied which policies.

Figure 6(a) shows that very few users follow Lin et al.'s policies most of the time. Whilst the AppPAL policy we used was a simplified version of Lin et al.'s policy, it suggests that there is a disconnect between user's privacy preferences and their behaviour (reminiscent of the *privacy paradox*); assuming the user population studied by Lin et al. behave similarly to data from the Carat study. A few users, however, did seem to be installing apps meeting these policies most of the time. This suggests that while users may have privacy preferences the majority are not attempting to enforce them. Policy enforcement tools, like AppPAL, can help users enforce their own policies which they cannot do easily using the current ad hoc, manual means available to them.

We found that 1 % of the users had a PUP or malicious app installed. Figure 6(b) shows that infection rates for PUPs and malware is low; though a user is 3 times more likely to have a PUP installed than malware. Users who were complying more than half the time with the conservative or advanced policies complied with the malware or PUP policies fully (Fig. 7(a)). This suggests that policy enforcement is worthwhile: users who can enforce policies about their apps experience less malware.

The *MalloDroid* tool [17] can scan apps for SSL misconfigurations. SSL misconfigurations are dangerous as they can undermine any privacy guarantees that SSL/TLS gives. MalloDroid distinguishes cases where the app is definitely misconfigured from those where there is some doubt. We set up AppPAL to use MalloDroid results as a constraint and measured the percentage of apps each Carat user had installed that did not have issues or suspected issues when scanned with MalloDroid. Users who were complied with the advanced policy were no better at avoiding apps with SSL errors than any other users, see Fig. 7(b). This emphasizes that AppPAL can help enforce complex policies that cannot be checked without additional tools.

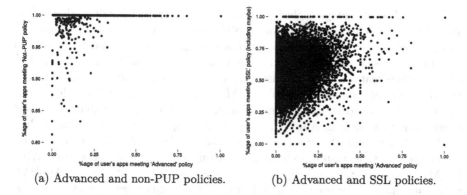

(a) Advanced and non-PUP policies.   (b) Advanced and SSL policies.

Fig. 7. Compliance with the advanced policy and the non-PUP and SSL policies. Each data-point represents a user. In (a) we see that users who followed the Advanced policy more than 50 % of the time did not install any malware. In (b) we see that even users who followed the Advanced policy were no better at avoiding apps with SSL problems than any other users.

There are limitations in this study: first, we do not have the full user purchase history, and we can only find out about apps whose names match those in available databases. So a user may have apps installed that break the policy without us knowing. Second, recently downloaded apps used for experiment may not be the same version that users had, in particular, their permissions may differ. Permissions tend to increase in apps over time [48]; so a user may be more conservative than our analysis suggests. Finally, as mentioned, we have compared a different set of users to the ones Lin et al. looked at. We plan to do a more comprehensive user study in the future that investigates AppPAL in use with different communities.

6   Related Work

Authorization logics have been used to enforce policies in several other domains. The earliest such logic, PolicyMaker [10], was general and undecidable. Logics that followed like KeyNote [9] and SPKI/SDSI [14] looked at public key infrastructure. The RT-languages [34] were designed for credential management. Cassandra [8] was used to model trust relationships in the British national health service.

SELinux is used to describe policies for Linux processes, and for access control (on top of the Linux discretionary controls). It was ported to Android [43] and is used in the implementation of the permissions system. SELinux describes the capabilities (in terms of system calls and file access) of processes, it cannot describe app installation policies or delegation relationships. Google also offer the *Device Policy for Android* app. This lets businesses configure company-owned devices to be trackable, remote lockable, set passwords and sync with their servers. It cannot be used to describe policies about apps, or describe trust relationships.

The SecPAL language is designed for access control in distributed systems. We picked SecPAL as the basis for AppPAL because it is readable, extensible, and is a good fit for the mobile ecosystem setting [26]. It has also been used to describe data usage policies [2] and inside Grid data systems [28]. Other work has added various features such as existential quantification [5] and extended to the DKAL family of languages [24,25]. DKAL contains more modalities than *says*, which lets policies describe actions principals carry out rather than just their opinions. For example in AppPAL a user might *say* an app is installable if they would install it (`"user" says App isInstallable`). In DKAL they can describe the conditions that would force them to install it (`"user" installs App`). With DKAL we can guarantee that the action was completed, whereas in AppPAL we do not know if the user actually installed a particular app. We chose to use SecPAL as the basis for AppPAL as we did not need the extra features DKAL added to express app installation policies for our initial applications.

Kirin [16] is a policy language and tool for enforcing app installation policies to prevent malware. Policy authors can specify combinations of permissions and broadcast events that should not appear together. For example, to stop malware sending premium rate text messages, we prevent an app having both the SEND_SMS and WRITE_SMS permissions one could write: `restrict permission [SEND_SMS] and permission [WRITE_SMS]`.

By analyzing apps which broke their policies Enck et al. found vulnerabilities in Android, but were ultimately limited by being restricted to permissions and broadcast events.

The Kirin approach has been shown to help identify malware, but it is less suitable for detecting PUPS. The behaviours and permissions PUP displays aren't necessarily malicious. One user may not want apps which need in-app-purchases to play, but another may enjoy them. With Kirin we are restricted to permitting or allowing apps. AppPAL can describe more scenarios than just allow or forbid, and use more app information than just permissions, such as constraints and static analysis results. By allowing delegation relationships we can understand the provenance and trust relationships in these rules.

7   Conclusions and Further Work

We have presented AppPAL: a language for describing app installation policies to help achieve security and privacy objectives but which can also *lock down* devices in other ways, e.g. restricting the use of certain apps while at work. We showed how static analysis tools can be connected to AppPAL to compose complex properties.

Further work is needed to tightly integrate AppPAL into Android. One way to integrate AppPAL on Android would be as a *required checker*: a program that checks all apps before installation. Google uses the required checker API to check for known malware and jailbreak apps. We would use AppPAL to check apps meet policies before installation. The API is protected, however, and it would require the phone to have a custom firmware. This is undesirable as it would make AppPAL difficult to install for most users, and negate the other security enhancements (such as timely updates and patches) provided by the standard Android system. AppPAL could be integrated as a service to reconfigure app permissions. Android Marshmallow has an iOS like permissions model where permissions can be granted and revoked at any time. These will be manually configurable by the user through the settings app. We can imagine AppPAL working to reconfigure these settings (and set the device's initial grant or deny states) based on a user's policy, as well as the time of day or the user's location. A policy could deny notifications while a user is driving, for example, by checking if they are using Android Auto [22] (an app to interact with a car's center console) or moving along a road at high speed.

Future work includes developing and testing, policies for users. Here we described a policy being specified by a user's employer. For most end-users writing a policy in a formal language unrealistic. With Ad-blocking software users subscribe to filter policies written by experts, such as EasyList [39]. We can imagine a similar scheme working well for app installation policies. Users subscribe to different policies by experts (examples could include no tracking apps, nothing with adult content, no in-app-purchase apps). Optionally the users could customize the policies further.

Policy composition raises further questions: what should happen when user's personal and work policies overlap or contradict? Future work will look at detecting these problems as well as integrating strategies to resolve them.

Another question might be whether we can use evidence to speed re-checking apps against a policy. Some static analysis tools, such as Evicheck [42], can create evidence that lets you check an app doesn't have certain behavior faster than it would be to infer the same property in the app without it, similar to proof-carrying code [36]. We can also imagine apps being distributed with evidence that proves the app meets an AppPAL policy but avoids the need to check the against the policy explicitly.

We might attempt to learn policies from existing user's behavior. Given app usage data, from a project like Carat [38], we could identify security conscious users. If we can infer these users policies we may be able to describe new policies that the less technical users may want. Given a set of apps one user has already installed, we could learn policies about what their personal security relevant installation policy is. This may help stores show users apps they're more likely to buy, and users apps that already behave as they want.

AppPAL gives us a framework for describing and evaluating policies for Android apps. The work provides new, rigorous, ways for machines to enforce user's and device-owner's rules about how apps should behave. These policies can be enforced more reliably, and with less interaction from the person operating the device.

Acknowledgements. Thanks to Igor Muttik at McAfee, and N Asokan at Aalto University and the University of Helsinki for discussions and providing us with data used in Sect. 5. Thanks also to the App Guarden project and colleagues at the University of Edinburgh for their comments, and the referees for their feedback.

References

1. Arzt, S., et al.: FlowDroid: precise context, flow, field, object-sensitive and lifecycle-aware taint analysis for Android apps. Program. Lang. Des. Implementation **49**(6), 259–269 (2014)
2. Aziz, B., Arenas, A., Wilson, M.: SecPAL4DSA. In: Cloud Computing and Intelligence Systems (2011)
3. Backes, M., Gerling, S., Hammer, C., Maffei, M., von Styp-Rekowsky, P.: App-Guard – enforcing user requirements on android apps. In: Piterman, N., Smolka, S.A. (eds.) TACAS 2013 (ETAPS 2013). LNCS, vol. 7795, pp. 543–548. Springer, Heidelberg (2013)
4. Barrera, D., Clark, J., McCarney, D., van Oorschot, P.C.: Understanding and improving app installation security mechanisms through empirical analysis of android. In: Security and Privacy in Smartphones and Mobile Devices, pp. 81–92, October 2012
5. Becker, M.Y.: Secpal formalization and extensions. Technical report, Microsoft Research (2009)
6. Becker, M.Y., Fournet, C., Gordon, A.D.: SecPAL: design and semantics of a decentralized authorization language. Comput. Secur. Found. (2006)
7. Becker, M.Y., Malkis, A., Bussard, L.: A framework for privacy preferences and data-handling policies. Technical report, Microsoft Research (2009)
8. Becker, M.Y., Sewell, P.: Cassandra: flexible trust management, applied to electronic health records. In: Computer Security Foundations, pp. 139–154 (2004)
9. Blaze, M., Feigenbaum, J., Keromytis, A.D.: KeyNote: trust management for public-key infrastructures. In: Christianson, B., Crispo, B., Harbison, W.S., Roe, M. (eds.) Security Protocols 1998. LNCS, vol. 1550, pp. 59–63. Springer, Heidelberg (1999)
10. Blaze, M., Feigenbaum, J., Lacy, J.: Decentralized trust management. In: Security and Privacy, pp. 164–173 (1996)
11. Bugiel, S., Davi, L., Dmitrienko, A.: Towards taming privilege-escalation attacks on Android. In: Network and Distributed System Security Symposium (2012)
12. Chia, P.H., Yamamoto, Y., Asokan, N.: Is this App Safe? World Wide Web, April 2012
13. Desnos, A.: Androguard. https://github.com/androguard/androguard
14. Ellison, C., Frantz, B., Lainpson, B., Rivest, R., Thomas, B.: RFC 2693: SPKI certificate theory. In: The Internet Society (1999)
15. Enck, W., Gilbert, P., Chun, B.G., Cox, L.P., Jung, J.: TaintDroid: an information-flow tracking system for realtime privacy monitoring on smartphones. In: Operating Systems Design and Implementation (2010)
16. Enck, W., Ongtang, M., McDaniel, P.: On lightweight mobile phone application certification. In: Computer and Communications Security, pp. 235–245, November 2009
17. Fahl, S., Harbach, M., Muders, T., Baumgärtner, L., Freisleben, B., Smith, M.: Why eve and mallory love Android. In: ASIA Computer and Communications Security, pp. 50–61, October 2012
18. Felt, A.P., Chin, E., Hanna, S., Song, D., Wagner, D.: Android permissions demystified. In: Computer and Communications Security, pp. 627–638, October 2011
19. Felt, A.P., Ha, E., Egelman, S., Haney, A., Chin, E., Wagner, D.: Android permissions: user attention, comprehension, and behavior. In: Symposium On Usable Privacy and Security, p. 3, July 2012

20. Feng, Y., Anand, S., Dillig, I., Aiken, A.: Apposcopy: semantics-based detection of Android malware through static analysis. In: Foundations of Software Engineering, pp. 576–587. ACM Request Permissions, New York, New York, USA, November 2014

21. Fuchs, A.P., Chaudhuri, A., Foster, J.S.: SCanDroid: automated security certification of Android applications. In: USENIX Security Symposium (2009)

22. Google: Android Auto. com.google.android.projection.gearhead

23. Google: Google Apps Device Policy. com.google.android.apps.enterprise.dmagent

24. Gurevich, Y., Neeman, I.: DKAL: distributed-knowledge authorization language. In: Computer Security Foundations, pp. 149–162 (2008)

25. Gurevich, Y., Neeman, I.: DKAL 2. Technical report, MSR-TR-2009-11, Microsoft Research, February 2009

26. Hallett, J., Aspinall, D.: Towards an authorization framework for app security checking. In: ESSoS Doctoral Symposium. University of Edinburgh, February 2014

27. Hornyack, P., Han, S., Jung, J., Schechter, S.: These aren't the droids you're looking for: retrofitting android to protect data from imperious applications. In: Computer and Communications Security (2011)

28. Humphrey, M., Park, S.M., Feng, J., Beekwilder, N., Wasson, G., Hogg, J., LaMacchia, B., Dillaway, B.: Fine-grained access control for GridFTP using SecPAL. In: Grid Computing (2007)

29. Jeon, J., Micinski, K.K., Vaughan, J.A., Fogel, A., Reddy, N., Foster, J.S., Millstein, T.: Dr. Android and Mr. Hide: fine-grained permissions in android applications. In: Security and Privacy in Smartphones and Mobile Devices, pp. 3–14, October 2012

30. Kelley, P.G., Consolvo, S., Cranor, L.F., Jung, J., Sadeh, N., Wetherall, D.: A conundrum of permissions. In: Useable Security, February 2012

31. Kim, J., Yoon, Y., Yi, K., Shin, J., Center, S.: ScanDal: static analyzer for detecting privacy leaks in android applications. In: Mobile Security Technologies (2012)

32. Krane, D., Light, L., Gravitch, D.: Privacy on and off the internet. Harris Interact. 18(5), 345–359 (2002)

33. Li, L., et al.: IccTA: detecting inter-component privacy leaks in Android apps. In: IEEE/ACM 37th IEEE International Conference on Software Engineering (2015)

34. Li, N., Mitchell, J.C.: Design of a role-based trust-management framework. In: Security and Privacy, pp. 114–130 (2002)

35. Lin, J., Liu, B., Sadeh, N., Hong, J.I.: Modeling users' mobile app privacy preferences. In: Symposium On Usable Privacy and Security (2014)

36. Necula, G.C., Lee, P.: Proof-carrying Code. Technical report, CMU-CS-96-165, Carniegie Mellon University (1996)

37. Oasis: eXtensible Access Control Markup Language (XACML) Version 3.0, January 2013

38. Oliner, A.J., Iyer, A.P., Stoica, I., Lagerspetz, E.: Carat: collaborative energy diagnosis for mobile devices. In: Embedded Network Sensor Systems (2013)

39. Petnel, R.: The Official EasyList Website. https://easylist.adblockplus.org/en/ (2016)

40. Poiesz, B.: Android M permissions. In: Google I/O (2015)

41. Scarfone, K., Hoffman, P., Souppaya, M.: NIST Special Publication 800–46: Guide to Enterprise Telework and Remote Access Security, June 2009

42. Seghir, M.N., Aspinall, D.: EviCheck: digital evidence for android. In: Finkbeiner, B., et al. (eds.) ATVA 2015. LNCS, vol. 9364, pp. 221–227. Springer, Heidelberg (2015). doi:10.1007/978-3-319-24953-7_17

43. Smalley, S., Craig, R.: Security enhanced (SE) android: bringing flexible MAC to Android. In: Network and Distributed System Security (2013)

44. Souppaya, M., Scarfone, K.: NIST Special Publication 800–124: Guidelines for Managing the Security of Mobile Devices in the Enterprise, June 2013

45. Svajcer, V., McDonald, S.: Classifying PUAs in the Mobile Environment, October 2013. sophos.com

46. Thompson, C., Johnson, M., Egelman, S., Wagner, D., King, J.: When it's better to ask forgiveness than get permission. In: The Ninth Symposium, p. 1, New York, USA. ACM, New York (2013)

47. Truong, H.T.T., Lagerspetz, E., Nurmi, P., Oliner, A.J., Tarkoma, S., Asokan, N., Bhattacharya, S.: The Company You Keep. In: World Wide Web, pp. 39–50, April 2014

48. Wei, X., Gomez, L., Neamtiu, I., Faloutsos, M.: Permission evolution in the Android ecosystem. In: Anual Computer Security Applications Conference, pp. 31–40. ACM Request Permissions, New York, New York, USA, December 2012

49. Zhou, Y., Wang, Z., Zhou, W., Jiang, X.: Hey, you, get off of my market: detecting malicious apps in official and alternative android markets. In: Network and Distributed System Security (2012)

Inferring Semantic Mapping Between Policies and Code: The Clue is in the Language

Pauline Anthonysamy[1,2(✉)], Matthew Edwards[2], Chris Weichel[2], and Awais Rashid[2]

[1] Google Switzerland, Zürich, Switzerland
anthonysp@google.com
[2] Security Lancaster, Lancaster University, Lancaster, UK
{p.anthonysamy,m.edwards7,c.weichel,a.rashid}@lancaster.ac.uk

Abstract. A common misstep in the development of security and privacy solutions is the failure to keep the demands resulting from high-level policies in line with the actual implementation that is supposed to operationalize those policies. This is especially problematic in the domain of social networks, where software typically predates policies and then evolves alongside its user base and any changes in policies that arise from their interactions with (and the demands that they place on) the system. Our contribution targets this specific problem, drawing together the assurances actually presented to users in the form of policies and the large codebases with which developers work. We demonstrate that a mapping between policies and code can be inferred from the semantics of the natural language. These semantics manifest not only in the policy statements but also coding conventions. Our technique, implemented in a tool (CASTOR), can infer semantic mappings with F1 accuracy of 70 % and 78 % for two social networks, Diaspora and Friendica respectively – as compared with a ground truth mapping established through manual examination of the policies and code.

1 Introduction

This paper addresses the problem of identifying areas of code that operationalize (or implement) one or more policy statement(s) from security or privacy policies. This problem is particularly challenging because information systems have grown not only in size and technical complexity but also in the volume of information they manage and process. The effort required to identify areas of code that implement relevant policies remains largely manual, at best aided by simple search techniques. Ideally, policies and code should be linked to ease processes such as compliance checks, verification, maintenance etc.; however this is not always the case for two main reasons:

(i) **Asynchronous Evolution of Policies and Code.** Policies describe organisations' actions on user data or personally identifiable information – and are often driven by regulatory and legal requirements. Program code, on the other hand, implements the various features and services provided by the

J. Caballero et al. (Eds.): ESSoS 2016, LNCS 9639, pp. 233–250, 2016.
DOI: 10.1007/978-3-319-30806-7_15

information system and must be compliant with the aforementioned policies. Modern information systems evolve rapidly as organisations continually update the system's functionality to provide a better quality of service and user experience. This is generally driven by factors such as changes in requirements, optimisation of code, fixes for bugs and security vulnerabilities, etc. Policies also change but such changes are less frequent and often driven by legislative requirements and regulatory frameworks or changes in business processes. This asynchronous evolution can often (unintentionally) lead to changes resulting in the code being non-compliant with the policy. A recent example is that of Facebook introducing a photo sync feature that allows users to sync their mobile photos with their Facebook account [5]. This feature introduced a vulnerability that allowed photos that had not been published on Facebook and should not have been visible to anyone be accessed by third-party applications; yet Facebook's terms of service continued to stipulate that private photos will stay private when connecting to external applications.

(ii) **Implementation Precedes Policies and Regulation.** In an ideal world, policies would be derived first, requirements established, and then passed on to software engineers for design and implementation. However, much modern software development does not follow this cycle. They also, almost always, out-pace the regulatory environment. Often, legal and regulatory requirements are not given full consideration during product development (requiring post-implementation compliance checks) or regulations come into existence *after* a system is in public use. For example, the European Commission only recently introduced regulation as part of its Data Protection Directive [4] requiring that users should have full export/download access to all of the data stored about them.

In this paper we present a technique and tool to infer and identify areas of code (aka functions) that implement particular policy statements described in natural language. Our inference technique is driven by the semantics of natural language and coding conventions, wherein verbs and nouns used in policy statements and source code (e.g., in function and parameter names) provide useful clues that enable a *semantic mapping* to be established between the two artefacts. Our use of naming conventions means that such mapping can be added to systems post-hoc – as we highlighted above, it is often impossible to attach security demands arising from high-level policies to methods at the time the code is written (e.g., because of codebases predating policies).

Contributions

We make the following novel contributions in this paper:

1. We describe a *semantic-mapping* approach to infer function specifications from natural language policies. The resulting technique aids developers in inferring and identifying relevant functions that implement one or more policy statement(s) to assist in compliance verification. Our technique demonstrates that the burden of identifying areas of code that operationalize relevant

policies can be reduced through inference using the semantic constructs of the natural language itself and coding conventions driven by such constructs.

2. An implementation of our technique in a tool called CASTOR is presented. It accepts as inputs policy statements and source code; and outputs a set of semantic mappings between policy statements and function specifications (methods). These mappings aid one to assess the completeness of an implementation with respect to stated policies. More importantly, the semantic mapping which is established *directly* between policy statements (as presented to users) and source code deem useful for organisations in quality assessment and compliance preservation.

3. We present an evaluation of our technique and tool on inferring mapping between privacy policies and the code implementing these policies for two open-source social networking sites, namely Diaspora and Friendica. Our evaluation shows that we can achieve a F1 accuracy of 70 % for Diaspora and 78 % for Friendica (for the balanced class experiment) in finding the semantic mappings required as compared with a ground truth mapping established by thorough manual examination of the policies and code.

2   Related Work

In this section we first contrast our work with techniques that automate mappings between textual documents and source code, followed by an approach to privacy leak detection using flow analysis. We then discuss techniques that automate mappings between policies and software requirements specifications.

Text Documents and Source Code: Pandita et al. [24] attempt to transform natural language descriptions of methods, as found in API documentation, into formal specifications for function behaviour, as described by *code contracts*. Their method involves parsing the API documentation through Part-Of-Speech (POS) tagging aided by domain-specific noun boosting and jargon handling, followed by the application of a shallow parser which attempts to classify the sentences of lexical tokens based on predefined semantic templates. The result of this process is a first-order logic expression which is then parsed for equivalences and redundancies and finally used to generate code contracts [3]. These contracts can then be inserted into the functions to which the corresponding API documents refer. While their work demonstrates the possibility of mapping between natural language and representations of source code, it differs from our work in two ways: firstly, we are interested in identifying implementations of policy statements in the source code, rather than in generating assertions for error–checking; and secondly, the mapping is done in a far more narrow domain than attempted in this paper, as API documentation is naturally more precisely connected to the source code it describes than user-facing texts such as policies.

Antoniol et al. [9,10] describe an approach to establish and maintain links between source code and free text documents such as requirements, design documents and user manuals, etc. Their work is based on the assumption that programmers use meaningful names for program concepts, such as functions, variables, types, classes, and methods; therefore, the analysis of these concepts (identifiers) can aid in associating high level concepts with program concepts, and vice-versa. The approach is based on a stochastic language model that assigns a probability value to every string of words taken from a prescribed vocabulary. The relevant documents are used to estimate the language models, one for each document or identifiable section. Then, a classifier – Bayesian classification – is used to compute the score of the sequence of mnemonics extracted from a selected area of code against the language models. A high score indicates a high probability that a particular sequence of mnemonics is extracted from the document, or a section, that generated the language model. This implies the existence of a semantic link between the document and the area of code from which the particular sequence of mnemonics is extracted. However, the approach is primarily applied to text that is likely to clearly express source code functionality (requirements specification documents). In contrast, our approach addresses the scenario where code and policies have either evolved independently or policies have come into existence post-development and deployment.

Privacy in Source Code: Jang et al. [18] approach the breach of privacy by user-facing websites through flow analysis of the Javascript code from several major websites. Their method involves the design and implementation of a language for specification of privacy-breaching information flows, and was trialled on a large sample of well-visited websites. Where their approach tests uniformly for four specific information breaching flows to identify violations, we aim to tie source code to the publicly expressed policies of social networking sites. Our approach also performs analysis of source code, but whereas their method analyses client-side code, we perform analysis of the server-side handling of information, which is arguably more critical for tracing potentially hidden violations.

Policies and Requirements: Massey et al. [21] evaluated the security and privacy requirements of an existing software system – the iTrust, open source electronic health record system – for legal compliance with a regulatory document (HIPAA). Their work mainly focuses on establishing trace links between software requirements and legal texts which, while an important initial step in legal compliance, does not fully complete the mapping between legal texts such as policies and the software code itself. Cleland-Huang et al. [14] proposed two machine learning methods to automatically generate links between regulatory codes (a subset of HIPAA) and product requirements. May et al. [22] present a framework that formalises regulatory rules, HIPAA, and exploit this formalisation to analyse the rules' conformance to a health-care system automatically. Fisler et al. [16] also attempt a model-checking based verification system, Margrave, for analysing role-based access control policies. However, these works focus on deriving software requirements from privacy policies and legal documents (primarily in the healthcare domains). In contrast, we aim to establish a semantic mapping between areas of code (functions) that implement particular policy statements described in natural language – and in situations where policies need to be mapped to code post-hoc implementation.

3   Semantic Inference

CASTOR's semantic inference mechanism (cf. Appendix A for CASTOR's architecture) presents a technique that enables software developers to infer and identify areas of code (aka functions) that implement particular policy statements described in natural language. This inference technique is driven by the semantics of natural language and coding conventions, wherein *verbs* and *nouns* used in policy statements and source code (e.g., in *function* and *parameter* names) provide useful clues that enable a *semantic mapping* to be established between the two artefacts.

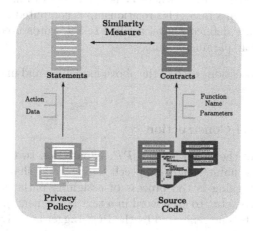

Fig. 1. An overview of our semantic mapping approach.

A premise of this work is that programmers use meaningful names for source code primitives, such as functions, parameters and classes. Much of the application domain knowledge that developers employ when writing code is often captured by these mnemonics for code primitives; thus these mnemonics aid in associating source code primitives with high-level concepts (e.g., policy statements) [10]. In this section, we provide basic definitions and concepts relating to policies, source code and the relationship between policy and source code primitives, wherein, we measure how close the code primitives, namely **functions** and **parameters**, are to the policy primitives of **actions** and **data**.

3.1   Definitions

The semantic mapping between policy and source code is drawn using a semantic relatedness measure between the primitives of these artefacts, namely, similarity between the words e.g., *data-parameter, action-function*, etc. As summarised below, we define a model for privacy policies, source code functions and followed by the *semantic relatedness* between the two. Herein,

- A policy, $\mathcal{PP}$, is considered to be a set of statements. Each statement $s \in \mathcal{PP}$ is modelled as the tuple $s = \langle a, D \rangle$, where $D = \{d_1, d_2, \ldots, d_n\} \subseteq \mathcal{D}$ is the data items referred to by the statement and $a \in A$ is the action verb (e.g., share, track, collect, etc.).
- $\mathcal{F}$ is the set of functions implemented in source code. A function $f = \langle c, n, P \rangle, f \in \mathcal{F}$ is modelled as a triple, where c is the class to which it belongs, n is the function's name and P is the set of its parameter names.
- Semantic relatedness, $\mathcal{R} : \mathcal{W} \times \mathcal{W} \rightarrow [0, 1]$ is the measure of semantic similarity between two words $w_0, w_1 \in \mathcal{W}$.
- $\mathcal{M}_F : \mathcal{S} \rightarrow \mathcal{W} \subseteq \mathcal{F}$ is a relationship between policy statements, $s \in \mathcal{S}$, and source code functions, $f \in \mathcal{W}$, where $\mathcal{W}$ is the subset of functions that map to one or more statements in $\mathcal{S}$. This relationship is computed using the semantic relatedness measure defined above, applied to the words used in policy statements and function/parameter names.

In the following subsections each of the above modelling and mapping techniques are elaborated.

3.2   Policy Model Construction

The construction of the policy model, $\mathcal{PP}$, is based on two common linguistic analysis techniques, namely part-of-speech tagging and shallow parsing. Part-of-Speech (POS) tagging is the process of assigning parts of speech, such as noun, verb, adjective, etc., to each word in a text (statements). A shallow parser accepts the lexical tokens generated by the POS tagger and divides those tokens into segments which correspond to certain syntactic units, such as noun phrases, verbs, verb phrases, etc. Figure 2 illustrates a simplified example of a parsed policy statement.

The annotated statements are mapped based on their grammatical functions to policy primitives of 'action', the activity that the actor performs and 'data', the data item to which an actor's action relates. In doing this, the fact that each grammatical function has a designated semantic role in natural language is exploited. Actions, for example, are expressed by any of the verbs or verb phrases (VP) in natural language, while data tends to be identified by nouns and noun phrases. For example, the tokens labelled [**VB:** post] and [**NN:** post] in Fig. 2 will be tagged as *action* and *data* respectively.

This grammatical mapping process is aided by a data dictionary to assist when mapping composite data primitives such as *'personally identifiable information'*. The data dictionary is used to associate and identify relevant noun phrases with pre-defined data classes. Without this, 'personally identifiable information' would be annotated as an adjective phrase by the POS tagger instead of as a noun phrase as required for this analysis. This association is essential to the

[WRB: Whenever] [PRP: you] [VB: post] [JJ: content]
[IN: like] [NN: status] [NNS: updates] [PRP: you] [MD: can]
[VB: select] [DT: a] [NN: privacy] [VBG: setting] [IN: for]
[DT: every] [NN: post].

Legend: **WRB:** Whadverb, **PRP:** Personal pronoun, **VB:** Verb, **JJ:** Adjective, **IN:** Preposition,
NN: Noun (singular), **NNS:** Noun(plural), **MD:** Modal, **DT:** Determiner, **VBG:** Verb

Fig. 2. An example of tagged policy statement.

semantic mapping step in which such composite data primitives are expanded
to obtain the individual data items that are grouped within that class such as
'gender', 'sexuality', 'relationship status', etc. Note, the POS Tagger used here
was adapted from the Stanford Parser [19].

To aid in retaining the core elements of the policy statements, i.e., verbs and
nouns, selected terms and grammatical constructs are removed. These include
stop words (e.g., *the, is, at, when*, etc.), personal and possessive pronouns. The
decision to retain only the core elements of the policy statement is to construct
an intermediate policy model that is easy to comprehend and allows for cohe-
sion with the original statement. Although, formalised policies like P3P [15]
and EPAL [11] have been proposed to make policies more readable and enforce-
able, they have several limitations, e.g., the P3P language does not have a clear
semantics and can therefore be interpreted and presented differently by different
user agents; and, an EPAL policy must be enforced at the time data is accessed
which causes significant performance overhead – every data access has to rely
on an external policy evaluation. Furthermore, the policy model proposed in
this work avoids the additional complexity that comes with formalisation and
utilises the semantics of natural language constructs which can be interpreted
and translated appropriately.

3.3   Source Code Model Construction

The source code model is constructed automatically using a (naive) static program analysis technique [26]. The analyser parses the code base of an online
social network and constructs a model based on class, functions and parameter names. This model is inspired by code contracts [23] which are a way of
abstractly expressing what a function accomplishes. Functions, $\mathcal{F}$, are modelled
as triples, $f = \langle c, n, P \rangle$, where c is the *class* to which the function belongs, n is
the *function's name* and P the set of its *parameter names*. Note: in this paper
the terms 'parameter' and 'variable' are used interchangeably. These code principles are extracted to ease the semantic mapping (described next) of policy and
source code primitives.

3.4 Semantic Mapping

The semantic mapping, $\mathcal{M}_F$, between $\mathcal{S}$ and $\mathcal{F}$ is based on the premise that policy statements are operationalized as functions at the source code level. The strategy for establishing this semantic mapping is based on a hybrid approach of Natural Language Processing (NLP) and machine learning applied to policy statements and source code. We use a lexical resource, namely WordNet[1] to discover the semantic relatedness, $\mathcal{R}$, (a measure of "similarity") between policy and source primitives. WordNet is a broad coverage lexical network of English words that contains around 100,000 terms, organised into taxonomic hierarchies. Nouns, adjectives, verbs and adverbs are organised into networks of synonym sets (synsets) that each represent one underlying concept and are interlinked with a variety of relations. For instance, a word that has multiple meanings (polysemous) will appear in one sysnset for each of its definitions. The measure of relatedness between two words, $w_0, w_1 \in \mathcal{W}$, in WordNet is computed using path length in the network graph: $\mathcal{R} : \mathcal{W} \times \mathcal{W} \to [0, 1]$. The shorter the path from one word to another, the more similar they are.

We then use a machine learning technique to map statements to functions using the computed similarity measures (input to the machine learning algorithm). The trained classifier can then distinguish between a correct and incorrect mapping when it is confronted with new similarity values by using the learned mapping model.

Examination of Naming Conventions: As previously mentioned, the semantic mapping approach is drawn from the concept of relating policy and source primitives. We measure how close the source primitives (variables/parameters or functions) are to the policy primitives of actions and data. Common programming practices tend to dictate that functions are named as verbs and variables are named as nouns [25]. These naming conventions are crucial to this approach, so we verified whether this practice held in the real world. A unigram POS tagger from the Python Natural Language Toolkit[2] was run across the source code from two social networks, Diaspora[3] and Friendica[4]. These two code bases are the datasets used for evaluation in this paper.

The tagger was trained on Brown corpus[5] (a general text collection that contains 500 samples of English-language text, totalling roughly to one million words), with a regular expression based backoff parser implementing a technical dictionary. We ran the tagger over a collection of function and variable names drawn from the source code of Diaspora and Friendica. As the common `camelCase` and `snake_case` coding conventions are likely to confuse a natural language tagger, such examples were split into their individual words (e.g., `camelCase` to `camel case`).

---

[1] http://wordnet.princeton.edu/.

[2] http://nltk.googlecode.com/svn/trunk/doc/howto/wordnet.html.

[3] https://github.com/diaspora.

[4] https://github.com/friendica/friendica.

[5] http://www.essex.ac.uk/linguistics/external/clmt/w3c/corpus_ling/content/corpora/list/private/brown/brown.html.

Table 1. Verb and Noun percentages of function names and variables.

	% Nouns	% Verbs	# Tagged
Parameters	77.50	6.82	21034
Parameters (split)	75.35	8.23	27967
Function name	68.04	25.68	5366
Function name (split)	56.48	27.89	11842
Function name (first token)	43.20	44.20	5366

As shown in Table 1, while function parameters mapped as expected to nouns (77.50 %), results vary for the function name mapping. Unsplit function names were mostly categorised as nouns by default, but splitting these names into constituent tokens revealed a modest increase in the proportion of tokens identified as verbs. Further examination showed that the first token after such splitting was in most cases a verb, as in **get_name** or similar constructs. 44.2 % of function names contained at least one verb token. The relatively high parameter-to-noun and function-to-verb semantic relatedness illustrates that the approach for data-to-parameter and action-to-function mappings is a viable measures in terms of drawing a similarity between policy and source primitives.

Mapping Inference: The problem of mapping policy statements to source code functions that operationalize those statements is formulated as a binary classification problem, because the mappings are either correct or incorrect. Our semantic inference is an application of the Random Forests [12] classifier, which is an effective approach to the problem of learning and classification [17,20]. We found that this classifier best fitted our mapping model and outperformed other standard classifiers such as naïve bayes [1] and support vector machine [2]. The classifier needs to be trained once per social network (domain-dependent), as random forests are a supervised learning technique. This is performed using manually created mappings. By confirming the manually mapped $s - f$ pairs, one can then provide more training data to the classifier and improve its prediction.

To infer the mapping, for each policy statement $s \in S$ the classifier predicts if a source code function $f \in \mathcal{F}$ maps to that statement, that is $\langle s, f \rangle \in \mathcal{M}_F$. And, to do this, labelled examples, i.e., a training dataset of correct and incorrect mappings, are required to estimate a 'target learning model' in the machine learning technique. This estimated learning model is then used to classify an input vector of features into classes. In CASTOR, the labelled examples are generated using manually created mappings. These manually created mappings are established based on a method that was derived in prior work [8]. The method provides a systematic means of studying the traceability (mapping) between privacy policies and controls in social networks, hence establishing the *degree of traceability* between the two. In [8], we define the degree of traceability as the level of certainty that we can have about the existence of an *externally observable relationship* measured using a qualitative 3-point scale.

By confirming the manually mapped statement/function pairs, one can then provide more training data to the classifier and improve its prediction (target learning model). We label such manual mapped $\langle s, f \rangle$ pairs as G (indicating correct mappings), while non-mapping pairs are labelled N (indicating incorrect mappings). For each statement, function pair $\langle s, f \rangle$ we extract a feature vector $\mathbf{v} = \langle dc, af, dp, pc \rangle$ for classification:

1. **data-class-similarity**, $dc = \mathcal{R}_{\max_{d \in \mathcal{D}_s}}(d, c_f)$, where $\mathcal{D}_s$ is the set of data items of the statement s, and c_f is the class name of the function f;
2. **action-function-similarity**, $af = \mathcal{R}_{\max_{a \in A_s}}(a, n_f)$, where A_s is the set of actions of the statement s, and n_f is the name of the function f;
3. **data-parameter-similarity**, $dp = \mathcal{R}_{\max_{d \in \mathcal{D}_s, p \in P_f}}(d, p)$, where $\mathcal{D}_s$ is the set of data items of the statement s, and P_f are the parameter names of the function f;
4. **parameter count**, $pc = |P_f|$ is the number of parameters of function f.

The WordNet path similarity is used as a measure for semantic relatedness of the feature variables dc, af, and dp. When actions, data items, parameter names or function names consist of multiple words W, the maximum similarity of these words were used as semantic relatedness: $\mathcal{R}(W, x) = \max_{w \in W}(w, x)$.

The measure of semantic relatedness as outlined above generates a set of vector of features for the learning method, which classifies each vector of features into the set of mapping classes, $V = \{G, N\}$. For example (for the training dataset), the feature vector for the statement–function pair $\langle s_1, f_6 \rangle$ shown below (see Statement s_1 & Listing. 1.1) is $\mathbf{v} = \langle 0.67, 0.00, 0.74, 5 \rangle$. Thus, in order to calculate the most probable class (G or N) for this vector, the features are run down all of the trees in the forest and the final class of the vector is decided by aggregating the votes (i.e., predicted class) of each tree – which is G in this case.

Statement, s_1: The default privacy setting for some of the information you post on Diaspora is set to "everyone".

```
1   def setDefault
2       #note :id references a postvisibility
3
4       @post = Post.where(:id => params[:post_id]).select("id, guid, author_id").first
5       @contact = current_user.contact_for(@post.author)
6
7       if @contact && @vis = PostVisibility.where(:contact_id => @contact.id,
8                                   :post_id => params[:post_id]).first
9           @vis.hidden =! @vis.hidden
10          if @vis.save
11              render 'update'
12              return
13          end
14      end
15      render :nothing => true, :status => 403
16  end
17  end
```

Listing 1.1. Snippet of function, f_6, setDefault from the Diaspora code base.

4   Evaluation

The data used in our experiments consists of privacy policies and source code of two social networks: Diaspora and Friendica. Both of these sites are decentralised social networks implemented using Ruby on Rails and PHP respectively. We selected these sites in accordance with the following constraints: availability of source code (open-source), at least 1000 function specifications, and the fact that they are implemented using different programming languages and frameworks. The motivation behind this selection is to test the coverage of our semantic mapping technique across different conventions used in real–world implementations.

Since the two social networks are decentralised open source networks, there were no publicly available privacy policies. This is a constraint that we faced since most popular social networks with a published privacy policy are closed source systems. We, therefore, synthesised policies drawing upon our earlier detailed investigation of privacy policies of 16 social networks [7], in which we showed that there exist a significant disconnect between policy statements and user-facing privacy controls. The synthesised privacy policies were representative of those that would be shown to users of these sites[6].

This section describes the different (independent) experiments conducted using machine learning techniques for the semantic inference. Recall that for each experiment the input to the classifier is the set of pairs $\langle s, f \rangle$ where each pair consists of the features $\mathbf{v} = \langle dc, af, dp, pc \rangle$. The results of these experiments and the conclusions drawn are then presented.

4.1   Experiment 1: Unbalanced Classes

There was a drastic imbalance of classes in our experimental datasets. Non-mapping statement–function pairs (class N) are far more common than mapping ones (class G) – see Table 2. This is due to the inherent nature of our input, there are significantly more contracts that are not relevant to policy statements compared to those that are relevant. In this unbalanced experiment we train the classifier on this unbalanced data, but adjust the weights of the class importance during learning, so that the equal error rate $EER = |FPR - FNR|$ is minimised (FPR is the false positive rate, FNR is the false negative rate).

Table 2. Class imbalance $\|N\| \cdot \|G\|^{-1}$ for all 2 datasets, with and without heuristics.

Dataset	No Heuristics	Heuristics	% Reduction
Diaspora	1347.01	601.98	44.69
Friendica	2195.99	700.81	31.91

---

[6] See example policies at http://www.paulineanthonysamy.com/myData.html.

For each dataset we manually created a ground truth mapping (based on the method in [8]). We trained network-dependent classifiers using an 80/20 training/test data split which we evaluated using a randomised cross validation. We report scores based on true positive rates (recall), *TPR*, false positive rates, *FPR*, precision, *PPV* and *F1* score. The recall score for each class, namely *G* and *N*, provides information on the number of semantic mappings that were successfully identified, while the precision score takes into account all identified mappings for each class and evaluates how many of them were actually relevant. Finally, the F1 score is the harmonic mean of precision and recall (see Appendix B).

4.2   Experiment 2: Balanced Classes

A common practice for dealing with imbalanced data sets is to rebalance them artificially. This is essential to evaluate the fundamental soundness of our semantic mapping approach. Over and under-sampling methodologies have received significant attention as a technique to rebalance classes [13]. Therefore, in the second experiment we trained and tested CASTOR's classifier on balanced datasets. For each dataset (one for each of the two social networks) we balance both classes (G, N), by randomly sampling an equal number of statement/function pairs. This random resampling method for balancing classes has been shown to be an effective technique when faced with an imbalance problem [13] as in our case.

$$\operatorname*{rand}_{s \in \mathcal{PP}, f \in \mathcal{F}} \langle s, f \rangle \text{ s.t. } |\langle s, f \rangle \in \mathcal{M}| = |\langle s, f \rangle \notin \mathcal{M}|.$$

4.3   Experiment 3: Introducing Heuristics

To alleviate the class imbalance, we introduce heuristics that exclude source code functions that are unlikely to map to policy statements. An expert would expect operationalizing functions to be located in specific places (i.e. packages and folders), depending on the programming language and framework that was used to implement the social network. We encode that knowledge and reject functions based on where in the source code they are defined. Below are some of the heuristics introduced:

– Global: Sources   within   the   'db/', 'spec/', 'config/', 'lib/', 'script/', 'markdown/' folders across our dataset were removed. These folders were selected as they contain database table descriptions, application wide configuration files, third-party library files, scripts and markdown files.
– Framework Specific: These were mainly to deal with the different terminologies and spellings among the folders.
  • PHP: Sources within the 'view/', 'util/', 'test/', 'mods/', 'library/' folders were removed.
  • Ruby: Sources within the 'presenters/', 'assets/', 'views/', 'mailers/', 'error_message', 'layout' folders were removed.

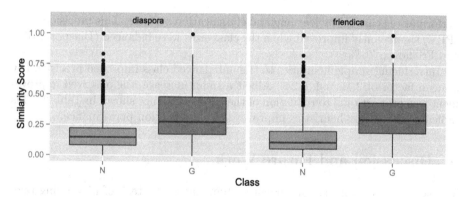

Fig. 3. Mean similarity scores of ground truth, G, and non-ground truth mappings, N.

Our heuristics do not reject functions that were manually labelled as ground truth. This way we reduce the class imbalance by 44.69 % for Diaspora and 31.91 % for Friendica respectively across the two social networks (see Table 2).

4.4   Results

Figure 3 depicts a box plot of mean similarity scores obtained from WordNet for ground truth (mapped), G, and non-mapping, N, statements respectively – indicated by the center horizontal line within each box. The outliers are represented by •. The scores are computed for each statement, function pair $\langle s, f \rangle$ with dc, af, dp, pc. As illustrated by Fig. 3 the mean scores for the two sites are higher for the ground truth (mapped) statements, namely 0.343 (s.d. 0.220) for Diaspora and 0.346 (s.d. 0.208) for Friendica. In comparison the non-mapping statements' means were 0.146 (s.d. 0.117) for Friendica and 0.166 (s.d. 0.127) for Diaspora. These values show that, although the overall similarity scores are small, WordNet consistently returned a higher similarity score for statements in G than statements in N, which warrants that our semantic mapping approach achieves its aim as to infer the mapping between policy statements and code.

The semantic mapping results are reported in Table 3 for all three experiments: unbalanced, balanced, and with heuristics. In all instances, the recall (TPR) rates were consistently high for Diaspora (between 0.69 and 0.78) and Friendica (between 0.79 and 0.80) indicating a high level of success in the identification of semantic mappings for each of our classes – G and N. These rates are crucial as it illustrates that our approach works in the non-optimal case, i.e., unbalanced classes, which is the norm in the real world. The consistent TPR and FPR rates shows that our approach generalises, and performs well, over different social networks. The EER (representing the number of false positive and false negative are equal) were also consistently low across all the experiments – at an average of 5 % and 6 % for Diaspora and Friendica.

We observe a very low precision in the unbalanced experiment (0.002). This is to be expected as it has been observed previously [13] that class imbalance (i.e., significant differences in class sizes) may produce a deterioration of the performance achieved by learning and classification systems. This precision score (PPV) significantly improved when the class sizes were balanced (Diaspora: 70 % and Friendica: 76 %).

Introducing simple heuristics to the unbalanced class improved precision (by a mean factor of 2.19, s.d. 0.24). Albeit a small increase, the observed improvement was proportional to reduction of the class imbalance shown in Table 2. This indicates that using heuristics improves the classification performance.

5   Discussion and Future Work

The scale and complexity of current systems make the task of identifying relevant sections of code (functions) that implement or realise a policy extremely challenging. Our technique demonstrates that this burden of identifying areas of code that operationalizes relevant policies can be reduced through inference (**F1 accuracy** of **70 %** and **78 %** for Diaspora and Friendica – balanced class experiment) using the semantic constructs of the natural language itself and coding conventions driven by such constructs. Though the functionality of a method is most critical in ensuring that requirements are upheld, this mandates that security demands arising from high-level policies are explicitly attached to methods at the time the code is written. This is infeasible nigh impossible in typical scenarios where code bases predate policies. Our approach allows this connection to be made based on well-established naming conventions. While this would never be as precise as a detailed semantic analysis of each method's code, the latter would be extremely expensive. Our usage of naming conventions means that such mapping can be easily added (post-hoc) to systems to highlight methods, which may need to be checked against security demands arising from policies. By identifying and short-listing the relevant methods, our approach not only benefits developers but potentially policy or compliance auditors for data sensitive systems such as Facebook and Google that are prone to accidental breaches.

Table 3. Table showing results from Random Forest classifier. The table labels are as follows:- Recall: TPR, False Positive Rate: FPR, Precision: PPV, F1 score: $F1$, Equal Error Rate: EER.

Dataset	TPR	FPR	PPV	F1	EER
Diaspora					
Balanced	0.693	0.296	0.700	0.696	0.011
Unbalanced	0.759	0.265	0.002	0.004	0.024
Heuristic	0.777	0.301	0.004	0.008	0.078
Friendica					
Balanced	0.788	0.245	0.762	0.775	0.033
Unbalanced	0.806	0.242	0.001	0.003	0.048
Heuristic	0.790	0.315	0.003	0.007	0.105

Table 4. Table showing results from Random Forest classifier with the verb synonym database. The table labels are as follows:- Recall: TPR, False Positive Rate: FPR, Precision: PPV, F1 score: $F1$, Equal Error Rate: EER.

Dataset	TPR	FPR	PPV	F1	EER
With verb synonym database					
Diaspora					
Balanced	0.785	0.251	0.757	0.771	0.031
Unbalanced	0.735	0.250	0.002	0.004	0.013
Heuristic	0.762	0.245	0.006	0.011	0.017
Friendica					
Balanced	0.797	0.209	0.792	0.795	0.006
Unbalanced	0.836	0.230	0.002	0.003	0.066
Heuristic	0.806	0.281	0.004	0.008	0.087

Limitations: Our semantic mapping approach relies on WordNet's similarity measures to compare policy and source code primitives. The overall WordNet similarity scores are low as it is designed as a dictionary based on psycholinguistic principles rather than a knowledge base. WordNet lacks contextual policy information. For example, in a social-networking policy, WordNet does not interpret 'track' as 'recording information' therefore we were compelled to take the most-related pair of synsets among the matched options. We hypothesized that these measures can be significantly increased if a verb-synonym database was available and later confirmed it [6]. The verb synonym database was built by extracting all the verbs from the privacy policies analysed in [8] and manually classifying them based on their semantic meanings. The semantic meanings of these verbs were determined using a lexical dictionary.

The results of the three experiments improved when conducted with a verb synonym database (cf. Table 4). In particular, the recall rates (TPR) increased for both datasets – Balanced: Diaspora: 78.5 % and Friendica: 79.7 %; Unbalanced: Diaspora 73.5 % and Friendica 83.6 %; and, Heuristic: Diaspora 76.2 % and Friendica 80.6 %. The precision also increased for both the datasets in the balanced class experiment, i.e., 77.1 % and 79.5 % accordingly. Although, the precision score was still relatively low in the unbalanced and heuristic experiments (due to the fact that the classes were still vastly disproportionate), there was a small hike in Friendica's PPV rates – 0.1 % rise – but there was no change in Diaspora. Whereas, the heuristic experiment improved the PPV rates for both datasets, i.e., about 0.2 % in Diaspora and 0.1 % in Friendica.

Acknowledgements. This research was funded by Lancaster University 40th Anniversary Research Studentship and has no ties to the first author's current employment at Google.

A   Implementation: CASTOR

We have implemented our technique in a tool called CASTOR. Figure 4 illustrates the architecture of CASTOR. CASTOR accepts as inputs policy statements and source code; and outputs a set of semantic mappings between policy statements and functions. Briefly, CASTOR works on the input as follows:

Policy Engine: CASTOR's policy engine is composed of a parser and a statement analyser which transforms the natural language policy into an intermediate representation (as described in Sect. 3.2). This intermediate representation maintains the relevant policy primitives of a statement, namely action (verbs) and data (nouns).

Code Engine: CASTOR's code engine is composed of a minimal recursive-descent parser that extracts a function's name, associated class and parameters, along with information identifying the source file and line number where the function can be found. This is inline with our source model construction in Sect. 3.3.

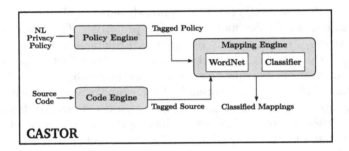

Fig. 4. CASTOR's architecture.

Mapping Engine: CASTOR's mapping engine infers the mapping between the privacy policy $\mathcal{PP}$ and source code functions $\mathcal{F}$ using its inbuilt WordNet corpora and classifier. The output of this engine is a set of semantic mappings between policy statement(s) and functions.

B   Formulae

$Recall$ (TPR) $= \frac{tp}{tp+fn}$; $False\text{-}Positive~Rate$ (FPR) $= \frac{fp}{fp+tn}$; $Precision$ (PPV) $= \frac{tp}{tp+fp}$; and $F1 = 2 \cdot \frac{Precision \cdot Recall}{Precision + Recall}$.

References

1. Naive bayes. http://www.nltk.org/_modules/nltk/classify/naivebayes.html
2. SVM. http://www.nltk.org/_modules/nltk/classify/svm.html
3. Code contracts (2010). http://research.microsoft.com/en-us/projects/contracts/
4. EU data directive 95/46/ec, February 2014. http://eur-lex.europa.eu/
5. Facebook photo leak flaw raises security concerns, March 2015. http://www.computerweekly.com/news/2240242708/Facebook-photo-leak-flaw-raises-security-concerns
6. Anthonysamy, P.: A framework to detect information asymmetries between privacy policies and controls of OSNs. Ph.D. thesis, Lancaster University (2014)
7. Anthonysamy, P., Greenwood, P., Rashid, A.: Social networking privacy: understanding the disconnect from policy to controls. IEEE Computer, June 2013
8. Anthonysamy, P., Greenwood, P., Rashid, A.: A method for analysing traceability between privacy policies and privacy controls of online social networks. In: Preneel, B., Ikonomou, D. (eds.) APF 2012. LNCS, vol. 8319, pp. 187–202. Springer, Heidelberg (2014)
9. Antoniol, G., Canfora, G., Casazza, G., De Lucia, A., Merlo, E.: Tracing object-oriented code into functional requirements. In: 8th International Workshop on Program Comprehension, 2000, Proceedings IWPC 2000, pp. 79–86 (2000)
10. Antoniol, G., Canfora, G., de Lucia, A., Casazza, G.: Information retrieval models for recovering traceability links between code and documentation. In: Proceedings of the International Conference on Software Maintenance (ICSM 2000). IEEE Computer Society, Washington, DC (2000)
11. Ashley, P., Hada, S., Karjoth, G., Powers, C., Schunter, M.: Enterprise Privacy Authorization Language (EPAL). Technical report, Rschlikon (2003)
12. Breiman, L.: Random forests. Mach. Learn. 45, 5–32 (2001). http://dx.doi.org/10.1023/A%3A1010933404324
13. Chawla, N.V., Japkowicz, N., Kotcz, A.: Editorial: special issue on learning from imbalanced data sets. SIGKDD Explor. Newsl. 6(1), 1–6 (2004)
14. Cleland-Huang, J., Czauderna, A., Gibiec, M., Emenecker, J.: A ML approach for tracing regulatory codes to product specific requirements. In: ICSE (2010)
15. Cranor, L., Langheinrich, M., Marchiori, M.: A P3P preference exchange language 1.0 (appel 1.0). World Wide Web Consortium, Working Draft WD-P3P-preferences-20020415, April 2002
16. Fisler, K., Krishnamurthi, S., Meyerovich, L.A., Tschantz, M.C.: Verification and change-impact analysis of access-control policies. In: Proceedings of the 27th International Conference on Software Engineering, ICSE 2005, pp. 196–205. ACM, New York (2005)
17. Haiduc, S., Bavota, G., Oliveto, R., De Lucia, A., Marcus, A.: Automatic query performance assessment during the retrieval of software artifacts. In: Proceedings of the 27th IEEE/ACM International Conference on Automated Software Engineering, ASE 2012, pp. 90–99. ACM, New York (2012)
18. Jang, D., Jhala, R., Lerner, S., Shacham, H.: An empirical study of privacy-violating information flows in javascript web applications. In: Proceedings of the 17th ACM Conference on Computer and Communications Security, CCS 2010, pp. 270–283. ACM, New York (2010)
19. Klein, D., Manning, C.D.: Accurate unlexicalized parsing. In: Proceedings of the 41st Annual Meeting on Association for Computational Linguistics (ACL 2003) - vol. 1. pp. 423–430, Stroudsburg, PA, USA (2003)

20. Ma, L., Torney, R., Watters, P., Brown, S.: Automatically generating classifier for phishing email prediction. In: 2009 10th International Symposium on Pervasive Systems, Algorithms, and Networks (ISPAN), pp. 779–783, December 2009

21. Massey, A., Otto, P., Hayward, L., Antn, A.: Evaluating existing security and privacy requirements for legal compliance. Requirements Engineering (2010)

22. May, M.J., Gunter, C.A., Lee, I.: Privacy APIs: access control techniques to analyze and verify legal privacy policies. In: Proceedings of the 19th IEEE Workshop on Computer Security Foundations, CSFW 2006, pp. 85–97. IEEE Computer Society, Washington, DC (2006)

23. Meyer, B.: Object-Oriented Software Construction, 1st edn. Prentice-Hall Inc, Upper Saddle River (1988)

24. Pandita, R., Xiao, X., Zhong, H., Xie, T., Oney, S., Paradkar, A.: Inferring method specifications from natural language api descriptions. In: Proceedings of the 34th International Conference on Software Engineering, ICSE 2012 (2012)

25. Rumbaugh, J., Blaha, M., Premerlani, W., Eddy, F., Lorensen, W.E., et al.: Object-Oriented Modeling and Design, vol. 199. Prentice Hall, Upper Saddle River (1991)

26. Wagner, D.: Static analysis and computer security: new techniques for software assurance. Ph.D. thesis, University of California at Berkeley, December 2000

Idea: Supporting Policy-Based Access Control on Database Systems

Jasper Bogaerts[✉], Bert Lagaisse, and Wouter Joosen

iMinds-DistriNet, KU Leuven, 3001 Leuven, Belgium
{Jasper.Bogaerts,Bert.Lagaisse,Wouter.Joosen}@cs.kuleuven.be

Abstract. Applications are increasingly operating on large data sets. This trend creates problems for access control, which in principle restricts the actions that subjects can perform on any item in that data set. Performance issues therefore emerge, typically for operations on entire data sets. Emerging access control models such as attribute-based access control do meet their limitations in this context. Worse, few solutions exist that addresses performance problems while supporting separation of concerns. In this paper, we present a first approach towards addressing this challenge. We propose a middleware architecture that performs policy transformations and query rewriting for externalized policies to optimize the access control process on the data set. We argue that this offers a promising approach for reducing the policy evaluation overhead for access control on large data sets.

Keywords: Access control · Policy-based access control · Databases · Attribute-based access control

1 Introduction

Applications are increasingly operating on large data sets. This is especially true for multi-tenant software-as-a-service (SaaS) applications, in which tenant organizations access a shared, typically web-based application hosted by a provider [12].

Such data must be protected. One important security measure to protect data is access control, which restricts actions performed by a subject (e.g., user) on an object (e.g., resource). A typical approach to realize this is to externalize an access control policy from the application and evaluate it each time a subject performs a request to the application [20, 22], a technique commonly referred to as Policy-Based Access Control (PBAC). This supports separation of concerns [7] and enables tenants in multi-tenant SaaS applications to specify their own policy without service interruption [8].

One challenge for policy-based access control is to enforce it for operations on a large data set. An *operation* comprises of the same action that is performed on each element of the data set. For example, when subjects perform a search on a database, only the elements (or *objects*) to which they are entitled should be returned. This involves a policy evaluation for the *view* action on the objects.

J. Caballero et al. (Eds.): ESSoS 2016, LNCS 9639, pp. 251–259, 2016.
DOI: 10.1007/978-3-319-30806-7_16

A naive approach to such issue is to serially evaluate the access control policy for each data item returned by the search query and filter the results. However, this can involve considerable evaluation overhead [21], especially for policies specified according to emerging access control models such as attribute-based access control [11]. On the other hand, while most database management systems support some form of efficient access control enforcement, they do not provide a solution to this problem that both scales in terms of the number of users and organizations, and meanwhile provides separation of concerns.

In this paper, we approach this issue by merging policy transformation and query rewriting techniques to optimize operations on large data sets. These techniques should be employable in a manner that is transparent to the application developer, thereby accomplishing a separation of concerns. This paper provides a first analysis of the approach that combines these techniques for this purpose, and discusses challenges that need to be addressed to realize such an approach.

Organization. Section 2 provides a scenario and motivates the requirements. This section also elaborates on related work. Section 3 describes the middleware that we propose to mitigate the problem. Section 4 discusses the proposed solution further, and elaborates on the challenges. Section 5 concludes the paper.

2   Motivation

This paper describes how policy transformation and query rewriting techniques can be combined to enforce access control policies for operations on large data sets. Although several such operations can be performed, such as batch insertions or updates, our main focus will be to enforce this for search operations.

Fig. 1. The document management application enables tenants to specify policies.

To illustrate this issue, consider a document management SaaS application, that manages millions of business documents for multiple tenants, as shown in Fig. 1. These tenants can each specify their own policies to restrict their respective affiliates. For example, they could permit read access to the creator of a document, and to all members of his/her department (e.g., accounting). The affiliates should also be restricted access based on policies that were specified by the application developers. In general, for example, such an affiliate cannot access documents of another tenant. A tenant can not be restricted access to its own documents through policies of another tenant [8].

When affiliates of a tenant search for documents in the document management application, they must only be returned the documents they are entitled to. This involves enforcement of application policies (e.g., subjects can only access documents of their tenant) and tenant policies (e.g., subjects of the accounting department can view paychecks).

2.1 Requirements

Besides functional requirements previously indicated in the scenario, the solution must support several additional characteristics:

- **Transparency.** The middleware must integrate seamlessly with the existing application, and respect the principle of separation of concerns. More particularly, the application developer should not take into account access control when writing queries.
- **Support for application-level policies.** The middleware must take into account policies that reason about application concepts. An access control policy is *safe* if it only refers to concepts that exist in the application or, alternatively, refers to the subject that performs actions on the application. Similarly, a query is safe if it corresponds to the underlying database schema.
- **Support for tenant policies.** The middleware must take into account policies specified by tenants. Since tenants provide untrusted input in the policies, we must ensure that they are *secure*, i.e., that they do not escalate privileges over provider policies, nor are they vulnerable to injection attacks.
- **Support for expressive policies.** The middleware should support expressive policies. This enables tenants to specify fine-grained access constraints.
- **Performance.** The middleware should reduce the access control evaluation overhead for operations on large data sets. The overhead that is introduced by the middleware itself must be minimized.

2.2 Related Work

A lot of related work has been performed in the domain of database access control. Many traditional database management systems employ views, stored procedures and access control lists to restrict access for individual subjects [3]. However, many such techniques assume a two-tier architecture, which has no use when the application performs a query on behalf of a subject, as is the case in multi-tier architectures that are common today [19]. As a result, such access control techniques cannot be effectively enforced at the database at a granularity level that exceeds the application. This is also the case for many techniques that employ query rewriting to optimize access control [5,10,15,18].

One technique that mitigates this issue is Virtual Private Database (VPD, [1]). VPD supports application identification, as a complement to subject identification, to enforce access control policies that are specified in the DBMS. While this supports the specification of views and queries that are aware of individual subjects, VPD requires policies to be specified at the database management system. This does not adhere to the principle of separation of concerns. Moreover, it requires the policies to restrict in terms of database operations (e.g., insertion and selection), whereas application actions may involve multiple such operations.

Opyrchal et al. [16] have addressed a similar issue to our goal by enforcing CPOL policies for databases. Their system first checks whether a query is permitted and evaluates the policy for each returned element if it is. However, their method involves only limited query rewriting and no policy transformations. Consequently, the system does not scale when large data sets are involved, especially if many elements are returned. By performing query rewriting and policy transformation, our approach is able to reduce data sets on which access control is performed, and optimizes access control evaluation for the remaining sets.

Axiomatics data access filter [2] also provides a solution that enforces policy-based access control on databases for attribute-based policies. However, it does not provide sufficient constraints to ensure safe and secure queries in the light of, among others, multi-tenant applications.

Cook et al. [6] focus on the safety of composing queries. They propose a method that effectively restricts developers from specifying queries that do not correspond to the application domain. This is complementary to our work, but it does not focus on the middleware that performs safe query rewriting.

This work relates to that of access control and usage control enforcement techniques. Notably, in [17], Pretschner et al. present an architecture that is capable of enforcing usage control policies in a distributed fashion. A similar approach is taken in [9] to ensure access control enforcement is decentralized. While this relates to our objective to speed up enforcement for access control, our approach does not perform distributed evaluation to achieve this.

3 Approach

In order to meet the requirements that were introduced in Sect. 2, we propose a middleware that transparently performs access control for operations on large data sets. This section outlines the middleware architecture. The next section discusses the challenges and motivates its feasibility.

The middleware is embedded in a database abstraction layer, and intercepts every query that is performed on the database. The database abstraction layer provides an abstraction over data access and integrates technologies such as JDBC[1] or object-relational mappers (ORM) such as JPA[2] to hide database-specific complexity. The middleware intercepts the query and determines the objects on which the operation can be performed before executing it on them.

In order to support expressive policies, the middleware supports a XACML-like policy language. XACML [14] provides a tree-structured, attribute-based

---

[1] Java Database Connectivity, see also http://www.oracle.com/technetwork/java/overview-141217.html.

[2] Java Persistence API, see also http://www.oracle.com/technetwork/java/javaee/tech/persistence-jsp-140049.html.

policy language in which policies can themselves contain other policies, thus forming a policy *tree*. Applicability of both policies and rules are determined during the evaluation, in which a *target* expression indicates applicability for a policy, and a *condition* expression indicates applicability for a rule. If a rule is applicable, its corresponding decision (i.e., permit or deny) is taken into account. Expressions compare attributes assigned to subjects, objects, actions and environment with each other and concrete values in order to determine the applicability. Combining algorithms (e.g., first applicable, permit overrides) provide conflict resolution when multiple policies or rules are applicable.

3.1 Design

The middleware combines policy transformation and query rewriting techniques to optimize access control evaluation for large data sets. Figure 2 outlines the architecture. The access control middleware introduces two components: the policy transformer and the query translator.

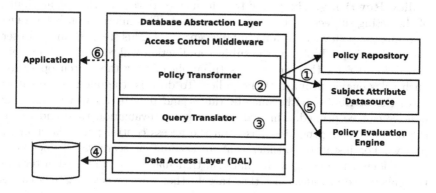

Fig. 2. The middleware employs policy transformations and query rewriting to reduce evaluation overhead for access control.

The policy transformer is responsible for retrieving, validating and transforming the policy. This results in a transformation of the original policy to a reduced form with the same semantics. The component achieves three goals. First, it substitutes all attributes that are not assigned to the object, and as a consequence, remain fixed for every element of the query that must be evaluated against the policy. Second, it prunes the policy by omitting the rules that are never applicable, thereby reducing evaluation time. Third, it reduces the policy and its tree structure in such a way that it simplifies the translation to a query.

The query translator takes as input the transformed policy, and translates it to a query that can be executed on the database. In order to support a wide range of underlying database systems, we do not require the policy to be incorporated in the resulting query entirely. Rather, the translator may select and translate a *partial* policy that can significantly reduce the data set on which the transformed policy must be evaluated. This enables the middleware to be used for database systems that have a constrained query language, which is common especially among NoSQL systems. Moreover, it supports translation to only queries that can be performed without significant overhead.

Figure 2 illustrates how the middleware handles the access control process. The access control process for search operations consists of six steps:

1. **Policy Retrieval and Validation.** The policy transformer retrieves and combines all policies that are relevant for the acting subject. Next, it validates that the concepts that are referred to in the policy apply to the application. This is especially a problem when attribute-based policies are involved, because attribute-based policies require an understanding of the properties associated with the subjects, objects, and actions of the application domain. Validation ensures that no errors can occur in the evaluation in this regard. Moreover, it can ensure that queries that are rewritten are safe (i.e., they refer only to application concepts) and secure (i.e., they do not lead to privilege escalation) because they are translated from a sanitized set of expressions.

2. **Policy Rewriting.** The policy transformer retrieves all relevant attributes of the acting subject, action and environment. This information is leveraged to prune the policy to contain only relevant rules. This stems from the observation that for each object, the subject and action will remain the same and may lead to rules that are always (in)applicable for the search operation. Such rules enable pruning of the policy. To do this, the middleware substitutes the attributes with concrete values and determines which rules and policies can be pruned. This can reduce policy evaluation time and reduce the query that is generated based on the access control policy later in the process. For instance, consider that an accountant searches all documents in the document management platform that match a certain search term. If the policy also includes rules that target other roles or actions, they can be pruned for the transformation.

3. **Policy to Query Translation.** In this step, the transformed policy is translated to a query. Note that the query should evaluate only concepts of the objects on which the search operation is performed, due to the substitution of attributes in the previous step. The query translator could select only a *partial* policy to translate to due to functional or performance constraints of the underlying database. Such a query must significantly reduce the size of the data set to optimize the access control evaluation process. For instance, consider again the example scenario of the previous step. Consider also that the policy states that accountants can only read financial documents that were created in the last year. The query translation could select only a part of this rule, e.g., that only financial documents may be read, to cope with constraints of the query language of the underlying database. This could already significantly reduce the resulting data set on which serial policy evaluation must be performed, because other types of documents for which the policy would evaluate to a deny decision are already filtered out.

4. **Query.** The query translator composes a query that takes into account both the original request parameters (e.g., search parameters) and the access control policy (translated in the previous step). Next, it retrieves all objects that satisfy the composed query from the underlying database.
5. **Policy Evaluation.** The policy transformer evaluates the previously transformed policy for each object that resulted from the query. This determines on which objects the subject is entitled to perform the search operation. For instance, consider the translation of a partial policy in the example of the third step. The previously transformed policy can be evaluated against the result set to enforce that the documents were created in the last year. Note that this step is redundant when the policy is fully translated to the query, and can be skipped in such a case.
6. **Result.** Finally, the resulting data set is returned to the subject.

4   Discussion

With the architecture presented in Sect. 3, we intend to significantly reduce the policy evaluation overhead for operations on large data sets. However, in order to do this, several challenges need to be addressed.

The middleware optimizes access control for large data sets through policy transformation and query rewriting. Both can introduce an overhead. On the one hand, policy transformation may introduce a processing overhead, and still requires considerable overhead when the transformed policy is evaluated against a large data set. On the other hand, performing a query that was rewritten according to a policy may also introduce an overhead when applied to a large data set. Consequently, a balance must be found in determining to what extent the query is rewritten. This is complicated by the variability of the underlying database schema and the query languages that are supported by the database system. This is a considerable challenge for future work, especially when NoSQL systems must be supported. In general, these systems are identified by constraints in their query language and a potentially large cost for performing certain types of queries. Whenever a policy is only translated partially into a query, this requires an evaluation of the transformed policy over the objects in the data set that results from this query. This is performed in the fifth step of the process. Because some of the expressions of the policy are already included as part of the query, the policy could be further transformed to omit these expressions and hence avoid redundant evaluation. When the policy can be translated fully into a query, this post-evaluation step can be omitted altogether.

While we focused on constraining search operations, a similar approach could be performed for write operations (e.g., batch updates). This would involve a step that filters out objects for which the operation is not permitted prior to performing the query, if the policy can not be translated fully to a query.

The strategy introduced in this paper requires that both database and middleware preside in the same security domain. Else, the solution would be subject to data leaking for objects that were not filtered by the query, but are withheld by the policy evaluation.

In order to determine the feasibility of the approach presented in this paper, we have induced a prototype that is capable of handling policy transformations for the STAPL [13] policy language (which closely resembles XACML [14]) and performs query rewriting for SQL-compliant database systems. To ensure the safety and security of the queries that are generated in our approach, validation techniques can be employed. Safety validation can be performed through matching the referred attributes to the application domain concepts [4]. This is done through a separate artifact that describes the properties of the subjects, objects and actions associated with the application domain and how they map on the database schema. This artifact can be extracted automatically from the application code. Security validation intends to prevent privilege escalation through queries generated from custom policies, and can also employ this artifact in combination with whitelisting techniques for the expressions of the policy to determine whether they can be translated to a query securely. The initial prototype indicates the feasibility of the approach, and a thorough evaluation of the performance will be presented in future work.

5   Conclusion

This paper has presented an initial step towards a middleware than can transparently enforce access control for search operations on large data sets. Evidently, many challenges remain. These include the way that the middleware handles variability of the underlying database schema and analyzing the performance issues that large policies may introduce on the translated query. However, we believe that such problems can be mitigated.

We are convinced that the middleware presented in this paper can significantly reduce the overhead introduced by performing access control for large data sets, and should be further researched in future work. This would enable policy-based access control to be enforced on both the application and the database, effectively supporting a separation of concerns.

Acknowledgments. This research is partially funded by the Research Fund KU Leuven, and by the EU FP7 project NESSoS. With the financial support from the Prevention of and Fight against Crime Programme of the European Union (B-CCENTRE).

References

1. Oracle Virtual Private Database (VPD). http://docs.oracle.com/cd/B28359_01/network.111/b28531/vpd.htm. (Accessed 02 September 2015)
2. Axiomatics. Data Access Filter (ADAF). http://www.axiomatics.com/solutions/products/authorization-for-databases/197-axiomatics-data-access-filter-adaf.html. (Accessed 2 October 2015)
3. Bertino, E., Sandhu, R.: Database security-concepts, approaches, and challenges. IEEE Trans. Dependable Secure Comput. **2**(1), 2–19 (2005)

4. Bogaerts, J., Decat, M., Lagaisse, B., Joosen, W.: Control, entity-based access: supporting more expressive access control policies. In: Proceedings of the 31st Annual Computer Security Applications Conference (2015)
5. Carminati, B., Ferrari, E., Cao, J., Tan, K.L.: A framework to enforce access control over data streams. In: ACM TISSEC (2010)
6. Cook, W.R., Rai, S., Safe query objects: statically typed objects as remotely executable queries. In: 27th International Conference on Software Engineering, ICSE, Proceedings, pp. 97–106. IEEE (2005)
7. De Win, B., Piessens, F., Joosen, W., Verhanneman, T.: On the importance of the separation-of-concerns principle in secure software engineering. In: Workshop on the Application of Engineering Principles to System Security Design (2002)
8. Decat, M., Bogaerts, J., Lagaisse, B., Joosen, W.: Amusa: middleware for efficient access control management of multi-tenant SaaS applications. In: Proceedings of the 30th Annual ACM Symposium on Applied Computing. ACM (2015)
9. Gay, R., Hu, J., Mantel, H.: CliSeAu: securing distributed java programs by cooperative dynamic enforcement. In: Prakash, A., Shyamasundar, R. (eds.) ICISS 2014. LNCS, vol. 8880, pp. 378–398. Springer, Heidelberg (2014)
10. Grummt, E., Müller, M.: Fine-grained access control for EPC information services. In: Floerkemeier, C., Langheinrich, M., Fleisch, E., Mattern, F., Sarma, S.E. (eds.) IOT 2008. LNCS, vol. 4952, pp. 35–49. Springer, Heidelberg (2008)
11. Hu, V., Ferraiolo, D., Kuhn, R., Schnitzer, A., Sandlin, K., Miller, R., Scarfone, K.: Guide to Attribute Based Access Control (ABAC) Definition and Considerations. NIST Special Publication (2014)
12. Mell, P., Grance, T.: The NIST definition of cloud computing. In: NIST (2009)
13. Moeys, J., Decat, M.: Simple Tree-structured Attribute-based Policy Language (STAPL). https://github.com/stapl-dsl. (Accessed 2 October 2015)
14. OASIS. eXtensible Access Control Markup Language (XACML) Standard v3.0 (2013). http://docs.oasis-open.org/xacml/3.0/xacml-3.0-core-spec-os-en.pdf
15. Olson, L.E., Gunter, C.A., Cook, W.R., Winslett, M.: Implementing reflective access control in SQL. In: Gudes, E., Vaidya, J. (eds.) Data and Applications Security XXIII. LNCS, vol. 5645, pp. 17–32. Springer, Heidelberg (2009)
16. Opyrchal, L., Cooper, J., Poyar, R., Lenahan, B., Zeinner, D.: Bouncer: policy-based fine grained access control in large databases. Int. J. Secur. Appl. 5(2), 1–16 (2011)
17. Pretschner, A., Hilty, M., Basin, D.: Distributed usage control. Commun. ACM 49(9), 39–44 (2006)
18. Rizvi, S., Mendelzon, A., Sudarshan, S., Roy, P.: Extending query rewriting techniques for fine-grained access control. In: SIGMOD Conference on Management of data. ACM (2004)
19. Roichman, A., Gudes, E.: Fine-grained access control to web databases. In: Symposium on Access Control Models and Technologies. ACM (2007)
20. Samarati, P., Vimercati, S., Control, A.: Policies, models, and mechanisms. In: Foundations of Security Analysis and Design, pp. 137–196 (2001)
21. Turkmen, F., Crispo, B.: Performance evaluation of XACML PDP implementations. In: Workshop on Secure Web Services. ACM (2008)
22. Vollbrecht, J., Calhoun, P., Farrell, S., Gommans, L., Gross, G., de Bruijn, B., de Laat, C., Holdrege, M., Spence, D.: RFC 2904: AAA Authorization Framework, August 2000

Idea: Enforcing Security Properties by Solving Behavioural Equations

Eric Rothstein Morris[✉] and Joachim Posegga

University of Passau, Passau, Germany
er@sec.uni-passau.de

Abstract. We present a novel theory of security property enforcement based on universal coalgebra and coinductive calculus. As an example, we show that it is possible to define sound and transparent runtime enforcers for noninterference using behavioural equations, and we preliminarily validate our approach by means of a Haskell implementation.

1 Introduction

Coalgebras [12,17] have arisen as a powerful category-theoretical framework for the specification and reasoning of models of computation. In general, coalgebras enable the uniform study of different systems, allowing the generalisation of well-known theorems in Computer Science; *e.g.*, Kleene's theorem [21]. In the context of systems security, Boreale *et al.* [3] lay a foundation for reasoning about information leakage for a variety of systems that builds on language-theoretic and coalgebraic concepts. In this work, we present a novel idea for the enforcement of security properties based on coalgebras.

We associate coalgebras with Haskell's *typeclasses*, which we use to implement them. For example, let I be an input alphabet and let $Bool = \{False, True\}$; consider a typeclass $\mathcal{DA}$ that defines the functions $accept: X \to Bool$ and $transition: X \to (I \to X)$. If the type Y implements the typeclass $\mathcal{DA}$, then we have a coalgebra $\langle Y, \langle accept, transition \rangle \rangle$ of the functor (explained below) $\mathcal{DA}(X) = Bool \times (I \to X)$, which defines coalgebras that model deterministic automata that recognise languages with alphabet I.

We work only with the category **Set** of sets and functions. Formally, a *functor* $\mathcal{F}: \mathbf{Set} \to \mathbf{Set}$ is a mapping from sets to sets and functions to functions that preserves identities and function composition. Given a functor $\mathcal{F}: \mathbf{Set} \to \mathbf{Set}$, an $\mathcal{F}$-*coalgebra* is a pair $\langle X, \alpha \rangle$ where X is a set and $\alpha: X \to \mathcal{F}(X)$ is a function.

Final coalgebras (see [17]) are associated with *denotational semantics* and notions of *behaviour*. An $\mathcal{F}$-coalgebra $\langle Z, \omega \rangle$ is final if and only if, for every $\mathcal{F}$-coalgebra $\langle X, \alpha \rangle$, there is *one and only one* function $[\![\cdot]\!]: X \to Z$, called the *semantic mapping*, such that $\mathcal{F}([\![\cdot]\!]) \circ \alpha = \omega \circ [\![\cdot]\!]$ (see also [12]). Every final $\mathcal{F}$-coalgebra $\langle Z, \omega \rangle$ satisfies two properties that are fundamental to our idea: first, the function ω is an isomorphism between Z and $\mathcal{F}(Z)$, and second, the set Z satisfies the principle of *coinduction*. In a nutshell, coinduction describes the *observation* and *dynamics* of the elements of Z; *i.e.*, how they may be observed and transformed.

J. Caballero et al. (Eds.): ESSoS 2016, LNCS 9639, pp. 260–268, 2016.
DOI: 10.1007/978-3-319-30806-7_17

Idea: Given a final $\mathcal{F}$-coalgebra $\langle Z, \omega\colon Z \to \mathcal{F}(Z)\rangle$ where $\omega = \langle f_1, \ldots, f_n\rangle$, we use a *system of behavioural equations* (see [19]) to define a property $P\colon Z \to \texttt{Bool}$; *i.e.*, for $z \in Z$, we say that $P(z)$ holds only if z satisfies a system of behavioural equations $f_1(z) = s_1, \ldots, f_n(z) = s_n$, where $\langle s_1, \ldots, s_n\rangle \in \mathcal{F}(Z)$. Using the system of behavioural equations, we then define a unary operator $enf_P\colon Z \to Z$, for $z \in Z$, by

$$enf_P(z) = \omega^{-1}(s_1, \ldots, s_n). \qquad (1)$$

Given that $\omega(enf_P(z)) = \langle s_1, \ldots, s_n\rangle$, we know that $enf_P(z)$ satisfies the property P, and we see that the operator enf_P models an enforcer for P. Finally, we extend enforcement of P to all $\mathcal{F}$-coalgebras by means of the semantic mapping.

The rest of the paper is organised as follows: Sect. 2 sets the notation for the rest of the paper and provides some preliminaries about systems modelled as coalgebras, and property enforcement. In Sect. 3, we take a popular security property, noninterference, and we show how to create an enforcer for it using behavioural equations. Section 4 discusses a preliminary Haskell implementation of the enforcement scheme. Section 5 explores related work, and we conclude in Sect. 6.

2 Preliminaries

We provide the following definitions related to enforcement mechanisms based on [1,7,13]. Let $\texttt{Sys}$ be a set of systems. We say that $\texttt{Sys} \to \texttt{Bool}$ is the set of *system properties*. Intuitively, a system σ satisfies a system property P if and only if $P(\sigma) = \texttt{True}$. Given a system property P, a *sound enforcer of* P is a mechanism $enf_P\colon \texttt{Sys} \to \texttt{Sys}$ such that $enf_P(\sigma)$ satisfies P, for all $\sigma \in \texttt{Sys}$. An enforcer enf_P is *transparent* if and only if whenever σ satisfies P, then $enf_P(\sigma)$ is *behaviourally equivalent* to σ.

Given a set X, the set of *sequences of elements of* X (either finite or infinite) is $[X]$. The *empty sequence* is denoted by $[\,]$. We denote the prepending of an element $x \in X$ to a sequence $w \in [X]$ by $x : w$.

3 Enforcing Noninterference via Behavioural Equations

In this section, we show how to enforce noninterference [9] using our behavioural equations method. Informally, noninterference is the notion of information flow security where the actions of a group of users does not affect what another group of users sees. A common formulation of noninterference uses a two-element lattice $\{L, H\}$ with $L \leq H$; *i.e.*, the actions of H-users must not affect what L-users see, but actions of L-users may affect what H-users see.

For the rest of the paper, let $\texttt{I}$ be a set of inputs and let $\texttt{O}$ be a set of outputs. We model reactive systems with only one output channel (of L-type) as coalgebras of the functor $\mathcal{F}(X) = (\texttt{I} \to \texttt{O}) \times (\texttt{I} \to X)$. We associate the functor $\mathcal{F}$ with a Haskell typeclass that requires the type X to implement two functions:

an *observation* function $\mathsf{obs}\colon X \to (I \to O)$ and a *dynamics/transition* function $\mathsf{trn}\colon X \to (I \to X)$. For $x \in X$ and $i \in I$, $\mathsf{obs}(x)(i)$ is the output written in the channel when i arrives at state x, and $\mathsf{trn}(x)(i)$ is the i-successor state of x; *i.e.*, the state that x makes a transition to when the input i arrives. Henceforth, we write x^i as a shorthand for $\mathsf{trn}(x)(i)$. We show how to model systems with two output channels in Sect. 3.3.

Let $[\![1_{\mathcal{F}}]\!] = [I] \to (I \to O)$. Based on [11, Lemma 6], we say that the set $[\![1_{\mathcal{F}}]\!]$ has a final $\mathcal{F}$-coalgebra structure if the functions $\mathsf{obs}\colon [\![1_{\mathcal{F}}]\!] \to (I \to O)$ and $\mathsf{trn}\colon [\![1_{\mathcal{F}}]\!] \to (I \to [\![1_{\mathcal{F}}]\!])$ are defined, for $\sigma \in [\![1_{\mathcal{F}}]\!]$, $i \in I$ and $w \in [I]$, by

$$\mathsf{obs}(\sigma) = \sigma([\,]), \tag{2}$$

$$\mathsf{trn}(\sigma)(i)(w) = \sigma(i : w). \tag{3}$$

For every $\mathcal{F}$-coalgebra $\langle X, \mathsf{obs}, \mathsf{trn} \rangle$, the *semantic mapping* $[\![\cdot]\!]\colon X \to [\![1_{\mathcal{F}}]\!]$ is defined, for $x \in X$, $i \in I$ and $w \in [I]$, by

$$[\![x]\!]([\,]) = \mathsf{obs}(x), \tag{4}$$

$$[\![x]\!](i : w) = [\![x^i]\!](w). \tag{5}$$

Given the existence of the semantic mapping, we refer to $[\![1_{\mathcal{F}}]\!]$ as the set of *behaviours* of $\mathcal{F}$-coalgebras.

Now that we have a final $\mathcal{F}$-coalgebra $\langle [\![1_{\mathcal{F}}]\!], \langle \mathsf{obs}, \mathsf{trn} \rangle \rangle$, we want to define a set of conditions $\mathsf{obs}(\sigma) = s_1$ and $\mathsf{trn}(\sigma) = s_2$ with $\langle s_1, s_2 \rangle \in \mathcal{F}([\![1_{\mathcal{F}}]\!])$ that characterises those elements $\sigma \in [\![1_{\mathcal{F}}]\!]$ that satisfy noninterference.

3.1   Defining Behavioural Equations for Noninterference

Before we define noninterference, we need to introduce a couple of concepts. Let $\mathsf{lvl}\colon I \to \{L, H\}$ be a function that classifies inputs into security levels, and let $\mathsf{trace}\colon [\![1_{\mathcal{F}}]\!] \to [I] \to [O]$ be the function defined by

$$\mathsf{trace}(\sigma, [\,]) = [\,], \tag{6}$$

$$\mathsf{trace}(\sigma, i : w) = \mathsf{obs}(\sigma)(i) : \mathsf{trace}(\sigma^i, w). \tag{7}$$

Now, let $\mathsf{filter}\colon [I] \to [I]$ be the function defined, for $w \in [I]$ and $i \in I$, by

$$\mathsf{filter}([\,]) = [\,], \tag{8}$$

$$\mathsf{filter}(i : w) = \begin{cases} i : \mathsf{filter}(w), & \text{if } \mathsf{lvl}(i) = L; \\ \mathsf{filter}(w), & \text{if } \mathsf{lvl}(i) = H. \end{cases} \tag{9}$$

Traditionally, *noninterference* [9] is the property $\mathsf{NI}\colon [\![1_{\mathcal{F}}]\!] \to \mathtt{bool}$ defined, for $\sigma \in [\![1_{\mathcal{F}}]\!]$, by

$$\mathsf{NI}(\sigma) = \forall w \in [I].\, \mathsf{trace}(\sigma, w) = \mathsf{trace}(\sigma, \mathsf{filter}(w)). \tag{10}$$

The property NI is defined in terms of traces and not in terms of "local conditions"; *i.e.*, conditions over obs and trn. To fit the method described in this work, we need to look for conditions over obs and trn that would imply NI, instead. Based on the *unwinding theorem* [10], we present the following two conditions: for all $i \in I$, in order for $\sigma \in [\![1_{\mathcal{F}}]\!]$ to satisfy NI, it is necessary that the following conditions hold: the condition

$$\mathsf{obs}(\sigma)(i) = \begin{cases} \mathsf{obs}(\sigma)(i), & \text{if } \mathsf{lvl}(i) = L; \\ [], & \text{if } \mathsf{lvl}(i) = H, \end{cases} \tag{11}$$

which requires H-level inputs to produce no visible L-output, the condition

$$\sigma^i = \begin{cases} \sigma^i, & \text{if } \mathsf{lvl}(i) = L; \\ \sigma, & \text{if } \mathsf{lvl}(i) = H, \end{cases} \tag{12}$$

which requires H-level inputs to cause no behavioural changes, and the condition

$$\mathrm{NI}(\sigma^i), \tag{13}$$

which requires the conditions (11) and (12) to hold invariantly.

We associate (11) with the notion of *local consistency* and (12) with the notion of *step consistency* from Ochoa *et al.*'s [16]. In [16], the authors show that those conditions are enough to build an unwinding relation for the testing of noninterference in UML state charts. Consequently, we assume that behaviours that satisfy Eqs. (11), (12) and (13) satisfy NI.

We now use the right hand side of Eqs. (11), (12) and (13) to define the behavioural equations that define our enforcer for noninterference.

3.2   Defining the Enforcer for Noninterference

Equations (11), (12) and (13) define the "local conditions" necessary for the satisfaction of NI, and we use them to determine the system of behavioural equations that defines the enforcer for NI. Let $enf_{\mathrm{NI}} : [\![1_{\mathcal{F}}]\!] \to [\![1_{\mathcal{F}}]\!]$ be the operator defined, for $\sigma \in [\![1_{\mathcal{F}}]\!]$ and $i \in I$, by the system of behavioural equations

$$\mathsf{obs}(enf_{\mathrm{NI}}(\sigma))(i) = \begin{cases} \mathsf{obs}(\sigma)(i), & \text{if } \mathsf{lvl}(i) = L; \\ [], & \text{if } \mathsf{lvl}(i) = H; \end{cases} \tag{14}$$

$$enf_{\mathrm{NI}}(\sigma)^i = \begin{cases} enf_{\mathrm{NI}}(\sigma^i), & \text{if } \mathsf{lvl}(i) = L; \\ enf_{\mathrm{NI}}(\sigma), & \text{if } \mathsf{lvl}(i) = H. \end{cases} \tag{15}$$

Implicitly, Eqs. (14) and (15) first map the behaviour $\sigma \in [\![1_{\mathcal{F}}]\!]$ to the pair $\langle \mathsf{obs}(enf_{\mathrm{NI}}(\sigma)), \mathsf{trn}(enf_{\mathrm{NI}}(\sigma)) \rangle \in \mathcal{F}([\![1_{\mathcal{F}}]\!])$. Then, Eqs. (14) and (15) use the inverse of the isomorphism $\langle \mathsf{obs}, \mathsf{trn} \rangle : [\![1_{\mathcal{F}}]\!] \to \mathcal{F}([\![1_{\mathcal{F}}]\!])$ to map the pair $\langle \mathsf{obs}(enf_{\mathrm{NI}}(\sigma)), \mathsf{trn}(enf_{\mathrm{NI}}(\sigma)) \rangle$ to the behaviour $enf_{\mathrm{NI}}(\sigma) \in [\![1_{\mathcal{F}}]\!]$, building a bridge between σ and its noninterferent version $enf_{\mathrm{NI}}(\sigma)$.

We prove that if σ satisfies Eqs. (11), (12) and (13), then $\sigma = enf_{\mathrm{NI}}(\sigma)$. We conduct the proof by *bisimulation*; i.e., we define a relation $R \subseteq [\![1_{\mathcal{F}}]\!] \times [\![1_{\mathcal{F}}]\!]$ that contains the pair $\langle \sigma, enf_{\mathrm{NI}}(\sigma) \rangle$ and we show that R is a *bisimulation relation*. More precisely, we show that $\mathsf{obs}(\sigma) = \mathsf{obs}(enf_{\mathrm{NI}}(\sigma))$ and that $\langle \sigma^i, enf_{\mathrm{NI}}(\sigma)^i \rangle \in R$, for all $i \in I$. For a comprehensive explanation on proofs by bisimulation, please refer to [12,17–19].

Theorem 1. *The relation* $R = \{ \langle \sigma, enf_{NI}(\sigma) \rangle \mid \sigma$ *satisfies* (11), (12) *and* (13) $\}$ *is a bisimulation.*

Proof. The equality $\mathsf{obs}(\sigma) = \mathsf{obs}(enf_{NI}(\sigma))$ holds because the right hand side of Eqs. (11) and (14) is the same.

To prove that $\langle \sigma^i, enf_{NI}(\sigma)^i \rangle \in R$, we split into two cases: one where $\mathsf{lvl}(i) = L$ and one where $\mathsf{lvl}(i) = H$. On the first case, since σ satisfies (13), we know that (11) and (12) also hold for σ^i. Consequently, we know that the pair $\langle \sigma^i, enf_{NI}(\sigma^i) \rangle \in R$. Additionally, we know that $enf_{NI}(\sigma)^i = enf_{NI}(\sigma^i)$ by Eq. (15), so $\langle \sigma^i, enf_{NI}(\sigma)^i \rangle \in R$ in this first case. On the second case, we have that $enf_{NI}(\sigma)^i = enf_{NI}(\sigma)$ by Eq. (15), and that $\sigma^i = \sigma$ by Eq. (12), so the pair $\langle \sigma^i, enf_{NI}(\sigma)^i \rangle$ is equal to $\langle \sigma, enf_{NI}(\sigma) \rangle$, which we know belongs to R. Thus, we conclude that R is a bisimulation relation. □

We derive the following

Corollary 1. *For all pairs* $\langle \sigma, enf_{NI}(\sigma) \rangle$ *that are members of the bisimulation relation* R, *we have that* $\sigma = enf_{NI}(\sigma)$. *This is because, in final coalgebras, all bisimulation relations are subsets of the equality relation (see [17, Theorem 9.2]).*

The operator enf_{NI} has the two properties that we look for in enforcement mechanisms: soundness and transparency. The function enf_{NI} is sound because, for all $\sigma \in [\![1_{\mathcal{F}}]\!]$, $enf_{NI}(\sigma)$ satisfies Eqs. (11), (12) and (13) by design; implying that $enf_{NI}(\sigma)$ satisfies NI. Additionally, the function enf_{NI} is transparent, because if σ satisfies NI, then $enf_{NI}(\sigma) = \sigma$.

To extend the enforcement of NI via the operator enf_{NI} to an arbitrary element $\mathsf{x} \in \mathsf{X}$ of an $\mathcal{F}$-coalgebra $\langle \mathsf{X}, \mathsf{obs}, \mathsf{trn} \rangle$, we first apply the semantic mapping, and then we apply enf_{NI}; *i.e.*, $enf_{NI}([\![\mathsf{x}]\!])$. This extension allows us to soundly and transparently enforce NI on any $\mathcal{F}$-coalgebra.

3.3 Noninterference with Multiple Channels

Arguably, the enforcement of NI in a system that has only one channel (of L-type) can be carried out by filtering all H-inputs so that they never reach the system. 'However, if a system has an H-channel that allows H-inputs to affect it, filtering H-inputs damages the transparency of the enforcer mechanism. In this section, we show how to adapt the behavioural equations for NI to fit a new model of computation with two output channels: one of H-level and one of L-level.

To model reactive systems with one H-channel and with one L-channel, we use coalgebras of the functor $\mathcal{G}(\mathsf{X}) = (\mathsf{I} \to \mathsf{O}) \times (\mathsf{I} \to \mathsf{O}) \times (\mathsf{I} \to \mathsf{X})$. $\mathcal{G}$-coalgebras have two observation operations instead of one: $\mathsf{obs}_H \colon \mathsf{X} \to (\mathsf{I} \to \mathsf{O})$ for the H-channel and $\mathsf{obs}_L \colon \mathsf{X} \to (\mathsf{I} \to \mathsf{O})$ for the L-channel. The dynamics function $\mathsf{trn} \colon \mathsf{X} \to (\mathsf{I} \to \mathsf{X})$ remains the same.
Consider the set $[\![1_{\mathcal{G}}]\!]$ defined by

$$[\![1_{\mathcal{G}}]\!] = [\![1_{\mathcal{F}}]\!] \times [\![1_{\mathcal{F}}]\!] = ([\![\mathsf{I}]\!] \to (\mathsf{I} \to \mathsf{O})) \times ([\![\mathsf{I}]\!] \to (\mathsf{I} \to \mathsf{O})), \tag{16}$$

The set $[\![1_{\mathcal{F}}]\!] \times [\![1_{\mathcal{F}}]\!]$ has a final coalgebra structure for the functor $\mathcal{G}$ (see [11, Lemma 6]) if $\mathbf{obs}_H$, $\mathbf{obs}_L$ and $\mathbf{trn}$ are defined, for $\langle \sigma_H, \sigma_L \rangle \in [\![1_{\mathcal{F}}]\!] \times [\![1_{\mathcal{F}}]\!]$ and $i \in I$, by

$$\mathbf{obs}_H(\sigma_H, \sigma_L) = \sigma_H([\,]) \tag{17}$$

$$\mathbf{obs}_L(\sigma_H, \sigma_L) = \sigma_L([\,]) \tag{18}$$

$$\mathbf{trn}(\sigma_H, \sigma_L)(i) = \langle \sigma_H^i, \sigma_L^i \rangle, \tag{19}$$

We want to allow L-inputs to affect the H-channel, but prevent H-inputs from affecting the L-channel. Thus, we propose the following system of equations: let $\langle \sigma_H, \sigma_L \rangle \in [\![1_{\mathcal{F}}]\!] \times [\![1_{\mathcal{F}}]\!]$ and $i \in I$ in

$$\mathbf{obs}_H(\mathit{enf}_{\mathrm{NI}}(\sigma_H, \sigma_L))(i) = \mathbf{obs}_H(\sigma_H, \sigma_L)(i) \tag{20}$$

$$\mathbf{obs}_L(\mathit{enf}_{\mathrm{NI}}(\sigma_H, \sigma_L))(i) = \begin{cases} \mathbf{obs}_L(\sigma_H, \sigma_L)(i), & \text{if } \mathtt{lvl}(i) = L; \\ [\,], & \text{if } \mathtt{lvl}(i) = H; \end{cases} \tag{21}$$

$$\mathbf{trn}(\mathit{enf}_{\mathrm{NI}}(\sigma_H, \sigma_L))(i) = \begin{cases} \mathit{enf}_{\mathrm{NI}}(\sigma_H^i, \sigma_L^i), & \text{if } \mathtt{lvl}(i) = L; \\ \mathit{enf}_{\mathrm{NI}}(\sigma_H^i, \sigma_L), & \text{if } \mathtt{lvl}(i) = H. \end{cases} \tag{22}$$

Equation (20) allows inputs of any level to affect the H-channel. Equation (21) matches Eq. (14), and prevents H-inputs from affecting the L-channel. Finally, Eq. (22) allows L-inputs to change the behaviour of the H-part of the system, but prevents H-inputs from changing the behaviour of the L-part of the system. These equations remind us of Devriese and Piessen's *secure multi-execution* [6], because we split systems into two "subsystems": one that provides outputs to the H-channel and one that provides outputs to the L-channel. However, our approach is different because secure multi-execution uses several instances of the same system, but using different inputs for each security levels We do not change the inputs of systems; we change the way systems respond to inputs, instead.

4   Implementation and Validation

Due to the functional nature of our definitions, Haskell offers a convenient way to implement the elements required for enforcement via behavioural equations. We test our enforcer with a very simple test case. We define $\mathtt{Input} = \mathtt{Int}$, $\mathtt{Output} = \mathtt{String}$, and we make $\mathtt{Int}$ our set of states. We define the functions $\mathtt{obs}\colon \mathtt{Int} \to \mathtt{Input} \to \mathtt{Output}$ and $\mathtt{trn}\colon \mathtt{Int} \to \mathtt{Input} \to \mathtt{Int}$ by $\mathtt{obs}(x, i) = \text{``}x\text{''}$ and $\mathtt{trn}(x, i) = x + i$. We use a simple policy $\mathtt{lvl}\colon \mathtt{Input} \to \{L, H\}$, defined for $i \in \mathtt{Input}$ by

$$\mathtt{lvl}(i) = \begin{cases} H, & \text{if } i \bmod_2 = 0; \\ L, & \text{otherwise .} \end{cases} \tag{23}$$

Our experiment consists in evaluating the expressions given in Table 1. The results suggest that noninterference is in fact being enforced, but that solving the behavioural equations at runtime heavily impacts performance.

Table 1. Expressions evaluated to test enforcement via behavioural equations.

Expression	Time	Result
`run 0 [1..1000]`	0.05 s	["0", "1", "3", "6", "10", "15", ...]
`run (enfNI (semanticMap 0)) [1..1000]`	0.72 s	["0", "1", "4", "9", "16", "25", ...]

5   Related Work

The only work we could find that combines behavioural equations and security is Boreale *et al.*'s [3]. They use behavioural equations to define the compositional semantics of their process calculus, and they also use Haskell to implement their calculus. However, their work focuses on quantification of information leakage, not on enforcement of security properties.

The concepts of final coalgebra, coinduction and bisimulation are tightly related, as shown in [17]. Sabelfeld's [20] explores notions of bisimulation to reason about noninterference, and Bohannon *et al.* [2] define variations of noninterference by coinduction. Unlike us, Bohannon *et al.* focus in coinduction for streams, and not in coinduction for final coalgebras in general. Consequently, we consider their definitions of behavioural properties to be traced-based, while ours are behaviour-based.

Clarkson and Schneider's hyperproperties [4] are a very general theory for the definition of systems and their properties. The recent extensions to LTL and CTL*; namely HyperLTL and HyperCTL*, allow us to express a wide range of hyperproperties as temporal formulas [5], and new algorithms enable the automatic verification of those formulas in finite state systems [8]. The main difference between hyperproperties and our approach is that hyperproperties is a trace-based approach to model systems, while ours is coalgebraic. Results are still too preliminary to conclude whether coalgebras and behavioural equations are more convenient than hyperproperties when it comes to enforcing complex security properties.

Finally, the work by Milushev and Clarke [14] uses coalgebras in order to provide an incremental approach to the verification of hyperproperties. However, they only study the verification of those hyperproperties, not their enforcement.

6   Conclusion and Future Work

Although our results are very preliminary, we believe that defining enforcers via behavioural equations is a promising method to enforce behavioural properties during runtime. We showed that it is possible to define sound and transparent enforcers for noninterference using behavioural equations, and we provided a preliminary validation of our results by means of a Haskell implementation.

Besides noninterference, it is also possible to capture and enforce notions of integrity using behavioural equations. For example, for $\sigma \in [\![1_{\mathcal{F}}]\!]$, we can define the operator $enf_{\text{INT}} \colon [\![1_{\mathcal{F}}]\!] \to [\![1_{\mathcal{F}}]\!]$ using the system behavioural equations

$$\mathsf{obs}(enf_{\text{INT}}(\sigma))(\mathtt{i}) = \mathsf{obs}(\sigma)(\mathtt{i}) \quad \text{and} \quad \mathsf{trn}(enf_{\text{INT}}(\sigma))(\mathtt{i}) = enf_{\text{INT}}(\sigma) \qquad (24)$$

The behaviour $enf_{\text{INT}}(\sigma)$ takes the first observation of σ and protects it from changes by inputs. In other words, enf_{INT} is an enforcer for the property that states: "the behaviour of the system must not change when inputs (of any kind) are received".

Concerning the functors that determine the coalgebras that model the systems we are interested in, we only imposed the restriction that a final coalgebra must exist. Given that polynomial functors (see [11]) always imply the existence of a final coalgebra, we believe that our method should work without problems for that class of functors.

Though the enforcement method is theoretically sound, the performance overhead caused by solving the behavioural equations during runtime needs to be mitigated; otherwise, the method becomes impractical.

There are several directions for future work. First, we need to study which security properties can be captured by means of behavioural equations. Defining well-known security properties as behavioural equations in order to test the expressivity of the method is an interesting line of work. Second, in order to improve performance, we are interested in finding optimisation methods for the solution of behavioural equations. Finally, we would like to study more complex systems, including non-deterministic and probabilistic systems, and describe behavioural equations for them. Finally, Ngo *et al.* [15] propose a generic construction of an enforcement mechanism for non-interference (among other properties) on black box reactive programs. We believe that the black box approach is definitely related to the use of Haskell typeclasses and coalgebraic modelling, so we see some similarities between our coalgebraic approach and theirs. Verifying how their framework compares to ours is an interesting direction for future work.

Acknowledgements. This work was partially supported by the European Commission funded project BIOMICS, Grant no. 318202.

References

1. Bielova, N.: A theory of constructive and predictable runtime enforcement mechanisms. Ph.D. thesis, University of Trento (2011)
2. Bohannon, A., Pierce, B.C., Sjöberg, V., Weirich, S., Zdancewic, S.: Reactive noninterference. In: Proceedings of the 2009 ACM Conference on Computer and Communications Security, CCS 2009, Chicago, Illinois, USA, 9–13 November 2009, pp. 79–90 (2009)
3. Boreale, M., Clark, D., Gorla, D.: A semiring-based trace semantics for processes with applications to information leakage analysis. Math. Struct. Comput. Sci. **25**(2), 259–291 (2015)

4. Clarkson, M.R., Schneider, F.B.: Hyperproperties. J. Comput. Secur. **18**(6), 1157–1210 (2010)
5. Clarkson, M.R., Finkbeiner, B., Koleini, M., Micinski, K.K., Rabe, M.N., Sánchez, C.: Temporal logics for hyperproperties. In: Abadi, M., Kremer, S. (eds.) POST 2014. LNCS, vol. 8414, pp. 265–284. Springer, Heidelberg (2014)
6. Devriese, D., Piessens, F.: Noninterference through secure multi-execution. In: 31st IEEE Symposium on Security and Privacy, S&P 2010, Berleley/Oakland, California, USA, 16–19 May 2010, pp. 109–124 (2010)
7. Falcone, Y., Fernandez, J., Mounier, L.: What can you verify and enforce at run-time? STTT **14**(3), 349–382 (2012)
8. Finkbeiner, B., Rabe, M.N., Sánchez, C.: Algorithms for model checking HyperLTL and HyperCTL*. In: Kroening, D., Păsăreanu, C.S. (eds.) CAV 2015. LNCS, vol. 9206, pp. 30–48. Springer, Heidelberg (2015)
9. Goguen, J.A., Meseguer, J.: Security policies and security models. In: 1982 IEEE Symposium on Security and Privacy, Oakland, CA, USA, 26–28 April 1982, pp. 11–20 (1982)
10. Goguen, J.A., Meseguer, J.: Unwinding and inference control. In: 1984 IEEE Symposium on Security and Privacy, p. 75, April 1984
11. Jacobs, B.: Objects and classes, coalgebraically. In: Object-Orientation with Parallelism and Persistence, pp. 83–103. Kluwer Academic Publishers (1995)
12. Jacobs, B.: Introduction to coalgebra. Towards mathematics of states and observations (2012). http://www.cs.ru.nl/B.Jacobs/CLG/JacobsCoalgebraIntro.pdf
13. Ligatti, J., Bauer, L., Walker, D.: Run-time enforcement of nonsafety policies. ACM Trans. Inf. Syst. Secur. **12**(3), 19:1–19:41 (2009)
14. Milushev, D., Clarke, D.: Towards incrementalization of holistic hyperproperties. In: Degano, P., Guttman, J.D. (eds.) POST 2012. LNCS, vol. 7215, pp. 329–348. Springer, Heidelberg (2012)
15. Ngo, M., Massacci, F., Milushev, D., Piessens, F.: Runtime enforcement of security policies on black box reactive programs. In: Proceedings of the 42nd Annual ACM SIGPLAN-SIGACT Symposium on Principles of Programming Languages, pp. 43–54. ACM (2015)
16. Ochoa, M., Cuéllar, J., Pretschner, A., Hallgren, P.: Idea: unwinding based model-checking and testing for non-interference on EFSMs. In: Piessens, F., Caballero, J., Bielova, N. (eds.) ESSoS 2015. LNCS, vol. 8978, pp. 34–42. Springer, Heidelberg (2015)
17. Rutten, J.: Universal coalgebra: a theory of systems. Theor. Comput. Sci. **249**(1), 3–80 (2000)
18. Rutten, J.: Behavioural differential equations: a coinductive calculus of streams, automata, and power series. Theor. Comput. Sci. **308**(13), 1–53 (2003)
19. Rutten, J.: A coinductive calculus of streams. Math. Struct. Comput. Sci. **15**(1), 93–147 (2005)
20. Sabelfeld, A.: Confidentiality for multithreaded programs via bisimulation. In: Broy, M., Zamulin, A.V. (eds.) PSI 2003. LNCS, vol. 2890, pp. 260–274. Springer, Heidelberg (2004)
21. Silva, A.: Kleene coalgebras. Ph.D. thesis, Radboud University Nijmegen (2010)

Author Index

Printed in the United States
By Bookmasters